Computational Molecular Biology

WILEY SERIES IN MATHEMATICAL AND COMPUTATIONAL BIOLOGY

BÜRGER—The Mathematical Theory of Selection, Recombination and Mutation

CHAPLAIN/SINGH/MCLACHLAN—On Growth and Form: Spatio-temporal Pattern Formation in Biology

CHRISTIANSEN—Population Genetics of Multiple Loci

CLOTE/BACKOFEN—Computational Molecular Biology: An Introduction

DIEKMANN/HEESTERBEEK—Mathematical Epidemiology of Infectious Diseases: Model Building, Analysis and Interpretation

Reflecting the rapidly growing interest and research in the field of mathematical biology, this outstanding new book series examines the integration of mathematical and computational methods into biological work. It also encourages the advancement of theoretical and quantitative approaches to biology, and the development of biological organisation and function.

The scope of the series is broad, ranging from molecular structure and processes to the dynamics of ecosystems and the biosphere, but unified through evolutionary and physical principles, and the interplay of processes across scales of biological organisation.

Topics to be covered in the series include:

- Cell and molecular biology
- Functional morphology and physiology
- Neurobiology and higher function
- Genetics
- Immunology
- Epidmiology
- Ecological and evolutionary dynamics of interacting populations

A fundamental research tool, the Wiley Series in Mathematical and Computational Biology provides essential and invaluable reading for biomathematicians and development biologists, as well as graduate students and researchers in mathematical biology and epidemiology.

Computational Molecular Biology
An Introduction

Peter Clote
Department of Computer Science and Department of Biology, Boston College, USA

Formerly
Ludwig-Maximilians-Universität München, Germany

Rolf Backofen
Ludwig-Maximilians-Universität München, Germany

JOHN WILEY & SONS, LTD

Chichester • New York • Weinheim • Brisbane • Singapore • Toronto

Baffins Lane, Chichester,
West Sussex, PO19 1UD, England

National 01243 779777
International (+44) 1243 779777

e-mail (for orders and customer service enquiries): cs-books@wiley.co.uk

Visit our Home Page on http://www.wiley.co.uk or http://www.wiley.com

Other Wiley Editorial Offices

John Wiley & Sons, Inc., 605 Third Avenue,
New York, NY 10158-0012, USA

Wiley-VCH Verlag GmbH
Pappelallee 3, D-69469 Weinheim, Germany

Jacaranda Wiley Ltd, 33 Park Road, Milton,
Queensland 4064, Australia

John Wiley & Sons (Asia) Pte Ltd, 2 Clementi Loop #02-01,
Jin Xing Distripark, Singapore 129809

John Wiley & Sons (Canada) Ltd, 22 Worcester Road,
Rexdale, Ontario, M9W 1L1, Canada

Library of Congress Cataloging-in-Publication Data

Clote, Peter.
Computational biology : a self contained approach to bioinformatics
/ Peter Clote, Rolf Backofen
p. cm – (Wiley series in mathematical and computational biology)
Includes bibliographical references (p.)
ISBN 0-471-87251-2 (alk. paper) – ISBN 0-471-87252-0 (pbk. : alk. paper)
1. Genetics—Mathematical Models. 2. Molecular biology—
Mathematical models. I. Backofen, Rolf. II. Title. III. Series.

QH438.4.M3 C565 2000
572.8'01'51187-dc21 00-038169

British Library Cataloguing in Publication Data

A catalogue record for this book is available from the British Library

ISBN 0-471-87251-2
ISBN 0-471-87252-0

Produced from PostScript files supplied by the authors.
Printed and bound in Great Britain by Antony Rowe Ltd, Chippenham, Wiltshire.
This book is printed on acid-free paper responsibly manufactured from sustainable forestry
in which at least two trees are planted for each one used for paper production.

To my wife, Marie, and to my son, Nicolas. (P.C.)
To my wife, Doris, and my children, Ina and Lara. (R.B.)

Contents

Series Preface

Theoretical biology is an old subject, tracing back centuries. At times, theoretical developments have represented little more than mathematical exercises, making scant contact with reality. At the other extreme have been those works, such as the writings of Charles Darwin, or the models of Watson and Crick, in which theory and fact are intertwined, mutually nourishing one another in inseparable symbiosis. Indeed, one of the most exciting developments in biology within the last quarter-century has been the integration of mathematical and theoretical reasoning into all branches of biology, from the molecule to the ecosystem. It is such a unified theoretical biology, blending theory and empiricism seamlessly, that has inspired the development of this series.

This series seeks to encourage the advancement of theoretical and quantitative approaches to biology, and to the development of unifying principles of biological organization and function, through the publication of significant monographs, textbooks and synthetic compendia in mathematical and computational biology. The scope of the series is broad, ranging from molecular structure and processes to the dynamics of ecosystems and the biosphere, but unified through evolutionary and physical principles, and the interplay of processes across scales of biological organization.

The principal criteria for publication beyond the intrinsic quality of the work are substantive biological content and import, and innovative development or application of mathematical or computational methods. Topics will include but not be limited to cell and molecular biology, functional morphology and physiology, neurobiology and higher function, immunology and epidemiology, and the ecological and evolutionary dynamics of interacting populations. The most successful contributions, however, will not be so easily categorized, crossing boundaries and providing integrative perspectives that unify diverse approaches; the study of infectious diseases, for example, ranges from the molecule to the ecosystem, involving mechanistic investigations at the level of the cell and the immune system, evolutionary perspectives as viewed through sequence analysis and population genetics, and demographic and epidemiological aspects at the level of the ecological community.

The objective of the series is the integration of mathematical and computational methods into biological work; the volumes published hence should be of interest both to fundamental biologists and to computational and mathematical scientists, as well as to the broad spectrum of interdisciplinary researchers that comprise the continuum connecting these diverse disciplines.

Simon Levin

Preface

In the early 1970s, gel electrophoresis was a novel technique in biochemistry labs. Now used as the work horse in genomic sequencing projects, molecular biology has undergone an incredibly rapid development, currently yielding so much raw data that efficient computer algorithms are mandatory for data analysis. The interdisciplinary field of *computational biology*, or *bioinformatics*, includes both theoretical and practical contributions from computer science, mathematics and biology and involves the development of mathematical models, statistical analysis, computer simulation, efficient algorithms, database systems, web interface, etc.

This text is meant to be a self-contained introduction to computational biology, where we provide the background mathematics required to understand *why* certain algorithms work. For instance, it seems to us that one must understand the underlying probability theory for development of hidden Markov models, rather than simply present the pseudocode for this important motif recognition algorithm. As a learning tool, many of the algorithms described in this book have prototype implementations in C/C++ and sometimes Java, whose source code can be found on the web pages for this book, found by following links from `http://www.wiley.co.uk/statistics`.

We would like to thank P. Baldi, J. Baglivo, C. Benham, E. Bornberg-Bauer, D. Chambers, S. A. Cook, S. Eddy, I. Hofacker, M. Kolmar, H.-P. Lenhof, R. Matthes, M. Nadel, K. Reinert, T. Santner, B. Steipe, A. von Haeseler, W. Thies, J. Timmer, J. Tromp, E. von Werner and S. Will for discussions, comments and suggestions and to Rob Calver, Ann-Marie Halligan and the staff of Wiley & Sons, Ltd. for their editorial assistance. Thanks are due as well to the students of the *Bioinformatik* courses taught every semester by the authors from 1996 through the present at the University of Munich. We have tried to provide a self-contained course for the typical computer science and biology students of our classes, in particular for those students in the new Bioinformatics Program at the University of Munich. Finally we would like to thank anonymous reviewers, whose comments were helpful in shaping this text. The authors nevertheless assume full responsibility for any remaining errors in the book.

The institutional affiliation of the first author (P. Clote) is now Boston College, Department of Computer Science and Department of Biology. This book was written during the period that the first author held the Gerhard-Gentzen Chair of Theoretical Computer Science at the University of Munich, which allowed him to build a new group in theoretical computer science, including the present group in computational biology. Were it not for initial funding by the Volkswagen Stiftung and the freedom allowed by Humboldt's vision of the German academic system, this book would not have been written.

Peter Clote and Rolf Backofen
April 30, 2001

1

Molecular Biology

> Survival machines began as passive receptacles for the genes, providing little more than walls to protect them from the chemical warfare of their rivals and the ravages of accidental molecular bombardment. In the early days they 'fed' on organic molecules freely available in the soup. (R. Dawkins, *The Selfish Gene* 1989 [Daw89])

In the past, living organisms were grouped into two distinct life forms or *domains*:

- *prokaryotes*, represented by *cyanobacteria* (blue-green algae) and common bacteria such as *Escherichia coli*, not having a nuclear membrane to separate genomic DNA from the cytoplasm; and
- *eukaryotes*, represented by unicellular organisms such as the flagellum *Trypanosoma brucei* (one species belonging to the *T. brucei* complex causes sleeping sickness in humans) and multicellular organisms, all of which have a nuclear membrane.

However, in August 1996, the genome[1] sequence of the methane-generating archaebacterium *Methanococcus jannaschii* was published by Bult et al. [BWO+96]. A small initial portion of the DNA sequence, consisting of over 1.6 million characters, is given as follows:

```
TACATTAGTGTTTATTACATTGAGAAACTTTATAATTAAAAAAGATTCATGTAAATTT
CTTATTTGTTTATTTAGAGGTTTTAAATTTAATTTCTAAGGGTTTGCTGGTTTGATTG
TTTAGAATATTTAACTTAATCAAATTATTTGAATTTTTGAAAATTAGGATTAATTAGG
TAAGTAAATAAAATTTCTCTAACAAATAAGTTAAATTTTTAAATTTAAGGAGATAAAA
ATACTCTGTTTTATTATGGAAAGAAAGATTTAAATACTAAAGGGTTTATATTATGAAG
TAGTTACTTACCCTTAGAAAAATATGGTATAGAAAAGCTTAAATATTAAGAGTGATGA
AGTATATTATGT...
```

Analysis of this sequence provided solid evidence for a startling hypothesis advanced two decades earlier by Carl Woese: there is a third domain of life called *Archaea*, which is distinct from *Prokarya* and *Eukarya*:

- *archaebacteria*, representing life domain *Archaea*, share cellular organization features with prokaryotes (such as lack of a nuclear membrane), but seem closer to eukaryotes in transcription and translation mechanism.

M. jannaschii was discovered in 1982 in a sediment sample collected by the research submarine *Alvin* from a *white smoker* on a hot spot of the sea floor of the Pacific Ocean

---

[1] The genome of an organism constitutes the entirety of its DNA.

(21° N on the East Pacific Rise) at a depth of 2600 meters. This methanogenic *archaeon* is *thermophilic*, living in a temperature range from 48°–94°C, with an optimum temperature around 85°C, and obtains its energy needs from high-energy bonds in certain inorganic compounds near thermal vents. Producing methane from carbon dioxide,[2] *M. jannaschii* represents a potentially non-polluting energy source.

How can one determine the (hypothetical) genes of *M. jannaschii* from its 1.66 megabase pair genome? How can one compute that 56% of its 1738 genes are completely new, unlike any genes found in prokaryotes or eukaryotes? *M. jannaschii* has only one kind of DNA polymerase,[3] while all other organisms studied have several kinds. Could this archaeon have an additional, completely different mechanism for DNA replication? If so, how can one begin to find the genes involved? How can one assert with confidence that *Archaea* is sufficiently different from *Protokarya* and *Eukarya* to warrant that it constitute a third domain of life? How can one devise a likely phylogenetic (ancestral) tree for life, which has *Protokarya* separating first from a common trunk, while *Archaea* separates somewhat later from the trunk leading to *Eukarya*? With genome sizes measuring into the millions and billions, only through the use of computers can one store and analyze genomic data. Computational biology, also called *bioinformatics*, is concerned with the development of efficient algorithms, statistical analysis, and mathematical modeling in order to answer such questions. In this chapter, we'll give a brief survey of those concepts of molecular biology, which are essential to computational biology.

The 'central dogma' of information flow in biology states that information flows from DNA to RNA to protein; since a protein's functionality is determined by its unique three-dimensional structure, it follows that the one-dimensional sequence information in DNA determines the three-dimensional structure of the corresponding protein.

> The central dogma states that once 'information' has passed into a protein it cannot get out again. The transfer of information from nucleic acid, or from nucleic acid to protein, may be possible, but transfer from protein to protein, or from protein to nucleic acid, is impossible. Information here means the precise determination of sequence, either of bases in the nucleic acid or of amino acid residues in the protein. (Francis Crick [Cri58])

Thus we have the following picture of the information flow in biology:

DNA ⟶ RNA ⟶ Protein

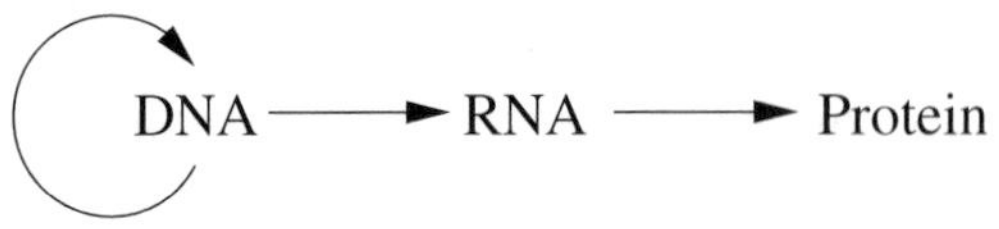

Following this diagram, the remainder of the chapter is organized as follows. We begin with an overview of organic chemistry needed in the sequel, then move on to DNA, RNA, amino acids and proteins. Finally, we describe the transcription and translation machinery that nature uses to process information in DNA into proteins.

---

[2] Note that $CO_2 + 8H = CH_4 + 2H_2O$. *M. jannaschii* uses a complicated pathway to perform this reduction to produce methane.

[3] As explained later, DNA polymerase is an enzyme necessary for DNA replication.

1.1 Some Organic Chemistry

Organic chemistry is centered around the chemistry of the carbon atom. The reason for this is that carbon has properties that distinguish it from most other elements.[4] These properties are

- carbon has a small size,
- carbon has 4 covalent bonds,
- several carbon atoms can form rings and chains.

Thus, it is possible to build large, stable molecules using carbon.

One distinguishes two types of bonds, namely covalent and non-covalent bonds. Non-covalent bonds are much weaker than covalent bonds (typically 30–300 times weaker in aqueous environment).[5] The average energy of a non-covalent bond is normally not much stronger than the energy of thermal movement (at room temperature).

A *covalent bond* is formed between two atoms that share electrons. Molecules are built using covalent bonds. *Non-covalent bonds* are subdivided into three classes, namely hydrogen bonds, ionic bonds, and van der Vaals interactions. Because of their importance, we concentrate on hydrogen bonds. Hydrogen bonds are used for base-pairing, which holds together the two strands of the DNA double helix. Hydrogen bonds also play an important role in RNA (which is normally single-stranded), as well as in stabilizing protein structure.

A *hydrogen bond* can be formed if there are two electronegative atoms (such as oxygen or nitrogen) that share a hydrogen atom, e.g.,

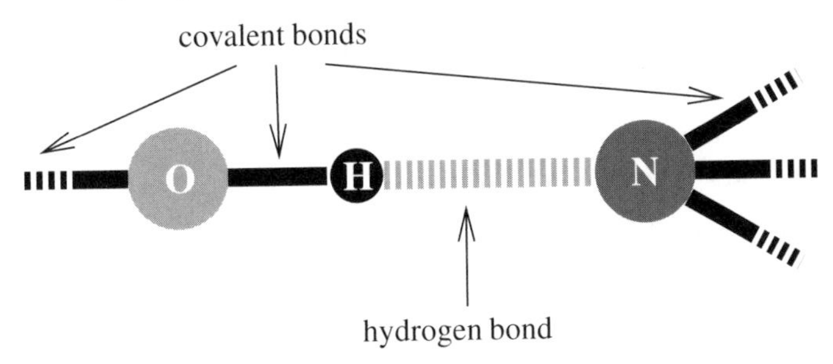

The hydrogen bond is caused by the polarity of the participating molecules. These bonds are in the range of 3–5 kJ/mole, and are easily disassociated by heat or certain chemicals. By comparison, a covalent C–C bond has 380 kJ/mole.

An important example of a polar molecule is the water molecule, which is shown in Figure 1.1. The electronegative oxygen attracts the electrons from both hydrogens, thus yielding electropositive and electronegative regions as depicted in the figure. Due to this polarity and the form of the region, water molecules are able to form hydrogen bonds with four other water molecules. At room temperature, 15% of the water molecules have (short-duration) hydrogen bonds with four other water molecules.

Other molecules that are polar can also form hydrogen bonds with water molecules. Since the formation of hydrogen bonds is energetically advantageous, polar molecules combine well with water and are therefore called *hydrophilic* molecules. On the other hand, non-polar

---

[4] A partial exception is silicon, which has some similar properties to carbon.

[5] Of course this does not hold for the vacuum, where the strength of an ionic bond is comparable to that of a covalent bond.

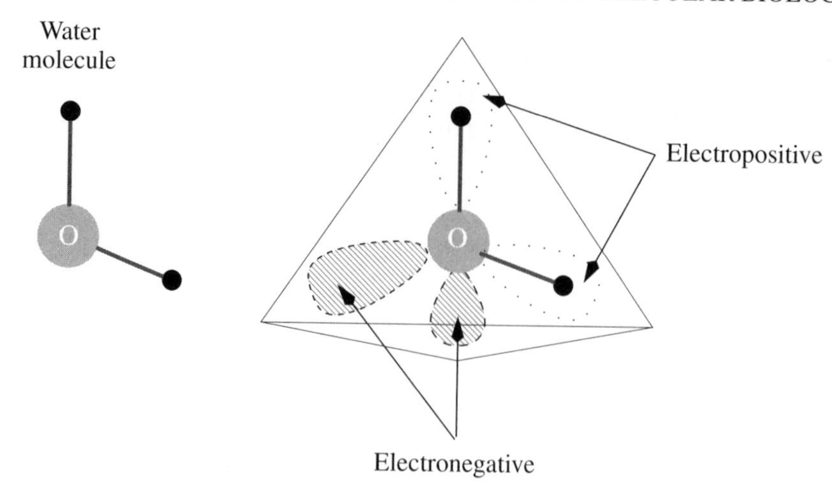

Figure 1.1 Polarity of the water molecule. The positive and negative charged regions, which span a tetrahedron, are shown in the left part.

groups cannot form hydrogen bonds. Since non-polar molecules disturb the network of water molecules connected via hydrogen bonds, such molecules do not combine well with water and hence are called *hydrophobic*. To minimize the effects of disturbing the (energetically favored) network of water molecules, hydrophobic molecules are forced together by the water. This force is called the *hydrophobic force*.

1.2 Small Molecules

A cell is made up of a small number of elements, where C, N, O and H make up 99% of the mass. A *hydrocarbon* molecule consists of only carbon and hydrogen. An *acid* (resp. *base*) has the property that in water it donates (resp. accepts) a free hydrogen ion, or proton, H^+. For instance, hydrochloric acid, HCl, disassociates in an aqueous solution to form H^+ and Cl^-, hence yielding the hydronium ion H_3O^+. The base sodium hydroxide, NaOH, disassociates in water to form Na^+ and OH^-.

Some specific simple groups such as the *hydroxyl group* $-OH$, the *carboxyl group* $-COOH$ (which characterizes an organic acid), and the *amino group* $-NH_2$ commonly occur in organic molecules (see Figure 1.2). Another important class of C–O compounds is generated by an *ester* linkage (see Figure 1.3). *Small molecules* are defined as organic molecules with up to 30 carbon atoms. A macromolecule, which is composed of a number of equal or similar smaller molecules (called *monomers*), is called a *polymer*. The most important polymers are DNA/RNA (composed of nucleotides) and proteins (composed of amino acids).

One distinguishes four groups of small molecules:

sugar: A sugar molecule is a small carbohydrate, where a *carbohydrate* has the generic formula $C_x(H_2O)_y$. Larger carbohydrates are called *polysaccharides*.

fatty acids: For example, components of lipid molecules, which make up the cell membrane.

nucleotides: Building blocks for DNA and RNA. The nucleotide ATP, adenosine

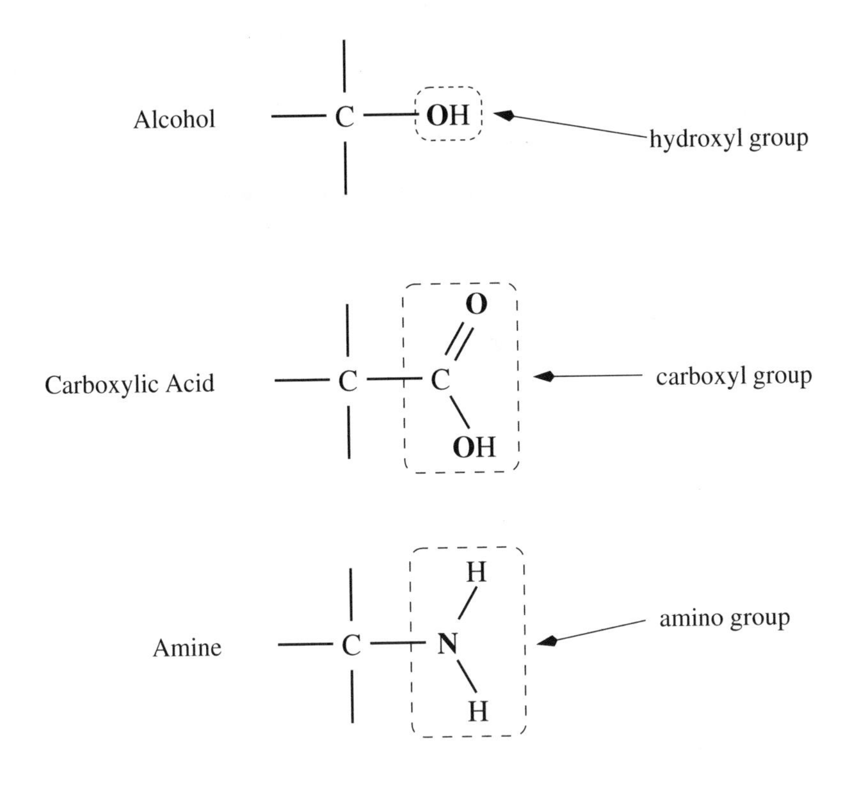

Figure 1.2 Groups of C–O and C–N compounds. Common groups are the hydroxyl group, the carboxyl group, and the amino group.

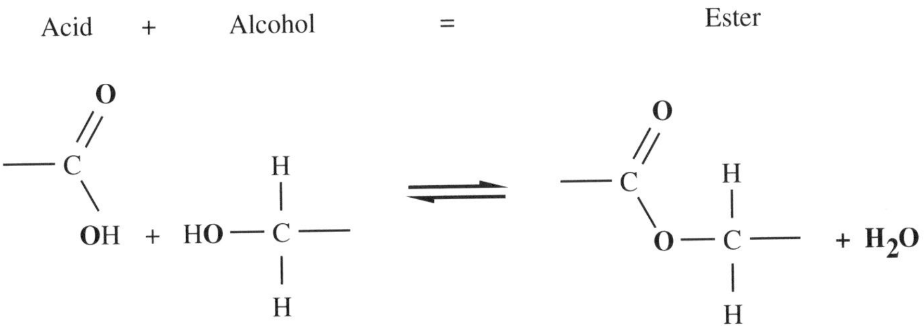

Acid + Alcohol = Ester

$$-C(=O)-OH + HO-CH_2- \rightleftharpoons -C(=O)-O-CH_2- + H_2O$$

Figure 1.3 An ester is the result of a reaction of an alcohol with an acid (such as carboxylic acid). The sign ⇌ indicates that the reaction can proceed in both direction. Proceeding from left to right, an ester and a water molecule are products, while from right to left, a water molecule is broken down, a process called *hydrolysis*.

triphosphate, is used for the storage and transportation of energy in cells, while GTP, guanine triphosphate, plays a role as neurotransmitter in brain.

amino acids: Proteins are composed of amino acids (where the amino acids are connected in a linear chain by a peptide bond).

1.3 Sugars

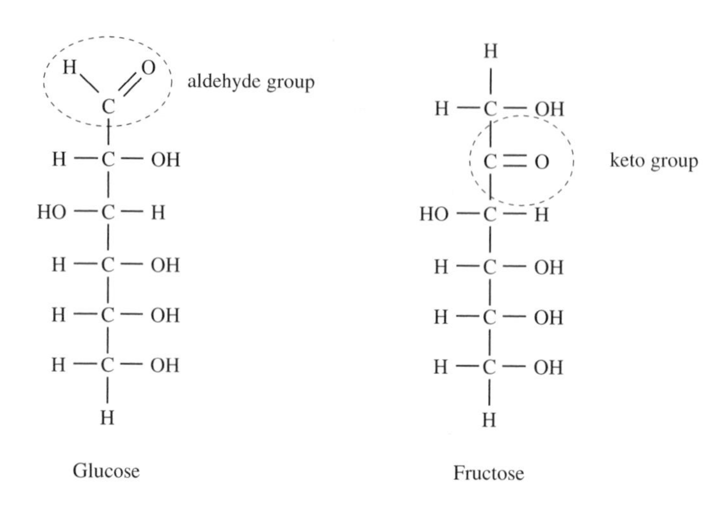

Figure 1.4 Two common 6-carbon sugars.

Sugars can be distinguished with respect to their number of carbon atoms *and* with respect to their molecular structure. Important classes with respect to the number of carbons are *pentoses* (having 5 Cs) and *hexoses* (having 6 Cs). With respect to structure, there are two forms of sugar, namely *aldehyde sugars* (also called *aldoses*) and *ketone sugars* (or *ketoses*). Figure 1.4 shows two common hexoses. Since the keto or aldehyde group can react with a hydroxyl group, a sugar can occur in cyclic form (see Figure 1.5). The carbons in a sugar are usually numbered, where one starts either with the aldehyde group, or with the end that is nearest to the keto group. Of particular significance for molecular biology is the 5-carbon sugar (or pentose) called *ribose*, which is a ketose. There are two different forms of ribose that occur, namely ordinary ribose, and *deoxyribose*, where one oxygen atom is missing (see Figure 1.6).

1.4 Nucleic Acids

1.4.1 Nucleotides

Both DNA and RNA are polymers, which are composed of nucleotides (nucleotide polymers are also called *oligonucleotides*). A *nucleotide* is a molecule consisting of a base, a ribose sugar (in DNA, deoxyribose), and a phosphate group. The base is bound to the $1'$ carbon, wherease the phosphate is bound to the $5'$ carbon (see Figure 1.7). A *nucleoside* consists of a base and a ribose sugar only. The *base* is a carbon ring molecule containing nitrogen atoms.

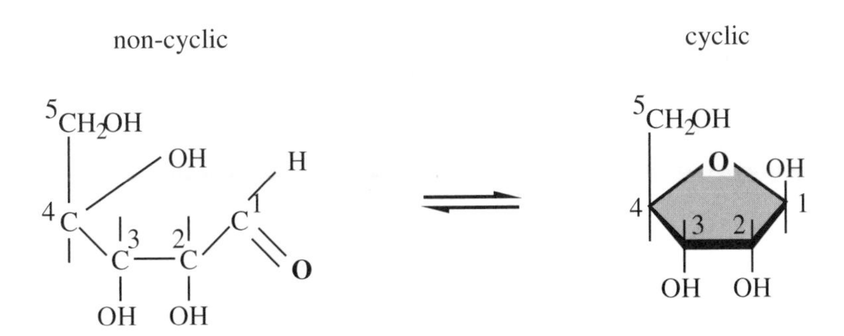

Figure 1.5 The cyclic and non-cyclic form of ribose. We have suppressed most of the H-atoms attached directly to C.

Figure 1.6 Ribose and deoxyribose.

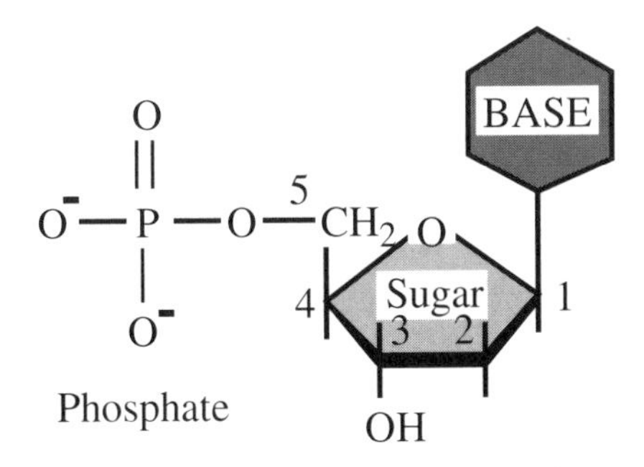

Figure 1.7 General picture of a nucleotide.

In DNA, we have the four bases *adenine* (A), *cytosine* (C), *guanine* (G), and *thymine* (T) (see Figure 1.8). In RNA, the thymine is replaced by *uracil* (U). The *purines* adenine and guanine have a 2-ring structure, and the *pyrimidines* cytosine, thymine and uracil have a 1-ring structure. Note that adenine and thymine have 2 hydrogen bond sites whereas cytosine and guanine have 3 hydrogen bond sites (see Figure 1.9).

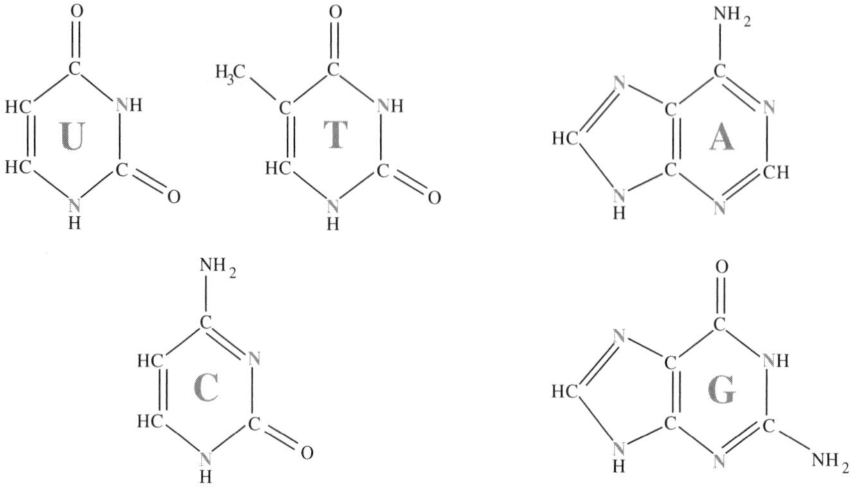

Figure 1.8 Chemical forms of the bases.

Nucleotides occur in three different forms, namely as *monophosphate*, *diphosphate*, and *triphosphate* nucleotides, having 1, 2, and 3 phosphate groups attached, respectively. The additional phosphate groups in diphosphate and triphosphate nulceotides can be split off, yielding energy that can be used for another process. Thus, the diphosphate and triphosphate nucleotides are used to transport energy in the cell.

1.4.2 DNA

At room temperature, DNA exists as a double-stranded molecule, formed by hydrogen bonds between *complementary* bases: *A with T, and C with G*, the so-called *Watson–Crick rules.*[6] These rules explain an experimental observation made by Chargaff in 1951 (two years before publication of the Watson–Crick model of DNA): adenine and thymine (resp. cytosine and guanine) appear in equal quantities in DNA.[7] In double-stranded B-DNA, the A-T and G-C base pairs are stacked in a planar fashion (with slight tilt) with about 3.4 Å per base pair.

A single strand of DNA is generated by chaining together nucleotides via a *phosphodiester bond.* In this bond, the phosphate molecule of the first nucleotide is attached to the hydroxyl group at the 3′ carbon of the next nucleotide (see Figure 1.10). The different ends of the DNA strand are numbered by the carbon atom position, where the next nucleotide can be attached.

[6] Francis H. Crick and James D. Watson received the 1962 Nobel Prize for Physiology or Medicine.

[7] According to [WHR+87], bacterial virus phage T2 lacks cytosine, but does contain a cytosine-like base that hydrogen-bonds to guanine.

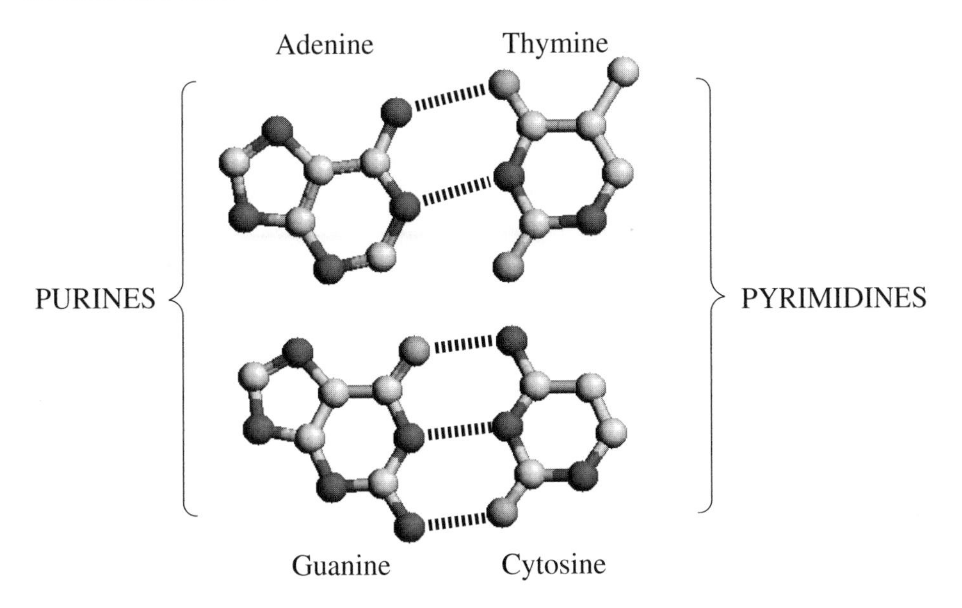

Figure 1.9 The four DNA bases together with their hydrogen bonds. The relative location of the bases is shown as they occur in DNA.

The $5'$ *end* contains a phosphate group, attached to the $5'$ carbon of the pentose sugar. The $3'$ *end* is characterized by the fact that the $3'$ position of the corresponding nucleotide is free for appending the next nucleotide.

Theoretically, there could be two directions for extending a single strand, namely by extending the $5'$ end, or by extending the $3'$ end. In nature, DNA is always extended at the $3'$ end, which implies that DNA grows in the $5' \rightarrow 3'$ direction. The reason is that the energy for the extension of the DNA is given by the new nucleotide containing a triphosphate group.

If the extension were in the $3' \rightarrow 5'$ direction, then the strand would have to provide the energy for extension (by having a triphosphate attached to the $5'$ end). But then it would be difficult to remove the last extended nucleotide from a strand, since this would remove the group providing the energy for extension. Thus, the error correction mechanism would be more difficult (which is one of the possible reasons why nature uses the $5' \rightarrow 3'$ direction for DNA).

Double-stranded DNA forms a *helix*, where one strand goes in the direction from $5'$ to $3'$, while the second goes from $3'$ to $5'$; thus the second strand is the *reverse complement* of the first strand. This mechanism allows one strand of DNA to serve as a template for producing the reverse complement strand, thus explaining how DNA can replicate.

In many viruses and bacteria, DNA exists in a circular form. For instance, the genome of *M. jannaschii* consists of one large circular chromosome consisting of 1 664 976 base pairs (bp), together with a large circular *extrachromosomal element*, or ECE, consisting of 58 407 bp, and one small circular ECE consisting of 16 550 bp. Locally, DNA is often tightly wound around *histone* proteins in a structure called a *nucleosome*, thus causing a bunching-up or banding, visible when using certain color stains under a light microscope – hence the name *chromosome*. Nucleosomes appear to be an essential structure, as manifested by the extreme conservation of the histone complex throughout organisms. On the order of 150–200 bp lie

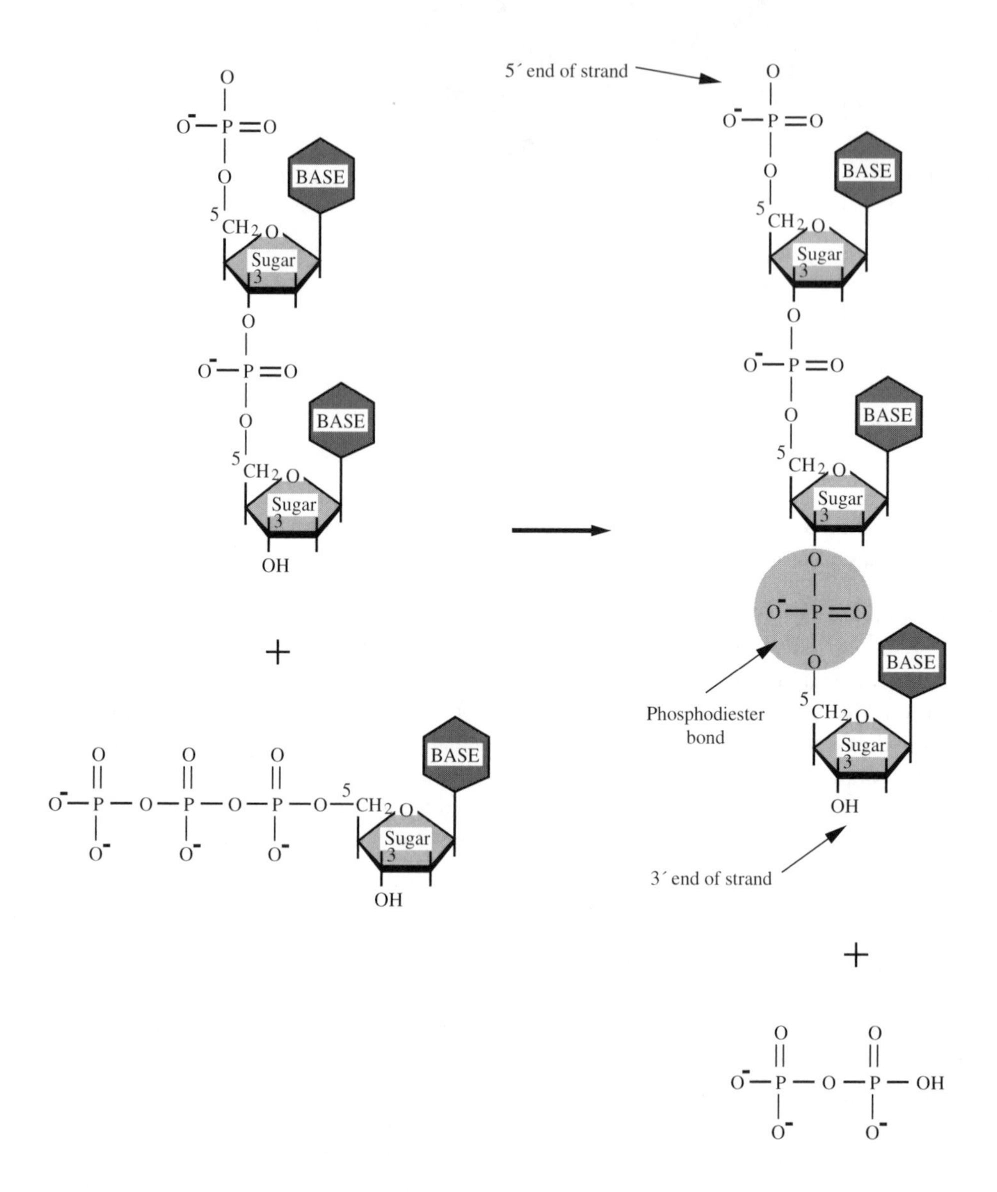

Figure 1.10 Extension of a single DNA strand, where the new nucleotides has a triphosphate group attached. The gray circle shows the $3' - 5'$ phosphodiester bond connecting two nucleotides.

Table 1.1 Nucleotide codes.

CODE	MEANING	COMPLEMENT
A	A	T
C	C	G
G	G	C
T	T	A
M	A or C	K
R	A or G	Y
W	A or T	W
S	C or G	S
Y	C or T	R
K	G or T	M
V	A or C or G	B
H	A or C or T	D
D	A or G or T	H
B	C or G or T	V
X/N	A or C or G or T	X
•	not A,C,G,T	•

on the nucleosome, forming about 2 superhelices. Moreover, the DNA within nucleosomes is more tightly wound – roughly 10 Å per helix, as contrasted with 10.4 Å per helix in B-DNA. In higher organisms, DNA may appear in linear chromosome pairs – one chromosome from each parent.

Circular DNA (and even linear DNA, since the latter can be locally considered circular in the region of a nucleosome) has *linking*, *winding*, and *twist* numbers. Think of an unbuckled belt that you twist several times and then buckle. Twist is the number of times you twisted the belt before buckling it. Now make a figure-of-eight with the belt, and notice how the twist number is reduced by increasing the writhe. Formal mathematical definitions of linking number (L), twist (T), and writhe (W) can be given, for which the important topological invariance

$$L = T + W \tag{1.1}$$

holds, a result due to J. White [Whi89, Whi95]. In a later chapter, we shall present a method due to C. Benham et al. [Ben90, SMFB95], which incorporates equation (1.1) to predict where double stranded DNA begins to separate in replication and transcription events. *In vivo*, DNA is usually negatively supercoiled, and it is known that certain enzymes, called *topoisomerases*, actually cut the DNA (topoisomerase I cuts a single strand while topoisomerase II cuts both strands).

When determining the entire nucleotide sequence, or *genome*, of an organism (such as *M. jannaschii* above), laboratory uncertainties lead at times only to a partial determination of certain bases – for instance, that either A or C occurs at a certain position. The nucleotide code displayed in Table 1.1 includes a symbol for each subset of $\{A, C, G, T\}$.

A segment of DNA that codes for a protein (as explained below) is called a *gene*. In *diploid* cells, where chromosomes donated from each parent are paired, the corresponding genes may be identical (*homozygous*) or non-identical (*heterozygous*). In the latter case, the corresponding, but slightly different genes are called *alleles*. Heterozygous alleles may be *dominant* or *recessive*.[8] Non-diploid germ cells are called *haploid*. These are formed by *meiosis*, where *crossing-over* between corresponding chromosomes happens randomly. In meiosis (germ cell formation), the cell's chromosomes are duplicated (with possible crossing-over), but the cell quadruples, leading to 4 haploid cells. This is contrasted with *mitosis*, where the cell's chromosomes are duplicated (without crossing-over) and the cell is then doubled.

While the human genome consists of 23 pairs of chromosomes amounting to a total of roughly 3×10^9 base pairs, biologists estimate that there are only 10^5 genes. From the genetic code, explained later, there are 3 bases that correspond to an amino acid, and most proteins consist of several hundred amino acids. Assuming that a protein consists of at most 1000 amino acids, a back-of-the-envelope calculation indicates that fewer than $3 \cdot 10^5 \cdot 10^3 = 3 \cdot 10^8$ bp of the genome correspond to genes, while $\frac{3\times10^9 - 3\times10^8}{3\times10^9} \approx 90\%$ of the genome is non-coding.

Coding regions encode proteins, as explained in Section 1.6.2. In general, the percentage of the genome consisting of coding regions *decreases* with increasing complexity of life. For instance, in the *polyoma* virus genome, consisting of 5297 base pairs, coding regions comprise over 90% of the genome [Ben93]. Certain viruses even encode two distinct proteins within roughly the same genomic region by a shift in the reading frame.[9] Indeed, for lower organisms, the reproduction rate depends essentially only on the DNA replication time, which in turn depends on genome length (bacteria in ideal conditions can reproduce every 20–30 minutes). However, in eukaryotes, genomic length seems to have little to do with either the complexity of the organism or time between mitosis.

The so-called TATA box is a consensus sequence consisting of TATA, or close relative thereof, which is part of the promotor sequence for a gene. The proximity of a TATA box upstream from the beginning of coding region is an indicator for a coding region, as well as a terminating hairpin structure and poly-A region at the end. A-T has two hydrogen bonds as compared with the three hydrogen bonds of G-C, so poly-A regions are less stable, and especially after a hairpin structure can lead to termination of transcription. An *open reading frame* or ORF is a region thought to encode a protein, but for which the functionality is not currently known.

Humans have 22 pairs of *autosomal*, or non-sex-related chromosomes, together with either an X,Y pair (in males) or a X,X pair (in females). DNA is a very thin (width about 10 Å) and long molecule (human DNA, if concatenated from all chromosomes and stretched out, would measure about 2 meters[10]). Generally, more complex organisms have longer DNA, though: for instance, the lungfish genome is almost 50 times as large as the human genome, and the *Amoeba dubia* genome is over 600 Mbp. Some sample genome sizes:

- Polyoma virus: one circular chromosome of size 5297 bp.
- *λ-phage virus*: one circular chromosome of size 48 502 bp.
- *E. coli*: one circular chromosome of size 4.6×10^6 bp.
- *S. cerevisiae* (yeast): 16 linear chromosomes of total size 13×10^6 bp.
- *Homo sapiens*: 23 pairs of linear chromosomes of total size 3×10^9 bp.

---

[8] See Exercise 14 at the end of Chapter 2, following work of P. Pudlák [PP97].

[9] Reading frame is explained later in Section 1.6.1.

[10] Multiplying 3×10^9 bp times 3.4 Å per bp gives a rough calculation of the length.

1.4.3 RNA

Ribonucleic acid (RNA) consists of the bases adenine, cytosine, guanine, and uracil (in place of thymine) attached in a similar manner to the pentose sugar ribose. As shown in Figure 1.6, ribose has an extra *hydroxyl* (OH) group at the $2'$ position, and so is able to form more hydrogen bonds than DNA. In deoxyribose, this oxygen atom is missing, hence the name *de*oxyribose. Because of the ability to easily form hydrogen bonds, RNA manifests a number of catalytic properties, including self-splicing activity in certain cases.[11] It is generally believed that an RNA world first existed [PJP98], where RNA played both an information storage and an enzymatic role.

It is known that cytosine spontaneously deaminates to uracil, and since there is no known RNA repair mechanism (unlike the various repair mechanisms of DNA, such as SOS, etc.), RNA cannot be reliably used as a carrier of genetic information. In replicating DNA, uracil is *always* removed. Moreover, it is speculated that this spontaneous deamination is the mechanism for one form of RNA editing, to be explained later. Thus it is believed that, once it evolved, the more stable DNA gradually replaced RNA in its information storage role, while enzymatic functions of RNA were gradually replaced by more efficient proteins. There are certain enzymatic functions still carried out by RNA. It is thought that such instances are *molecular fossils*.

In contrast to DNA, RNA is single-stranded, though it can can form Watson–Crick hydrogen bond pairs (A-T, C-G) and even the weaker pairs A-G with itself, thus forming hairpin loops and more complicated structures.

Though DNA serves only the function of information storage, RNA serves certain catalytic functions through its complex 3-dimensional forms. Before describing transcription and translation, here is an overview of the different kinds of RNA (undefined concepts are explained in the next section):

1. *mRNA*: messenger RNA, transcribed from DNA where introns have been spliced out and pointwise editing performed.
2. *tRNA*: transfer RNA generally consisting of 70–80 bases having a cloverleaf secondary struction, and an L-shaped tertiary structure. The anticodon occurs at the end of a hairpin loop. The corresponding amino acid is bonded to the tRNA by a reaction involving aminoacyl-tRNA synthetase.
3. *rRNA*: ribosomal RNA, main component of the ribosomes. Mammals have on the order of 10^6 ribosomes per cell. Sometimes rRNA is described using Svedberg units (denoted by S), which are sedimentation rates in an ultracentrifuge; for instance, 16 S rRNA.
4. *snRNA*: small nuclear RNA, a molecular fossil [WHR$^+$87] which assists in splicing nuclear mRNA.
5. *gRNA*: guide RNA, used to control the editing of mRNA (pointwise nucleotide insertions or deletions in RNA).

---

[11] In 1981, *Tetrahymena* r-RNA self-splicing was first observed.

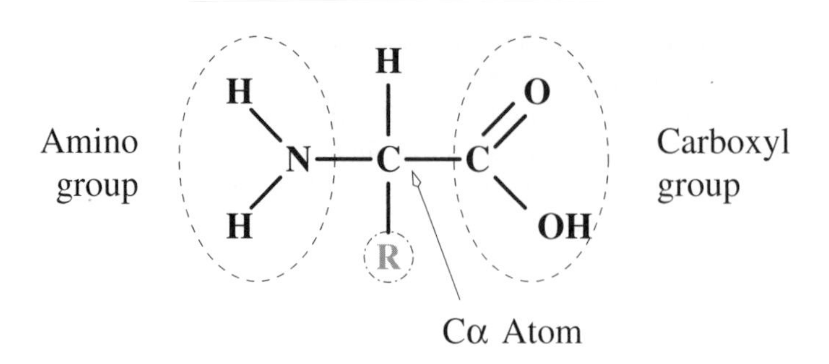

Figure 1.11 General form of an amino acid.

1.5 Proteins

1.5.1 Amino Acids

Proteins are sequences composed of an alphabet of 20 amino acids. An amino acid is a chemical group of the form given in Figure 1.11, where R is a chemical group (called *chain residue*) specifying the type of the amino acid. The central carbon atom is the *α carbon* (Cα), the left NH_2 group is the *amino group*, and the right COOH the *carboxyl group*. There are 20 different chain residues, which have different chemical properties. Residues can be hydrophobic or hydrophilic, small or large, charged or uncharged.

Two amino acids can be connected via a *peptide bond*, where the carboxyl group of the first amino acid reacts with the amino group of the second. The result is a group of the form

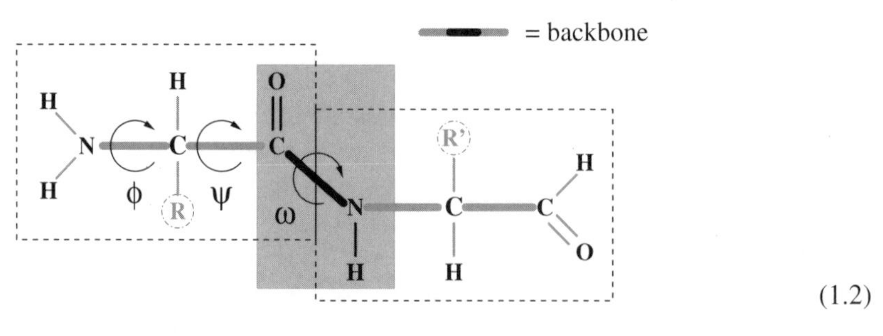

(1.2)

Using the peptide bond, long linear chains of amino acids (i.e., proteins) can be generated. The ends are characterized by the fact that they have an amino group (resp. carboxyl group) that is not part of the peptide bond. Thus, one speaks of the *N terminus* (resp. *C terminus*) of the amino acid chain.

The peptide bond itself (indicated with a gray rectangle in (1.2)) is usually planar, which means that there is no free rotation around this bond.[12] There is more flexibility for rotation around the N-Cα-bond (called the ϕ-angle), and around the Cα-C bond (called the ψ-

[12] There are two possible conformation rotation angles ω of the peptide bond, namely *trans* (corresponding to a rotation angle of 180°) and *cis* (corresponding to a rotation angle of 0°). The *cis* conformation is rare and occurs usually in combination with a specific amino acid, namely proline (which is in fact an imino acid, i.e., both ends of proline are connected with the backbone).

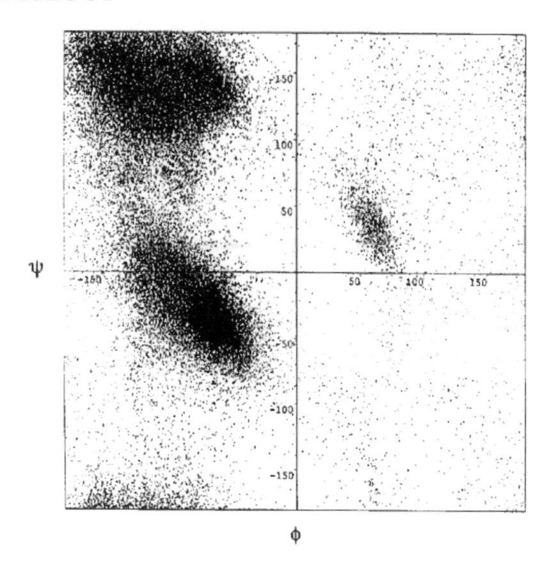

Figure 1.12 Ramachandran plot for 310 proteins. Every dot corresponds to a (ϕ, ψ) pair of one amino acid (except proline and glycine amino acids). Picture taken from [Cre92], page 60.

angle). Nevertheless, the allowed values of combinations of ϕ and ψ angles are restricted to small regions in natural proteins (which are displayed on so-called Ramachandran plots, see Figure 1.12). Using this freedom of rotation, the protein can *fold* into a specific three-dimensional structure (called conformation). In natural proteins, the final structure that is achieved is uniquely determined by the sequence of amino acids. For this reason, one speaks of the *native structure* of a given amino acid sequence. The native structure of a protein is believed to be the conformation that is a free energy global minimum. Terms in the energy function include van der Waals attraction, electrostatic (Coulomb) forces, hydrogen bonds between different residues (e.g. between H bound to N and the O in polypeptides; such bonds are responsible for the secondary structures α helices and β strands), salt bridges (ionic attractions), disulfide bridges (disulfide bond between two cysteines), and hydrogen bonds with the solvent water.

1.5.2 Protein Structure

The function of a protein is determined by its 3-dimensional structure – for instance, antibodies in the human immune system recognize antigens by having a complementary surface to that of the antigen (*key and lock* paradigm). Enzymes are proteins, which, by forming a substate to lower the reaction energy, facilitate or accelerate chemical reactions, which would otherwise might never take place. Because of the importance of the immune system, immunoglobulins have been widely studied, and hence form a large portion of current protein databases.[13]

---

[13] It is sometimes necessary, as in Sippl's polypeptide folding algorithm discussed in a later chapter, to obtain an unbiased sampling of the protein database, by requiring that any two proteins from the sample have, for instance, at most 25% homology. Such unbiased sampling is supported by *PDB select*, where for instance `PDBselect 30` produces a sample of proteins that pairwise share no more than 30% homology.

Proteins may roughly be divided into three classes: *globular* (enzymes), *fibril* (collagen, elastin), and *membrane* proteins. Transmembrane proteins, responsible for pores/channels in cell membranes, have in part been much less studied because of technical difficulties in their X-ray crystallography. The Brookhaven Protein Database (PDB) contains the 3-dimensional coordinates of 10 310 proteins[14] and 788 nucleic acids, as determined by X-ray crystallography or nuclear magnetic resonance (NMR).[15] See Appendix B.2 for a more detailed description of an example PDB file.

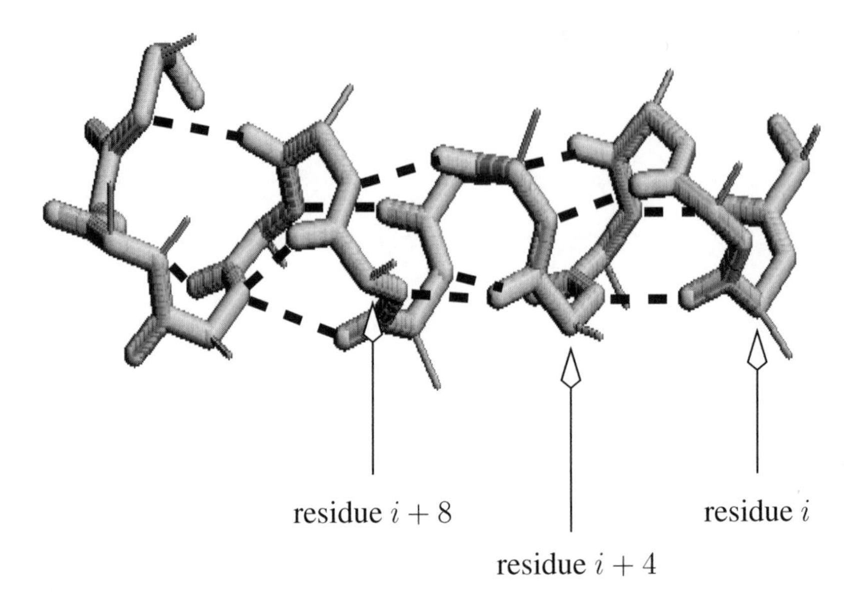

Figure 1.13 An α helix. The stabilizing hydrogen bonds are shown with dashed lines.

Concerning the 3-dimensional structure of proteins, there are some architectural features that are commonly used in natural proteins. These structural pieces are called *secondary structure elements*. The two main motifs are α helices and β sheets. An α helix is generated by stacking amino acids in a helix, forming a fixed cylinder (see Figure 1.13). Note that residues $1, 5, 9, \ldots$ appear aligned, with residue $n+4$ stacked above residue n in the enclosing cylinder. A β *sheet* is a structure where parts of the amino acid chain are stacked onto each other in a linear manner (see Figure 1.14). Since a protein has a direction (from the N terminus to the C terminus), one distinguishes *parallel* β *sheets* and *antiparallel* β *sheets*. In the parallel form, the folded parts have the same direction, whereas in the antiparallel form, they have alternating directions.

Certain combinations of α helices and β sheets have been identified as *motifs* (sometimes called *supersecondary structures*) common to several proteins. An example is the helix–turn–helix (H-T-H) motif, often signaling a DNA-binding site.

PDB select files can be retrieved by anonymous FTP at `ftp.embl-heidelberg.de` in the directory `/pub/databases/protein_extras/pdb_select`.

[14] This figure includes proteins, peptides and viruses.

[15] Figures as of February 2000.

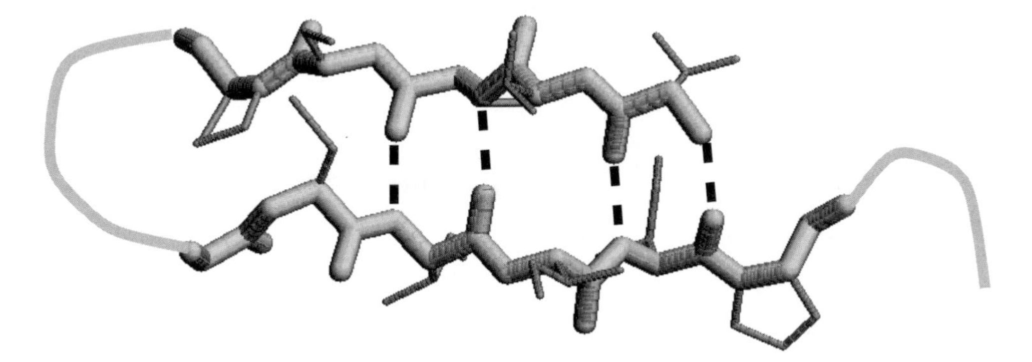

Figure 1.14 An antiparallel β sheet. The stabilizing hydrogen bonds are shown with dashed lines.

The structure of a protein is described on different levels:

primary structure: the amino acid sequence of the protein.

secondary structure: describes the regions in the primary structure where secondary structure elements (such that α helix, β sheets, etc.) occur.

tertiary structure: the 3-dimensional structure of a protein domain in the native structure. If a protein is dimeric (i.e., is composed of several subproteins), then the tertiary structure describes the structure of a subprotein.

quarternary structure: the 3-dimensional, native structure of the fully functional protein.

1.6 From DNA to Proteins

1.6.1 Amino Acids and Proteins

Amino acids have numerous different chemical properties, including molecular size, electric charge, etc. It is held that one of the most important driving forces in protein folding is the *hydrophobic force*, or tendency for hydrophobic residues to avoid contact with water molecules and hence form a compact inner core in the protein. *Salt bridges* (attractions between oppositely charged residues) and *disulfide bonds* between the sulfur atoms of cysteine residues also play a role in the conformation determination. The 20 amino acids found in proteins are listed in Table 1.2, along with their 1-letter and 3-letter codes, together with a hydrophobicity value technically called *polar requirement* [WDD$^+$66] and the designation 'H' (hydrophobic) or 'P' (polar, i.e., hydrophilic).

The genetic code is now known to be a triplet, non-overlapping, *comma-free*[16] code, where successive *codons* consisting of 3 successive RNA nucleotides encode one of the 20 amino acids or the signal to stop translation. For instance, using Table 1.3, the oligonucleotide

```
1  2  3  4  5  6  7  8  9  10 11 12
G  U  U  U  U  A  A  G  U  C  C  U
```

[16] Comma-free means that there is no code for a punctuation mark between two codons. Thus 6 successive nucleotides code for 2 successive amino acids.

Table 1.2 Amino acid codes.

AMINO ACID	CODE (1 ch)	CODE (3 ch)	POLAR REQ	H/P
Alanine	A	Ala	7.0	H
Arginine	R	Arg	9.1	P
Asparagine	N	Asn	10.0	P
Aspartic acid	D	Asp	13.0	P
Asparagine or aspartic acid	B	Asx		P
Cysteine	C	Cys	4.8	P
Glutamine	Q	Gln	8.6	P
Glutamic acid	E	Glu	12.5	P
Glutamine or glutamic acid	Z	Glx		P
Glycine	G	Gly	7.9	P
Histidine	H	His	8.4	P
Isoleucine	I	Ile	4.9	H
Leucine	L	Leu	4.9	H
Lysine	K	Lys	10.1	P
Methionine	M	Met	5.3	H
Phenylalanine	F	Phe	5.0	H
Proline	P	Pro	6.6	H
Serine	S	Ser	7.5	P
Threonine	T	Thr	6.6	P
Tryptophan	W	Trp	5.2	H
Tyrosine	Y	Tyr	5.4	P
Valine	V	Val	5.6	H

codes for the amino acid sequence `Val Leu Ser Pro` if the *reading frame* begins at position 1, while it codes for `Phe Stop` if the reading frame begins at position 2, and `Phe Lys Ser` if the reading frame begins at position 3. To complicate matters, in double-stranded DNA there are three additional reading frames possible for the reverse complement strand. In prokaryotes and eukaryotes, the reading frames of different proteins do not overlap. This is different in some viruses, where the need to compactify the genetic information may lead to a situation where different reading frames for viral proteins overlap greatly.

There is much interesting speculation about the origin of the genetic code (see [HH91, FH98, KSH95, ELT+89]). It is thought that in the protobiotic RNA world, different genetic codes may have existed, in which case pressure from natural selection would have optimized both the codes and translation machinery, until, as F. Crick has suggested, a *frozen accident* occurred with the emergence of life, meaning that further random mutations would most likely be detrimental to living organisms. Two rather different views for the emergence of the genetic code have been proposed. In the late 1960s, Crick suggested that few amino acids were initially encoded, that with time, more amino acids were encoded using similar codes to those that then existed, and that tRNA and aminoacyl-tRNA synthetases were derived from those already existent. In 1973, C. Woese *et al.* [WDD+66] proposed that stereochemical affinities between codons and respective amino acids exist, that early translation mechanisms were imprecise, and that ancestors of tRNA were capable of recognizing only classes of codons, rather than unique codons. In particular, Woese *et al.* introduced a measure of hydrophobicity of amino acids, and noted that amino acids having A (resp. U) as middle codon are hydrophilic

Table 1.3 Genetic code.

	U	C	A	G	
U	Phe	Ser	Tyr	Cys	U
	Phe	Ser	Tyr	Cys	C
	Leu	Ser	Stop	Stop	A
	Leu	Ser	Stop	Trp	G
C	Leu	Pro	His	Arg	U
	Leu	Pro	His	Arg	C
	Leu	Pro	Gln	Arg	A
	Leu	Pro	Gln	Arg	G
A	Ile	Thr	Asn	Ser	U
	Ile	Thr	Asn	Ser	C
	Ile	Thr	Lys	Arg	A
	Met	Thr	Lys	Arg	G
G	Val	Ala	Asp	Gly	U
	Val	Ala	Asp	Gly	C
	Val	Ala	Glu	Gly	A
	Val	Ala	Glu	Gly	G

(resp. hydrophobic), while the nucleotide adenine (resp. uracil) is chemically hydrophobic (resp. hydrophilic). As later explicitly stated in [HH91], middle anti-codon positions with A (resp. U) are thus associated with hydrophobic (resp. hydrophilic) amino acids. It has also been suggested by M. Eigen and P. Schuster that the original code was of the form RNY, meaning purine, followed by any nucleotide, followed by pyrimidine.

1.6.2 Transcription and Translation

A segment of a chromosome, corresponding to a *coding region* (a region that gives rise to a protein), is often preceded by a *promotor* sequence, usually consisting of a rough consensus pattern of TATAA ... (called a TATA box) as well as other sequence signals. RNA polymerase is capable of recognizing the promotor sequence, and after DNA strand separation in the coding region occurs, RNA polymerase assists in the *transcription* of DNA into a pre-edited form of messenger RNA called the RNA *transcript*. The transcription occurs in the $5'$ to $3'$ direction, where the DNA nucleotides $\mathrm{A}, \mathrm{C}, \mathrm{G}, \mathrm{T}$ are respectively transcribed into RNA nucleotides $\mathrm{U}, \mathrm{G}, \mathrm{C}, \mathrm{A}$.

There are different models for the transcription and translations in prokaryotes (i.e., bacteria), and eukaryotes (non-bacteria; e.g. unicellular paramecia, and all plants and animals). *Prokaryotes* are cells that have no compartment to separate genomic DNA from the cytoplasm[17] and no membraneous organelles, and have one circular chromosome. The reading

[17] The cytoplasm is the cell contents (including organelles) within the plasma membrane, without the nucleus in the case of eukaryotes.

frame for the transcription of a proteins is continuous, i.e., not interupted. *Eukaryotes* on the other hand, do have a compartment, called a *nucleus*, to separate DNA from the cytoplasm, have membraneous organelles (e.g. mitochondria), and generally have more than one linear chromosome. The reading frame for a protein is not continuous, but rather is disrupted by *introns* (or *intervening sequences*). Phillip A. Sharp and Richard J. Roberts discovered introns in 1977, for which they later later received the Nobel Prize. The remaining segments between the introns are called *exons*. Either by self-catalysis, or facilitated by enzymes, the introns in the RNA transcript are excised and removed (see Figure 1.15). In proteins, a similar splicing occurs, where the removed segments are called *inteins* and the remaining segments *exteins*. Moreover, it is an important algorithmic problem in bioinformatics to determine intron/exon splice sites.

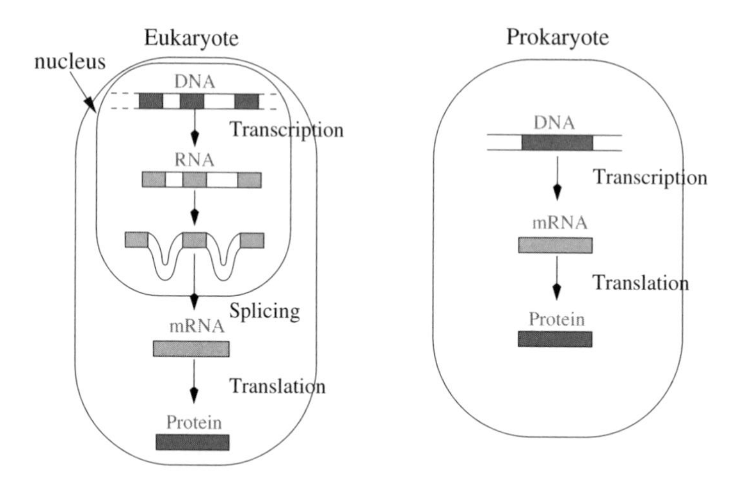

Figure 1.15 The different models of transcription and translation in prokaryotes and eukaryotes.

After removal of introns,[18] it may occur that nucleotides are inserted or deleted (pointwise, non-consecutive insertions and deletions). This phenomenon is called *RNA editing*, and, at least in the case of kinetoplastid protozoa [vHBS$^+$92], is directed by *guide RNA* (gRNA). That editing occurs in most living organisms is truly remarkable, since the protein coded by mRNA after editing may have nothing to do with the protein that would have been coded by mRNA before editing occurred. This is because the genetic code is a triplet, non-overlapping code, and an insertion of one nucleotide effectively shifts the original reading frame, producing codes for possibly completely unrelated amino acids.

Transfer RNA (tRNA), whose existence was first postulated by F. Crick in his *adapter molecule* hypothesis, is a small (around 70 bp) RNA, having a cloverleaf secondary structure and an L-shaped tertiary structure. At the extremity of the middle lobe of the cloverleaf lies an *anticodon*, i.e. three nucleotides, whose Watson–Crick reverse complement is a triplet codon for an amino acid, as given in Table 1.3. *Aminoacyl-tRNA synthetase* assists in the attachment

[18] Mammalian mitochondrial DNA and generally prokaryotic DNA have no introns.

of the *correct* amino acid to tRNA, i.e. the amino acid whose codon is the reverse complement of the anticodon in the tRNA. Despite the fact that, with the exception of mitochondria and certain bacteria, the genetic code is essentially universal, different organisms may prefer the usage of certain codons for particular amino acids. In particular, an organism may have no tRNA for particular codon, or may produce no aminoacyl-tRNA synthetase for that codon. For instance, leucine aminoacyl-tRNA synthetase was not found in *M. jannaschii*, as reported in [BWO$^+$96].

Ribosomes are cytoplasmic complexes constituted from ribosomal RNA (rRNA) together with certain proteins, and form a surface for mRNA (after intron removal and possible editing) and tRNA to assemble amino acids together in a linear chain, thus forming a protein. This process is called *translation*. Eukaryotic translation usually begins with the codon for the amino acid *methionine*, which serves as a start codon. Methionine may later be removed from a functional protein.

It is known that certain proteins, called *chaperones*, are found in the vicinity of ribosomes, and are responsible for refolding incorrectly folded proteins, or ensuring their destruction, when correct refolding does not occur. Chaperones recognize incorrectly folded proteins, because of hydrophobic[19] residues on the protein's surface (rather than in the core). At least for one particularly well-studied class, chaperones are known to have a cylindrical 3-dimensional structure with hydrophobic-rich interior walls. When a protein with a hydrophobic surface lies within the chaperone's walls, the chaperone undergoes a conformational change causing the inner walls to become hydrophilic and the 'lid' of the cylinder to clap shut, thus providing an inert 'test tube' medium, allowing the protein to refold correctly.

DNA replication occurs in a similar fashion to RNA transcription, but facilitated with DNA polymerase, as opposed to RNA polymerase. After DNA strand separation begins to occur (i.e. as the 'zipper' proceeds to unzip), the adjoining of complementary nucleotides to the template strand always occurs in the $5'$ to $3'$ direction, because of the high-energy phosphate group at the $5'$ end of the nucleotide. This occurs without major difficulty along one template strand, but is more complicated along the other template strand, since the $5'$ direction lies at the border between the 'zipped' and 'unzipped' portion of the partially separated double strands. The replication error rate along this latter strand is significantly higher. Various correction mechanisms for DNA replication, including the SOS system, reduce the final replication error rate to less than 1 in 10^9.

The above description of DNA replication holds for *mitosis* (cell division) as well as *meiosis* (germ cell formation). In mitosis, though DNA can in principle be replicated forever (eternal life), the accumulation of replication errors is thought to lead to dysfunctional cells, cancerous cells, etc. In meiosis, in addition to *pointwise mutation* caused by DNA replication errors as just explained, another source of mutation, essential for evolution, consists of *inversions* and *transpositions* as well as *crossover*.

1.7 Exercises

1. Use the software *RasMol* to view DNA and certain proteins and protein motifs. Procure a PDB file of a TATA box in the promotor sequence in DNA. Note the bend

---

[19] Hydrophobic amino acid residues, as contrasted with hydrophilic (polar) amino acid residues, avoid contact with water molecules and tend to lie buried in a protein's core.

in the DNA at this point, thus allowing RNA polymerase to recognize transcription origin.

2. Write an efficient program to recognize *palindromes* within a genome, and test it on the genome of *M. jannaschii*. In computer science, a palindrome is a word of even or odd length in a finite alphabet, which reads the same forwards as backwards. In biology, a palindrome generally means a word of even length in the alphabet A,C,G,T such that, the ith character from the beginning can form a Watson–Crick pair with the ith character from the end of the word.
3. Go to the Protein Data Base (PDB, `http://www.rcsb.org/pdb/`). Search for a structure of the human hemoglobin molecule. What is the PDB code? What is the classification of this protein in the CATH (Class, Architecture, Topology, and Homologous superfamily) hierarchy?

   Retrieve the structure in RasMol format. View the heme group (which is the group enclosing the iron atom) in red.

Acknowledgments and References

The background in molecular biology presented in this chapter was drawn from Linus Pauling's classic treatise [Pau70] on general chemistry, along with the classic textbook on molecular biology and genetics by J.D. Watson *et al.* [WHR+87, WTK83], the classic textbooks on molecular biology of the cell by B. Alberts *et al.* [ABL+94] and the biochemistry text by Moran *et al.* [MSH+94]. Some facts were drawn from [Nei87] as well. The pictures of the secondary structure elements were generated using the RasMol software package [SMW95].

2

Math Primer

Probability theory is nothing but 'common sense reduced to calculation'.
Laplace[1]

An important aspect of computational biology is the development of mathematical models for biological processes, together with the design of efficient and sensitive algorithms. Biological processes depend on random events, diffusion, etc. so that mathematical models are often stochastic. In this chapter, we review fundamental mathematical concepts necessary for bioinformatics – namely concepts from probability theory, information theory (entropy), and combinatorial optimization techniques.

2.1 Probability

Ants and wasps diverged from a common ancestor roughly 100 million years ago (100 My). Despite physical similarities such as abdominal poison glands, ants are social insects living in a hierarchically organized manner in nests of 200 000 to millions of individuals, whereas certain wasps (such as *Cynipidae* and *Sphecoidae*) are solitary. Ants seem to display a cooperative, altruistic behavior, putting the welfare of the group above that of self. Why? A commonly advanced theory is that, on average, worker ants share $\frac{3}{4}$ of their sisters' genes, whereas wasps, on average, share only half their siblings' genes. Let us use this example as a starting point in recalling some basic concepts of elementary probability theory.

Suppose that a female wasp has the alleles A and B of a given gene, while a male wasp has the alleles a and b. When male and female wasps mate, the possible allele combinations of female AB and male ab are Aa, Ab, Ba, and Bb. Suppose that one offspring (Fred) has allele combination Aa (all other cases are analogous). A second offspring (Ethel) has $\frac{1}{4}$ chance of being any one of the four previously listed combinations. The *average* number of shared genes is $\frac{2+1+1+0}{4} = 1$, where in each of the four possible cases we have the following number of shared genes: Ethel is Aa (2), Ethel is Ab (1), Ethel is Ba (1), Ethel is Bb (0). Since on average two sibling wasps share 1 gene, out of a total genome of 2 genes, they share 50% of their genes.

Consider, in contrast, what happens when a queen ant Aa mates with a drone BB', where B' is identical to B.[2] The possible allele combinations of *worker ants* are AB, aB, AB', and aB'. Suppose one offspring (Lucy) has allele combination AB (the other cases are

---

[1] Pierre Simon Marquis de Laplace, French mathematician and physicist, 1749–1827.

[2] An unfertilized ant egg develops into a male drone ant.

analogous). The average number of genes shared between Lucy and a second offspring (Ethel) is $\frac{2+2+1+1}{4} = 1.5$, where in each of the four possible cases we have the following number of shared genes: Ethel is AB (2), Ethel is AB' (2), Ethel is aB (1), Ethel is aB' (1). Thus on average two sibling worker ants have $\frac{6}{4} = 1.5$ genes in common, and hence share $\frac{1.5}{2} = 75\%$ of their genes.

Let us now recall standard notions from probability theory. The *frequency* of event A in an experiment consisting of n trials is $n(A)/n$, where $n(A)$ is the number of times A occurred in the n trials. A *sample space* Ω is a non-empty set of mutually exclusive *elementary events* e; for instance, in $\Omega = \{H, T\}$, the elementary events are 'heads' and 'tails', while in $\Omega = \{1, 2, 3, 4, 5, 6\}$, the elementary events are the outcome of the roll of a die.

Suppose that Ω is a *discrete* sample space, i.e., Ω is either finite or countably infinite (in one-to-one correspondence with the set $\mathbb{N} = \{0, 1, 2, \dots\}$ of natural numbers). In this case, an *elementary probability function* p on Ω is a map from Ω into the unit interval $[0, 1]$, such that $\sum_{e \in \Omega} p(e) = 1$. For $A \subseteq \Omega$, define $Pr[A] = \sum_{e \in A} p(e)$. Clearly, the following properties are satisfied:

1. $Pr[\emptyset] = 0$, $Pr[\Omega] = 1$.
2. $Pr[\Omega - A] = 1 - Pr[A]$.
3. $Pr[A \cup B] = Pr[A] + Pr[B] - Pr[A \cap B]$, for all $A, B \subseteq \Omega$. This is called the *addition law*.

The probability of event $e \in \Omega$ is then $Pr[\{e\}] = p(e)$.

In the general case, where Ω may be a discrete or continuous sample space, a *probability function* Pr is a mapping from the power set $\mathcal{P}(\Omega)$ of all subsets of Ω into the unit interval $0, 1]$ that satisfies the previous three conditions.

The expression $\binom{n}{m} = \frac{n!}{m!(n-m)!}$ counts the number of ways of choosing m element subsets of an n element set. These are the *binomial coefficients* ; i.e. $\binom{n}{m}$ is the coefficient of x^m in the expansion of $(1 + x)^n$. Taking $x = 1$ in this expression yields

$$\sum_{i=0}^{n} \binom{n}{i} = 2^n.$$

Moreover, for $\lambda \leq \frac{1}{2}$,

$$\sum_{i=0}^{\lambda n} \binom{n}{i} \leq 2^{h(\lambda)n},$$

where $h(\lambda) = -\lambda \log_2(\lambda) - (1 - \lambda) \log_2(1 - \lambda)$ is the *entropy* of the probability distribution $(\lambda, 1 - \lambda)$. The *multinomial coefficients* generalize this, in that there are

$$\frac{n!}{n_1! n_2! \cdots n_k!} \tag{2.1}$$

many ways of partitioning n into k classes, where the ith class has n_i elements, so that $n = n_1 + \cdots + n_k$.

THEOREM 2.1 (STIRLING'S FORMULA)

$$n! \sim \sqrt{2\pi n} \left(\frac{n}{e}\right)^n,$$

where $f \sim g$ *means* $\lim_{n \to \infty} \frac{f(n)}{g(n)} = 1$.

For $A, B \subseteq \Omega$, we often write $Pr[A \cap B]$ as $Pr[A, B]$.

DEFINITION 2.2 (CONDITIONAL PROBABILITY)
$Pr[A|B] = \frac{Pr[A,B]}{Pr[B]}$. Events A, B are independent if and only $Pr[A, B] = Pr[A]\,Pr[B]$, i.e., if and only if $Pr[A] = Pr[A|B]$.

DEFINITION 2.3 (TOTAL PROBABILITY FORMULA)
Assume that $B_1, \ldots, B_n$ are mutually exclusive and exhaustive; i.e. $B_i \cap B_j = \emptyset$, for all distinct i, j, and $\Omega = B_1 \cup \cdots \cup B_n$. Then

$$Pr[A] = \sum_k Pr[A|B_k]Pr[B_k].$$

Note that $Pr[A|B] = \frac{Pr[A,B]}{Pr[B]}$ and $Pr[B|A] = \frac{Pr[A,B]}{Pr[A]}$, so that $Pr[A|B]\,Pr[B] = Pr[B|A]\,Pr[A]$; hence

$$Pr[B|A] = \frac{Pr[A|B]\,Pr[B]}{Pr[A]}.$$

This and its generalization are known as *Bayes' rule.*

THEOREM 2.4 (BAYES' RULE)
Suppose that the hypotheses $B_1, \ldots, B_n$ are mutually exclusive and exhaustive events. Then

$$\begin{aligned} Pr[B_k|A] &= \frac{Pr[B_k]\,Pr[A|B_k]}{Pr[A]} \\ &= \frac{Pr[B_k]\,Pr[A|B_k]}{\sum_{k=1}^{n} Pr[A|B_k]\,Pr[B_k]}. \end{aligned}$$

In Bayes' rule, the B_i are considered intuitively to be the hypotheses responsible for event A. This has applications in hidden Markov models, discussed in a later chapter.

2.1.1 Random Variables

DEFINITION 2.5
A random variable X is a function $X : \Omega \to \mathbb{R}$.

We shall usually consider *discrete* random variables; in the continuous case, one requires a random variable to be a measurable real function.

In the discrete case, we write $Pr[X = x]$ to mean $Pr[\{e \in \Omega \,|\, X(e) = x\}]$ or $\sum_{e \in \Omega | X(e) = x} p(e)$. Then $\sum_{x=-\infty}^{\infty} Pr[X = x] = 1$. In the continuous case, for given $x \in \mathbb{R}$, it is usually the case that $Pr[X = x] = 0$, so a *probability density* function $p_X : \mathbb{R} \to \mathbb{R}$ is defined for a continuous random variable X to satisfy

$$\begin{aligned} Pr[a \le X \le b] &= Pr[\{e \in \Omega \mid a \le X(e) \le b\}] \\ &= \int_a^b p_X(x)\,dx. \end{aligned}$$

Note that $\int_{-\infty}^{\infty} p_X(x)\,dx = 1$. The *distribution function* for a continuous random variable X is

$$\Phi_X(x) = \int_{-\infty}^{x} p_X(y)\,dy,$$

so that

$$Pr[a \leq X \leq b] = \Phi_X(b) - \Phi_X(a) = \int_a^b p_X(x)\,dx.$$

The *expectation* (also called the *mean* or *average*) of discrete random variable X, denoted by $E[X]$ or sometimes $\langle X \rangle$, is defined by

$$E[X] = \sum_{i=-\infty}^{\infty} i\,Pr[X = i]$$

and in the continuous case

$$E[X] = \int_{-\infty}^{\infty} x\,p_X(x)\,dx.$$

From the definition, it is easy to see that expectation is *linear* in the sense that $E[aX + bY] = aE[X] + bE[Y]$, for constants a, b. Moreover for *independent* random variables X_1, X_2,

$$E[X_1 X_2] = E[X_1]E[X_2].$$

Note that $X_1 \leq X_2$ implies $E[X_1] \leq E[X_2]$.

The *mean square* or *second moment* of X is defined by

$$E[X^2] = \sum_{-\infty}^{\infty} x^2\,Pr[X = x]$$

for discrete random variables and

$$E[X^2] = \int_{-\infty}^{\infty} x^2 p_X(x)\,dx$$

in the continuous case. In general, the rth *moment* is defined by

$$E[X^r] = \sum_{-\infty}^{\infty} x^r\,Pr[X = x]$$

respectively

$$E[X^r] = \int_{-\infty}^{\infty} x^r p_X(x)\,dx.$$

The *variance* $V[X]$ is defined as the second moment of $X - \mu$, where $\mu = E[X]$. In other words,

$$\begin{aligned} V[X] &= E[X - \mu]^2 \\ &= E[X^2] - 2\mu E[X] + \mu^2 \\ &= E[X^2] - 2\mu^2 + \mu^2 \\ &= E[X^2] - \mu^2. \end{aligned}$$

The *standard deviation* σ satisfies $\sigma(X) = \sqrt{V[X]}$. From the above, $V[cX] = c^2 V[X]$ and, in the case that X, Y are *independent* random variables, the variance is additive:

$$V[X + Y] = V[X] + V[Y].$$

2.1.2 Some Important Probability Distributions

The *uniform* distribution satisfies

$$Pr[X = \omega] = \frac{1}{|\Omega|}$$

in the discrete case, while in the continuous case, X is distributed uniformly on the real interval $[a, b]$ if

$$Pr[c \leq X \leq d] = \int_c^d \frac{dx}{b-a} = \frac{d-c}{b-a}$$

where the density function

$$p_X(x) = \begin{cases} \frac{1}{b-a} & \text{if } a \leq x \leq b \\ 0 & \text{otherwise.} \end{cases}$$

In the continuous case, the expectation is

$$\int_a^b \frac{x}{b-a}\,dx = \frac{b^2-a^2}{2(b-a)} = \frac{a+b}{2}$$

and the variance is

$$\begin{aligned} \int_a^b \frac{x^2}{b-a}\,dx - \left(\frac{a+b}{2}\right)^2 &= \frac{b^3-a^3}{3(b-a)} - \left(\frac{a+b}{2}\right)^2 \\ &= \frac{b^2+ab+a^2}{3} - \frac{a^2+2ab+b^2}{4} \\ &= \frac{b^2-2ab+a^2}{12}. \end{aligned}$$

Binomial Distribution

A *Bernouilli* trial is an experiment with probability p of success; i.e. a Bernouilli random variable $Y : \Omega \rightarrow \mathbb{R}$ where Y takes only the values of $0, 1$ (1 for success, 0 for failure). Trials are independent, so, letting X be a random variable counting the number of successes in n trials,

$$b(n;k) = Pr[X = k] = \binom{n}{k} p^k (1-p)^{n-k}.$$

The random variable X is said to be binomially distributed. Note that $X = Y_1 + \cdots + Y_n$, where each Y_i is an independent Bernouilli random variable. Since $E[Y_i] = p$, and $V[Y_i] = E[Y_i^2] - E[Y_i]^2 = p - p^2 = p(1-p)$, it follows by additivity of expectation and variance (the latter requires independence) that

$$E[X] = np$$

and

$$V[X] = np(1-p).$$

If the number of trials n is large and the number k of successes is small, then a good approximation is

$$b(n;k) = Pr[X = k] \approx \frac{(np)^k}{k!} e^{-(np)},$$

where np is the expected number of successes. The latter is the Poisson distribution, which we shall soon discuss.

The *multinomial distribution* generalizes the binomial distribution to the case where there are m possible outcomes of an experiment. If p_i is the probability of outcome i in one trial, then the probability $P[X = (n_1, \ldots, n_m)]$ that in n successive trials, there are n_i outcomes of outcome i is defined by

$$M(n; n_1, \ldots, n_m) = P[X = (n_1, \ldots, n_m)] = \frac{n!}{n_1! \cdots n_m!} p_1^{n_1} \cdots p_m^{n_m}.$$

The multinomial distribution naturally models repeated rolls of a die.

Geometric Distribution

Consider the previously defined *Bernouilli* random variable X with probability p of success. The random variable X has the *geometric distribution* if for all $k \in \mathbb{N}$,

$$Pr[X = k] = (1 - p)^{k-1} p.$$

In this situation, X can be interpreted to count the number of experiments performed before a successful outcome is realized.

Suppose that $p < 1$, and denote $1 - p$ by q. Then a computation shows that the expectation satisfies

$$\begin{aligned}
E[X] &= \sum_{k=0}^{\infty} k q^{k-1} p \\
&= p \sum_{k=0}^{\infty} \frac{d}{dq}(q^k) \\
&= p \frac{d}{dq} \left(\sum_{k=0}^{\infty} q^k \right) \\
&= p \frac{d}{dq} \left(\frac{1}{1-q} \right) \\
&= \frac{p}{(1-q)^2} \\
&= \frac{1}{p}.
\end{aligned}$$

A similar computation yields the variance $V[X] = E[X^2] - E[X]^2$, as follows:

$$\begin{aligned} E[X^2] &= \sum_{k=0}^{\infty} k^2 q^{k-1} p \\ &= p \sum_{k=0}^{\infty} \frac{d}{dq} \left(kq^k\right) \\ &= p \frac{d}{dq} \left(\sum_{k=0}^{\infty} kq^k \right) \\ &= p \frac{d}{dq} \left(\frac{q}{(1-q)^2} \right) \\ &= p \frac{1+q}{(1-q)^3} \\ &= \frac{1+q}{p^2}, \end{aligned}$$

so that

$$V[X] = E[X^2] - E[X]^2 = \frac{1+q}{p^2} - \frac{1}{p^2} = \frac{q}{p^2}.$$

Poisson Distribution

A random variable has the *Poisson distribution* with parameter λ if

$$p(k;\lambda) = Pr[X = k] = \frac{\lambda^k}{k!} e^{-\lambda}$$

for $k \in \mathbb{N}$. Recall that the kth term in the Taylor expansion of e^x is $\frac{x^k}{k!}$, so that $\sum_{k=0}^{\infty} \frac{\lambda^k}{k!} e^{-\lambda}$ is $e^{-\lambda} \sum_{k=0}^{\infty} \frac{\lambda^k}{k!} = e^{-\lambda} e^{\lambda} = 1$. It turns out that both the expectation and variance of X are given by λ. The expectation is given by

$$\begin{aligned} E[X] &= \sum_{k=0}^{\infty} k \frac{\lambda^k}{k!} e^{-\lambda} \\ &= \lambda e^{-\lambda} \frac{d}{d\lambda} \left(\sum_{k=0}^{\infty} \frac{\lambda^k}{k!} \right) \\ &= \lambda e^{-\lambda} \frac{d}{d\lambda} \left(e^{\lambda}\right) \\ &= \lambda e^{-\lambda} e^{\lambda} = \lambda. \end{aligned}$$

Note for future reference that it follows that $\sum_{k=0}^{\infty} \frac{k\lambda^k}{k!} = \lambda e^{\lambda}$. The second moment is given by

$$\begin{aligned} E[X^2] &= \sum_{k=0}^{\infty} k^2 \frac{\lambda^k}{k!} e^{-\lambda} \\ &= \lambda e^{-\lambda} \frac{d}{d\lambda} \left(\sum_{k=0}^{\infty} \frac{k\lambda^k}{k!} \right) \\ &= \lambda e^{-\lambda} \frac{d}{d\lambda} (\lambda e^{\lambda}) \\ &= \lambda e^{-\lambda} (e^{\lambda} + \lambda e^{\lambda}) \\ &= \lambda + \lambda^2. \end{aligned}$$

Thus

$$V[X] = E[X^2] - E[X]^2 = \lambda + \lambda^2 - \lambda^2 = \lambda.$$

The Poisson distribution is often used in modeling situations in biology where events occur infrequently. In particular, in Section 4.2.1 of Chapter 4, the substitution of nucleotide bases is modeled as a Poisson process. While the Poisson distribution models the *number* of events that occurred within a given time interval, the *exponential* distribution models the *interarrival time* between two occurrences of events.

Normal Distribution

The continuous random variable X with density

$$p_X(x) = \frac{1}{\sqrt{2\pi}} e^{-x^2/2}$$

is said to have a *normal* or *Gaussian* distribution. The graph of density p_X is the familiar bell-shaped curve. The integral defining the distribution,

$$\Phi_X(x) = \frac{1}{\sqrt{2\pi}} \int_{-\infty}^{x} e^{-u^2/2}\, du,$$

cannot be evaluated in closed form, and so must be approximated numerically. By symmetry of the bell-shaped curve,

$$E[X] = \frac{1}{\sqrt{2\pi}} \int_{-\infty}^{\infty} x e^{-x^2/2}\, dx = 0$$

and integration by parts,

$$\begin{aligned} V[X] = E[X^2] - E[X]^2 &= \int_{-\infty}^{\infty} x^2 e^{-x^2/2}\, dx - 0^2 \\ &= \int_{-\infty}^{\infty} (-x) \frac{-x e^{-x^2/2}}{\sqrt{2\pi}}\, dx \\ &= \left. \frac{-x e^{-x^2/2}}{\sqrt{2\pi}} \right|_{-\infty}^{\infty} + \int_{-\infty}^{\infty} \frac{e^{-x^2/2}}{\sqrt{2\pi}}\, dx \\ &= 0 + 1 = 1, \end{aligned}$$

shows that $V[X] = 1$. In general, a random variable X with density

$$p(x) = \frac{1}{\sqrt{2\pi}\sigma} e^{-(x-\mu)^2/2\sigma^2}$$

is normally distributed with mean μ and variance σ^2. In this case, we shall say that X has distribution $\mathcal{N}(\mu, \sigma)$. It can be computed that about 68% (resp. 95%) of the area under the curve given by the normal density function $p(x)$ lies within 1 (resp. 2) standard deviations σ (resp. 2σ) to either side of the mean. In biological applications, it is sometimes assumed that data is normally distributed, so that if the data average is μ and standard deviation is σ, then one states that the value from the data is $\mu \pm \sigma$ (resp. $\mu \pm 2\sigma$), where σ (resp. 2σ) is called the *standard error* (this often appears in [Nei87]).

The central limit theorem states that in the limit, the distribution of the sum of n independent identically distributed random variables having finite mean and variance is normally distributed.

THEOREM 2.6 (CENTRAL LIMIT THEOREM)
Let $X_1, X_2, \ldots$ be independent, identically distributed random variables, having finite mean μ and finite variance σ^2. Consider the sum

$$S_n = \sum_{i=1}^{n} X_i$$

of the first n random variables, and consider the normalized sum

$$S_n^* = \frac{S_n - E[S_n]}{\sqrt{V[S_n]}} = \frac{S_n - n\mu}{\sigma\sqrt{n}}.$$

Then

$$\lim_{n \to \infty} Pr[a \leq S_n^* \leq b] = \frac{1}{\sqrt{2\pi}} \int_a^b e^{-x^2/2}.$$

Since convergence is fast, even for $n = 10$, $S_n^* = \frac{S_n - n/2}{\sqrt{n/12}}$ is close to normal. This observation furnishes an easy algorithm to generate random values with a normal distribution, given a uniform random generator. For X a uniformly distributed continuous random variable with values between 0 and 1, by previous discussion, $E[X] = 1/2$ and $V[X] = 1/12$. (We shall assume that a computer's pseudorandom real generator yields such values.) By the central limit theorem,

$$S_n^* = \frac{S_n - n/2}{\sqrt{n/12}}$$

is approximately normally distributed with mean 0 and variance 1. By earlier remarks about additivity of expectation and variance, it follows that $E[\mu + \sigma X] = \mu + \sigma E[X]$ and $V[\mu + \sigma X] = \sigma^2 V[X]$. Thus $\mu + \sigma S_n^*$ has mean μ and variance σ^2.

These remarks justify the following code, where we renormalize S_n^* to have a given mean and variance, so the output real values are normally distributed with *mean* and *variance* given by the function parameters:

```
double normal ( double mean, double variance ) {
  const int N = 20;
  int i; double x=0;
  for (i=0;i<N;i++) x += (double) rand()/ RAND_MAX;
  return (mean+sqrt(variance)*(x-N/2.0)/sqrt(N/12.0));
}
```

A more efficient algorithm is due to Box–Muller (see [Knu81]):

```
double boxMuller ( double mean, double variance ) {
  double u,v,x,y;
  x = (double) rand()/RAND_MAX;
  y = (double) rand()/RAND_MAX;
  u = sqrt(-2*log(x))*cos(2*M_PI*y);
      // M_PI is approx to PI = 3.14159...
  return(u);
}
```

Hypergeometric Distribution

Suppose there are n balls in an urn, r of which are red and $b = n - r$ of which are black, and that $m \leq n$ balls are chosen at random without replacement. The discrete random variable X has the hypergeometric distribution if

$$h(n, r; m, k) = Pr[X = k] = \frac{\binom{r}{k}\binom{n-r}{m-k}}{\binom{n}{m}},$$

where $Pr[X = k]$ is the probability that out of m randomly chosen balls, exactly k are red.

If p, q are probabilities with $p+q = 1$, and there are $r = pn$ red balls (resp. $b = n-r = qn$ black balls), then with probability

$$\binom{m}{k} p^k q^{m-k}$$

there are exactly k red balls when choosing m balls *with replacement* from an urn containing n balls. For large n, the hypergeometric distribution can be approximated by the binomial distribution: $h(n, r; m, k) \approx b(m; k)$ (see p. 50 of [Fel68a]). Specifically, for $p = \frac{r}{n}$ we have

$$\binom{m}{k}\left(p - \frac{k}{n}\right)^k \left(q - \frac{m-k}{n}\right)^{m-k} < h(n, k; m, k) < \binom{m}{k} p^k q^{m-k} \left(1 - \frac{m}{n}\right)^{-m}.$$

It follows that for large populations, there is hardly a difference between sampling with and without replacement. Later we shall see an application of the hypergeometric distribution in a *segmentation* algorithm for finding regions of the genome with approximately the same entropy.

Exponential Distribution

Fix $\alpha > 0$. Then

$$f(t) = \begin{cases} \alpha e^{-\alpha t} & \text{if } t \geq 0, \\ 0 & \text{otherwise} \end{cases}$$

is the density function for the continuous random variable X with exponential distribution, and

$$F(t) = \begin{cases} 1 - e^{-\alpha t} & \text{if } t \geq 0, \\ 0 & \text{otherwise} \end{cases}$$

is its distribution. Straightforward integration by parts shows that $E[X] = \frac{1}{\alpha}$ and $V[X] = \frac{1}{\alpha^2}$. The exponential distribution models interarrival time between events (such as mutation of a nucleotide), where there is no *memory* of previous events having occurred – for instance, a nucleotide mutation is just as likely to occur at a mutation site as an non-mutation site of a previous mutation.

Following [Gar86], we derive the exponential distribution from first principles concerning interarrival times for successive events. Let μ be the mean interarrival time between successive events, let Δt be a small, fixed interval of time. Divide the time interval t into $\frac{t}{\Delta t}$ many subintervals. For very small Δt,

$$Pr[\text{there is an arrival within } \Delta t \text{ time}] \approx \frac{\Delta t}{\mu},$$

so that

$$Pr[\text{there is no arrival within } \Delta t \text{ time}] \approx 1 - \frac{\Delta t}{\mu}.$$

Assuming that the probability of an arrival in any one of the $\frac{t}{\Delta t}$ subintervals is independent of whether an arrival occurred in any other subinterval, we have

$$Pr[\text{there is no arrival within time } t] \approx \left(1 - \frac{\Delta t}{\mu}\right)^{\frac{t}{\Delta t}}.$$

so by taking the limit as $\Delta t \to 0$, it follows that

$$Pr[\text{there is no arrival within time } t] = e^{-\frac{t}{\mu}};$$

hence

$$Pr[\text{there is an arrival within time } t] = 1 - e^{-\frac{t}{\mu}}.$$

This is a derivation of the exponential distribution

$$F(t) = \begin{cases} 1 - e^{-\alpha t} & \text{if } t \geq 0, \\ 0 & \text{otherwise,} \end{cases}$$

where the mean interarrival time μ is $\frac{1}{\alpha}$.

For computer simulations, one can produce an exponentially distributed random variable for a given mean interarrival time μ as follows. Let X be a uniformly distributed continuous random variable with $0 \leq X \leq 1$. Note that $Pr[X > x] = 1 - x$. Then

$$\begin{aligned} Pr[\text{ there is an arrival in time } t] &= 1 - e^{-\frac{t}{\mu}} \\ &= Pr[X > e^{-\frac{t}{\mu}}] \\ &= Pr\left[\ln X > -\frac{t}{\mu}\right] \\ &= Pr[-\mu \ln X < t]. \end{aligned}$$

Thus repeatedly evaluating $-\mu \ln X$ for uniformly distributed random real $0 \leq X \leq 1$ yields a sequence of interarrival times with mean interarrival time of μ. This could be used, for instance, to simulate the interarrival time for nucleotide substitutions, explained in Chapter 4. Here is C-code fragment for the exponential distribution, whose mean is given by the function parameter:

```
double exponential ( double mean ) {
    double x;

    x = (double) rand()/RAND_MAX;
    return (- mean*log(x) );
}
```

The following theorem states a connection between the exponential and Poisson distributions.

THEOREM 2.7 (FELLER [FEL68A])
Let $X_1, X_2, \ldots$ be independent, identically distributed random variables with exponential distribution

$$F(t) = \begin{cases} 1 - e^{-\alpha t} & \text{if } t \geq 0 \\ 0 & \text{otherwise.} \end{cases}$$

Consider the sum $S_n = \sum_{i=1}^{n} X_i$ of the first n random variables, and define the random variable $N(t)$ so that $N(t) = n$ holds exactly when $S_n \leq t$ and $S_{n+1} > t$. Then

$$Pr[N(t) = n] = e^{-\alpha t} \frac{(\alpha t)^n}{n!},$$

so that $N(t)$ has a Poisson distribution with parameter αt.

It follows that given $\frac{1}{\alpha}$ as the mean interarrival time, the probability that within time t there are exactly n arrivals is approximately $p(n; \alpha) = \frac{(\alpha t)^n}{n!} e^{-\alpha t}$. Thus the exponential and Poisson distributions are closely related.

Boltzmann Distribution

In the latter part of the 19th century, L. Boltzmann[3] reasoned that in an ideal gas having N molecules, the number N_i of molecules having energy E_i satisfies

$$N_i = N\frac{e^{-E_i/kT}}{Z},$$

where k is the *Boltzmann constant* having value 13.805×10^{-24} J deg^{-1}, T is the absolute temperature in degrees Kelvin, and Z, called the *partition function*, satisfies

$$Z = \sum_{s \in Q} e^{-E_s/kT},$$

where Q is the finite set of possible energy states. For applications in computer science, such as in the Metropolis–Hastings algorithm, the Boltzmann constant k is often dropped, and the Boltzmann distribution, sometimes called the Gibbs distribution, is defined by

$$p_T(i) = \frac{e^{-E_i/T}}{Z}. \tag{2.2}$$

The Boltzmann distribution appears almost magically as that probability distribution $p_1, \ldots, p_n$ having maximal entropy $H(p_1, \ldots, p_n) = -\sum_{i=1}^{n} p_i \log_2(p_i)$ for which a given average energy $\langle E \rangle$ is the expected value

$$\langle E \rangle = \sum_{i=1}^{n} p_i E_i.$$

This will be proved later, but is the basis for a remarkable analogy between a collection of ideal gas molecules and a representative database of protein conformations, a topic we shall pursue later when concerned with protein folding. In particular, following [Sip90, Krö96] we shall see later how to apply Boltzmann's law to determine energy functions, called *amino acid pair potentials*, computed from frequencies of amino acid pairs, as measured in from a representative sampling of known conformations in protein databases.

At high temperature T, the Boltzmann distribution is close to the uniform distribution, while at positive temperature close to 0, the Boltzmann distribution is concentrated on the global energy minimum. In particular, if the energy function takes on a unique minimum in state $i_0 \in S$, then for positive $T \approx 0$,

$$p_T(i_0) \approx \delta_{i_0}(i) = \begin{cases} 1 & \text{if } i = i_0, \\ 0 & \text{otherwise.} \end{cases}$$

A computation illustrates this point well. For example, the following table gives sample energies E_i for 10 states, along with Boltzmann probabilities $p_T(i) = \frac{e^{-E_i/T}}{Z}$ for temperature $T = 10^{-3}$:

---

[3] Ludwig Boltzmann, Austrian physicist, 1844–1906.

i	$E(i)$	$p_T(i)$
1	0.53522821	3.3666×10^{-15}
2	0.50190333	0.99999979
3	0.53884837	9.0152×10^{-17}
4	0.54406764	4.8783×10^{-19}
5	0.53822782	1.6768×10^{-16}
6	0.56446537	6.7554×10^{-28}
7	0.57723159	1.9291×10^{-33}
8	0.55286341	7.3845×10^{-23}
9	0.51729184	2.0742×10^{-7}
10	0.52294047	7.3061×10^{-10}

To prove that at high temperature, the Boltzmann distribution is approximately the uniform distribution, while at low temperature, the Boltzmann distribution is concentrated on the global energy minimum (or minima), following [Wat95], we proceed as follows.

Let Q be a finite set of states, and assume a given energy function $E : Q \to \mathbb{R}$. Suppose that the global energy minimum $m_0 = \min\{E(v) \mid v \in Q\}$ is achieved exactly at the values

$$Q_0 = \{v \in Q \mid E(v) = \min_{u \in Q} E(u)\} = \{v_1, \dots, v_k\}.$$

For a fixed temperature T, define

$$\pi_T(v) = \frac{e^{-E(v)/T}}{\sum_{w \in Q} e^{-E(w)/T}};$$

$\pi_T(v)$ is the probability that system is in state v. Clearly $\sum_{v \in Q} \pi_T(v) = 1$. Two limiting distributions are when the temperature is 0 and infinite; i.e. π_0, π_∞, where

$$\pi_0(v) = \lim_{T \to 0^+} \pi_T(v),$$

$$\pi_\infty(v) = \lim_{T \to \infty} \pi_T(v).$$

CLAIM 2.8

$$\pi_0(v) = \begin{cases} \frac{1}{k} & \text{if } m_0 = E(v) \\ 0 & \text{otherwise.} \end{cases}$$

PROOF

$$\begin{aligned} \pi_0(v) &= \lim_{T \to 0+} \frac{e^{-E(v)/T}}{\sum_{w \in Q} e^{-E(w)/T}} \\ &= \lim_{T \to 0+} \frac{1}{e^{E(v)/T} \sum_{w \in Q} e^{-E(w)/T}} \\ &= \lim_{T \to 0+} \frac{1}{\sum_{w \in Q} e^{(E(v)-E(w))/T}}. \end{aligned}$$

CASE 1: $v \in \{v_1, \dots, v_k\}$
Then $w \in \{v_1, \dots, v_k\}$ implies that $E(w) = E(v)$ and so

$$\lim_{T \to 0^+} e^{[E(v)-E(w)]/T} = 1,$$

while $w \notin \{v_1, \ldots, v_k\}$ implies that $E(w) > E(v)$ and so

$$\lim_{T \to 0^+} e^{[E(v)-E(w)]/T} = \lim_{T \to 0^+} \frac{1}{e^{[E(w)-E(v)]/T}} = \lim_{n \to \infty} \frac{1}{e^{10^n [E(w)-E(v)]}} = 0.$$

So altogether

$$\pi_0(v) = \frac{1}{\sum_{w \in Q} \lim_{T \to 0^+} e^{[E(v)-E(w)]/T}} = \frac{1}{k}.$$

CASE 2: $v \notin \{v_1, \ldots, v_k\}$
Let w_0 be a global minimum for E; i.e. $w_0 \in \{v_1, \ldots, v_k\}$. Then $E(v) - E(w_0) > 0$, so

$$\begin{aligned} \lim_{T \to 0^+} \frac{1}{\sum_{w \in Q} e^{[E(v)-E(w)]/T}} &\leq \lim_{T \to 0^+} \frac{1}{e^{[E(v)-E(w_0)]/T}} \\ &= \lim_{n \to \infty} \frac{1}{e^{10^n [E(v)-E(w_0)]}} \\ &= \frac{1}{\infty} = 0. \end{aligned}$$

Thus

$$\pi_0(v) = \begin{cases} \frac{1}{k}, & \text{if } v \text{ is a global minimum of } E \\ 0 & \text{otherwise.} \end{cases}$$

■

CLAIM 2.9

$$\pi_\infty(v) = \frac{1}{|Q|}.$$

PROOF

$$\begin{aligned} \pi_\infty(v) &= \lim_{T \to \infty} \frac{e^{-E(v)/T}}{\sum_{w \in Q} e^{-E(w)/T}} \\ &= \lim_{T \to \infty} \frac{1}{\sum_{w \in Q} e^{(E(v)-E(w))/T}} \\ &= \frac{1}{\sum_{w \in Q} e^0} \\ &= \frac{1}{|Q|}. \end{aligned}$$

■

In summary, π_0 is uniformly concentrated on global energy minima, while π_∞ is uniformly distributed on Q.

We will soon see that the concentration of the Boltzmann distribution on the global energy minima can be exploited to devise an algorithm to solve combinatorial optimization problems. Specifically, the Metropolis–Hastings algorithm (also called MCMC or Markov chain Monte Carlo algorithm) performs a walk on the state space for the given combinatorial optimization problem, where a random neighbor of the current position is chosen with uniform probability, and a move to that neighbor is allowed (i.e. an update of the current position) if either the neighbor has lower energy (greedy step) or with Boltzmann probability when the neighbor has higher energy. Here, temperature plays only a formal role, in regulating the likelihood of performing an update of the current position to an energy-unfavorable neighbor. While unnecessary from our computational standpoint, we make a few closing remarks concerning the role of the Boltzmann distribution in chemistry [Pau70].

The *Gibbs free energy* G in a system (of gas molecules at equilibrium, or a solution of molecules, etc.) satisfies

$$G = U - TS \tag{2.3}$$

where U is the *enthalpy* or heat content (in the case of DNA in solvent under physiological conditions, enthalpy comprises the energy from the ionic, hydrogen and covalent bonds, etc.), T is the absolute temperature in degrees Kelvin and S the *entropy* or measure of disorder. Thus equation (2.3) states that free energy is the difference of *ordering* forces minus *disordering* forces. Since inner energy and entropy cannot be measured directly, in laboratory experiments, one usually measures the changes, i.e.

$$\Delta G = \Delta U - T\Delta S.$$

For an ideal gas at temperature T, the difference ΔG in free energy between the gas at pressure P and at pressure P_0 of 1 atmosphere satisfies

$$\Delta G = -RT \ln\left(\frac{P}{P_0}\right),$$

where R is the gas constant having value 8.3146 J deg^{-1} mole^{-1}. Moreover, *Boltzmann's constant* $k = R/N$, where N is Avogadro's number 6.0229×10^{23}. It follows that

$$P = P_0 e^{\frac{-\Delta G}{RT}} \tag{2.4}$$

and it is exactly the term $e^{\frac{-\Delta G}{T}}$ which appears in the Metropolis step of MCMC, as we shall see later.

2.1.3 Markov Chains

Consider a physical process having discrete observable states, which when we monitor over time, yields the sequence $q_0, q_1, q_2, \ldots$ of observed states. A (first-order, time-homogeneous) *Markov chain* is a stochastic model of this system, whose main property is that the state at time $t + 1$ depends only on the state at time t. For instance, protein folding can be modeled by such a Markov chain, where the states are possible conformations on a lattice.

To formalize this concept, we need a few definitions. Throughout, only finite Markov chains are considered. A *stochastic* matrix is an $n \times n$ matrix of non-negative values, each of whose rows sums to 1. A *doubly stochastic* matrix has the same property for its columns, and a *substochastic* matrix is a non-negative matrix, each of whose rows sums to a value less than or equal to 1.

Let $Q = \{1, \ldots, n\}$ be a finite set of states, and consider the initial probability distribution $\pi = (p_1, \ldots, p_n)$, considered as a row vector, and the $n \times n$ stochastic matrix $P = (p_{i,j})$. A (first-order, time-homogeneous) *Markov chain* $M = (Q, \pi, P)$ is a stochastic process, whose state q_t at time t is a random variable determined by

$$\begin{aligned} Pr[q_0 = i] &= \pi_i, \\ Pr[q_{t+1} = j | q_t = i] &= p_{i,j}. \end{aligned}$$

Define $p_i(t) = Pr[q_t = i]$ and $p_{i,j}^{(t)} = Pr[q_t = j | q_0 = i]$. Clearly, the (i, j)th entry of the tth power P^t of P equals $p_{i,j}^{(t)}$; moreover, by time-homogeneity it follows that $p_{i,j}^{(t)} = Pr[q_{t_0+t} = j | q_{t_0} = i]$, for all t_0.

DEFINITION 2.10 (PERSISTENCE)
A state i of a Markov chain is persistent if

$$\sum_{n=0}^{\infty} p_{i,i}^{(n)} = \infty.$$

If i is not persistent, then it is transient.

THEOREM 2.11
If the initial state is persistent, then with probability 1, the system returns infinitely often to this state. If the initial state is transient, then with probability 1 the system returns only finitely many times to this state.

State j can be reached (or is *accessible*) from state i if there is an N for which $p_{i,j}^{(N)} > 0$.

DEFINITION 2.12 (IRREDUCIBLE AND STATIONARY MARKOV CHAINS)
A Markov chain is irreducible if every state can be reached from every other state:

$$(\forall i, j \in Q)(\exists N \geq 0)[p_{i,j}^{(N)} > 0].$$

A Markov chain is stationary if for each state i, the probabilities $p_i(n)$ are constant for $n = 0, 1, \ldots$.

The following theorem is a classic result for Markov chains, whose proof is adapted from [KS60]. First, we need some notation. A matrix $P = (p_{i,j})$ is defined to be *non-negative*, denoted $P \geq 0$, if every entry $p_{i,j} \geq 0$, while P is *positive*, denoted $P > 0$, if $P \geq 0$ and there exists at least one entry $p_{i,j} > 0$. Finally, P is *strictly positive* if every entry $p_{i,j} > 0$. Similarly, define $P \geq \epsilon$ for any real number ϵ. Recall that a superscript 'T' denotes the transpose of a vector or matrix.

THEOREM 2.13
Let $P = (p_{i,j})$ be a strictly positive stochastic $n \times n$ matrix, and $P \geq \epsilon > 0$. Fix $1 \leq j \leq n$, and let $\vec{r} = (p_{1,j}, \cdots, p_{n,j})^{\mathrm{T}}$ be the jth column vector in matrix P, and let $\vec{s} = (s_1, \cdots, s_n)^{\mathrm{T}} = P \cdot \vec{r}$. Consider the maximum and minimum components of vectors $\vec{r}$,

$\vec{s}$, given by

$$\begin{aligned} M_0 &= \max_{1 \le i \le n} r_i, \\ m_0 &= \min_{1 \le i \le n} r_i, \\ M_1 &= \max_{1 \le i \le n} s_i, \\ m_1 &= \min_{1 \le i \le n} s_i. \end{aligned}$$

Then $m_0 \le m_1 \le M_1 \le M_0$ and $(M_1 - m_1) \le (M_0 - m_0)(1 - 2\epsilon)$.

PROOF Suppose that $1 \le i_0, i_1 \le n$ are such that $p_{i_0,j} = m_0$ and $p_{i_1,j} = M_0$. For $1 \le i \le n$, define column vectors $\vec{a}$, $\vec{b}$ by

$$a_i = \begin{cases} m_0 & \text{if } i = i_0, \\ M_0 & \text{otherwise} \end{cases}$$

and

$$b_i = \begin{cases} M_0 & \text{if } i = i_1, \\ m_0 & \text{otherwise.} \end{cases}$$

Clearly $\vec{b} \le \vec{r} = (p_{1,j}, \cdots, p_{n,j})^{\mathrm{T}} \le \vec{a}$, so by linearity, $P \cdot \vec{b} \le \vec{s} = P \cdot \vec{r} \le P \cdot \vec{a}$, where inequality is defined componentwise. Moreover, the ith component in $P \cdot \vec{a}$ is $M_0 \sum_{j=1}^{n} p_{i,j} - p_{i,i_0}(M_0 - m_0)$, since

$$P \cdot \vec{a} = P \cdot \begin{pmatrix} M_0 \\ \vdots \\ M_0 \end{pmatrix} - P \cdot \begin{pmatrix} 0 \\ \vdots \\ M_0 - m_0 \\ \vdots \\ 0 \end{pmatrix},$$

where the term $M_0 - m_0$ appears in the i_0th row, and hence equals $M_0 - p_{i,i_0}(M_0 - m_0)$, since P is a stochastic matrix and so $\sum_{j=1}^{n} p_{i,j} = 1$. Now $\epsilon \le p_{i,j}$ for all $1 \le i, j \le n$, and so $M_1 \le M_0 - \epsilon(M_0 - m_0)$.

In a similar manner, the ith entry of $P \cdot \vec{b}$ is equal to $m_0 + p_{i,i_1}(M_0 - m_0) \ge m_0 + \epsilon(M_0 - m_0)$, and hence

$$M_1 - m_1 \le M_0 - \epsilon(M_0 - m_0) - [m_0 + \epsilon(M_0 - m_0)] = (M_0 - m_0)(1 - 2\epsilon).$$

■

For $1 \le j \le n$, let $m_0(j)$ and $M_0(j)$ represent the minimum and maximum, respectively, of the jth column of matrix P, and let $m_{t+1}(j)$ and $M_{t+1}(j)$ represent the minimum and maximum, respectively, of the jth column of matrix $P^t \cdot (p_{1,j}, \cdots, p_{n,j})^{\mathrm{T}}$. From the previous argument, it follows that $\lim_{t \to \infty} P^t = P^* = (p^*_{i,j})$ exists, and that each entry in the jth column of P^* has the same value. Denoting this common value of the jth column by p^*_j, we have that for $1 \le i \le n$,

$$p^*_{i,j} = p^*_j = \lim_{t \to \infty} m_t(j) = \lim_{t \to \infty} M_t(j).$$

Since $m_0(j) \leq m_1(j) \leq \cdots \leq p_j^*$, the matrix P^* must have strictly positive entries. Letting $d_t(j) = M_t(j) - m_t(j)$, it follows that $d_t(j) \leq (M_0 - m_0)(1 - 2\epsilon)^t$, so that convergence is exponentially fast. Now

$$P^* = \lim_t P^t = \lim_t P^{t+1} = (\lim_t P^t) \cdot P = P^* \cdot P;$$

hence it follows that $(p_1^*, \ldots, p_n^*) \cdot P = (p_1^*, \ldots, p_n^*)$. In other words, 1 is a *left eigenvalue* of P, whose eigenvector is the row vector $(p_1^*, \ldots, p_n^*)$.

COROLLARY 2.14
Let $M = (Q, \pi, P)$ be a finite Markov chain, whose transition matrix P satisfies $P \geq \epsilon > 0$. Then $\lim_t p_i(t) = p_i^*$.

PROOF $p_i(t)$ is the ith coordinate of $\pi \cdot P^{(t)}$, so equals $\sum_{j=1}^n \pi_j p_{j,i}^{(t)}$. Recalling that $\sum_{j=1}^n \pi_j = 1$, we have

$$\begin{aligned}
|p_i(t) - p_i^*| &= \left| \sum_{j=1}^n \pi_j p_{j,i}^{(t)} - p_i^* \right| \\
&= \left| \sum_{j=1}^n \pi_j p_{j,i}^{(t)} - \sum_{j=1}^n \pi_j p_i^* \right| \\
&= \left| \sum_{j=1}^n \pi_j (p_{j,i}^{(t)} - p_i^*) \right| \\
&\leq \sum_{j=1}^n \pi_j (M_0 - m_0)(1 - 2\epsilon)^t \\
&\leq (M_0 - m_0)(1 - 2\epsilon)^t.
\end{aligned}$$

It follows that $\lim_t (p_i(t) - p_i^*) = 0$. ■

We now claim that $(p_1^*, \ldots, p_n^*)$ is the unique left eigenvector of the transition matrix P having eigenvalue 1. Indeed, suppose that $\vec{q} = (q_1, \ldots, q_n)$ is a probability distribution satisfying $q_j = \sum_{i=1}^n q_i p_{i,j}$ for $1 \leq j \leq n$. Then $\vec{q} \cdot P = \vec{q}$, $\vec{q} \cdot P^2 = \vec{q}$, ..., $\vec{q} \cdot P^t = \vec{q}$, etc. and so $\vec{q} \cdot \lim_t P^t = \vec{q} \cdot P^* = \vec{q}$. But it follows from Corollary 2.14 that $\pi \cdot P^* = \eta \cdot P^* = (p_1^*, \ldots, p_n^*)$ for any initial distributions π, η, and so $(q_1, \ldots, q_n) = (p_1^*, \ldots, p_n^*)$, and uniqueness follows.

DEFINITION 2.15
The period of a state $i \in Q$ of finite Markov chain $M = (Q, \pi, P)$ is the greatest common denominator of $\{t > 0 \mid p_{i,i}^t > 0\}$. A Markov chain is aperiodic if the period of each state $i \in Q$ is 1.

I.e, if P is an $n \times n$ 'anti-diagonal' matrix having 1s along the diagonal from the bottom left corner to the top right corner and 0s elsewhere (i.e. $p_{i,j} = 1$ if $j = n + 1 - i$, otherwise 0), then the period for every state is 2 (except $\frac{n+1}{2}$ when n odd).

The following technical lemma is proved in the appendix to this chapter (Section 2.5).

LEMMA 2.16 (POSITIVE TRANSITION MATRIX)
If $M = (Q, \pi, P)$ is a finite, aperiodic, irreducible Markov chain, then some power of P is strictly positive.

Putting things together, we have the following.

THEOREM 2.17 (CONVERGENCE TO STATIONARY MARKOV CHAIN)
Given a finite, aperiodic, irreducible Markov chain $M = (Q, \pi, P)$, where Q consists of n states, there exist stationary probabilities

$$\lim_{t \to \infty} p_i(t) = p_i^*,$$

where the p_i^* form a unique solution to the conditions

- $p_i^* \geq 0$,
- $\sum_{i=1}^n p_i^* = 1$,
- $p_j^* = \sum_{i=1}^n p_i^* p_{i,j}$.

The distribution $(p_1^*, \ldots, p_n^*)$ from Theorem 2.17 is also called the *equilibrium distribution.*

The following notion of *reversible* Markov chain plays an important role in Monte Carlo algorithms, where equation (2.5) is called the *detailed balance equation.*

DEFINITION 2.18
Suppose that $M = (Q, \pi, P)$ is a Markov chain having stationary probabilities $p_1^*, \ldots, p_n^*$. Then M is reversible if for all $i, j \in Q$,

$$p_i^* p_{i,j} = p_j^* p_{j,i}. \tag{2.5}$$

Because of the exponentially fast convergence, it is simple to compute the stationary probabilities for small Markov chains by repeatedly computing the powers P^t of transition matrix P, until the desired numerical convergence has occurred. The common value p_i^* of the ith column is the stationary probability that the system is in state i. Moreover, the mean recurrent time μ_i, to go from state i to state i, satisfies $\mu_i = 1/p_i^*$. In contrast, for the case of substochastic matrix P, the iterated powers of P converge to the zero matrix. For example, for the stochastic matrix

$$P = \begin{pmatrix} 0.8 & 0.1 & 0.1 \\ 0.33 & 0.33 & 0.34 \\ 0.25 & 0.5 & 0.25 \end{pmatrix},$$

after 17 iterations (i.e. P^{17}), the matrix

$$\begin{pmatrix} 0.5953 & 0.2238 & 0.1808 \\ 0.5953 & 0.2238 & 0.1808 \\ 0.5953 & 0.2238 & 0.1808 \end{pmatrix}$$

is obtained, with stationary probabilities $p_1^* = 0.5953$, $p_2^* = 0.2238$, and $p_3^* = 0.1808$. As well, for the substochastic matrix

$$P = \begin{pmatrix} 0 & 0.1 & 0.2 \\ 0 & 0.25 & 0.75 \\ 0.25 & 0.75 & 0 \end{pmatrix},$$

after 114 iterations (i.e. P^{114}), the zero matrix was obtained (using 4-place decimal precision).

The existence of stationary probabilities is not just an interesting mathematical theorem, but rather provides the justification for convergence of the Markov chain Monte Carlo algorithm, where a *sampling* is made on an underlying aperiodic, irreducible Markov chain $M = (Q, \pi, P)$, whose limiting stationary probabilities are given by the Boltzmann distribution: $p_i^* = \frac{e^{-E(i)/T}}{Z}$, where $E(i)$ is the energy of state i, T is the temperature, and Z is the *partition function*, satisfying $Z = \sum_{i \in Q} e^{-E(i)/T}$.

We terminate this section by an important approach due to A. Sinclair [Sin93] in bounding the time for convergence to the stationary probabilities. This has applications to expected runtime for Markov chain Monte Carlo algorithms. First, we define some new concepts.

Let $M = (Q, \pi, P)$ be a finite Markov chain. The *relative pointwise distance* is defined as

$$\Delta(t) = \max_{i,j \in Q} \frac{|p_{i,j}^{(t)} - p_j^*|}{p_j^*}.$$

Fix a subset $X \subseteq Q$ and define *capacity* to be $C_X = \sum_{i \in X} p_i^*$. Define the *ergodic flow* out of X to be $F_X = \sum_{i \in X, j \notin X} p_{i,j} p_i^*$. Since $0 < F_X \leq C_X < 1$, the quotient $\Phi_X = F_X / C_X$, may be considered to be the conditional flow out of X, provided the system is in X. The *conductance* $\Phi = \min_{C_X \leq 1/2} \Phi_X$, where the minimum is take over all $X \subseteq Q$.

THEOREM 2.19 (A. SINCLAIR[SIN93])
Let $M = (Q, \pi, P)$ be a finite reversible, irreducible, aperiodic Markov chain, for which all left eigenvalues of P are non-negative. Then

$$\Delta(t) \leq \frac{\lambda_1^t}{\min_{i \in Q} p_i^*}$$

where $\lambda_1 < 1$ is the second largest eigenvalue. Moreover, $\Delta(t) \leq \frac{(1-\phi^2/2)^t}{\min_{i \in Q} p_i^*}$ and if $\phi \leq 1/2$ then $\Delta(t) \geq (1 - 2\phi)^t$.

2.1.4 Metropolis–Hastings Algorithm

The Monte Carlo algorithm, proposed by N. Metropolis *et al.* [MRR+53], is a modification of the greedy algorithm, where with small probability (given by the Boltzmann distribution), occasionally unfavorable, non-greedy moves are allowed, with the intent to avoid becoming trapped in local minima of the energy function. Later, Kirkpatrick *et al.* [KGV83] introduced a temperature cooling schedule to implement *simulated annealing*, an idea borrowed from metallurgy, where in order to produce small crystals, a metal is repeatedly quenched and reheated, so that large, irregular crystals can be removed in favor of small, regular crystals. This heuristic was given mathematical rigor only in 1984, when S. Geman and D. Geman [GG84] proved that for exponentially slow temperature schedules ($T_n \geq \frac{cT_{hi}}{\ln(n+1)}$), simulated annealing correctly computes the energy minimum.

Actually, this statement is not quite correct – all that can be shown is that with probability tending to 1, simulated annealing correctly determines the energy minimum. Moreover, using brute force, in exponential time, one could search the entire state space, so the analytical result of Geman–Geman concerning exponentially slow temperature schedules is only of theoretical interest. As given in the pseudocode below, the temperature is usually reduced by a factor of something like 0.9, so that the number of temperature reductions is on the order of $\frac{\ln(9 \cdot T_{hi})}{\ln 10}$, i.e. logarithmic in n, rather than exponential in n.

Applications of Monte Carlo in computational biology include protein folding on lattice models [ŠSK94b, ŠSK94a, VKBS95], optimality of the genetic code [SC97], DNA strand separation in replication and transcription events [SMFB95], the double digest problem [Wat95], etc.

Suppose that Q is a finite set of states of a physical system and we have an energy function $E : Q \to \mathbb{R}$. Define a *neighborhood system* satisfying the following properties, where $i \in N_j$ denotes that i is a neighbor of j:

1. $i \notin N_i$.
2. $i \in N_j \Leftrightarrow j \in N_i$.
3. $|N_i| = |N_j|$, for all $i, j \in Q$.
4. For $i, j \in Q$, if $i \neq j$, then there exist $m \geq 0$ and $i_1, \ldots, i_m \in Q$ such that

$$(i \in N_{i_1}, i_1 \in N_{i_2}, \ldots, i_{m-1} \in N_m, i_m \in N_j).$$

Let $T_{hi} > T_{lo} > 0$ be given high and low temperatures. Pseudocode for Monte Carlo with simulated annealing is given in Algorithm 2.1. Monte Carlo without annealing is obtained removing lines 3,11,12 and usually optimizing with respect to temperature.

Algorithm 2.1 Monte Carlo with simulated annealing (practical version)

```
1   T = T_hi
2   i = initial
3   while (T > T_lo) {
4       repeat M times {
5           choose random j ∈ N_i
6           if (E(j) ≤ E(i)) then
7               i = j
8           else
9               x = random(0,1)
10              if (x < e^{-[E(j)-E(i)]/T} then i = j       }
11      T = T * 0.9
12  }
13  return i and E(i)
```

As soon to be seen in our proof of its convergence, the Monte Carlo algorithm requires that the move set be *ergodic* and *balanced*. Here, *ergodic* means that from any two distinct states, there is a succession of moves from one to the other, while *balanced* means that the detailed balance equation (2.5) holds for the underlying Markov chain. It suffices that the state transition matrix be a symmetric matrix, all of whose entries are positive – i.e. for all states i, j the probability of moving from i to j is positive and equals the probability of moving from j to i.

Following [Wat95], we prove the convergence of Monte Carlo without annealing. For temperature T, define the Markov chain $M_T = (Q, \pi, P_T)$, where if Q consists of n states, then $\pi = (1/n, \ldots, 1/n)$ is the uniform distribution on Q, and $P_T = (p_T(i,j))$ is defined by

$$p_T(i,j) = \begin{cases} \frac{1}{|N_i|} & \text{if } j \in N_i \text{ and } E(j) \leq E(i), \\ \frac{e^{-[E(j)-E(i)]/T}}{|N_i|} & \text{if } j \in N_i \text{ and } E(j) > E(i), \\ 0 & \text{if } j \notin N_i \text{ and } i \neq j, \\ 1 - \sum_{k \neq i} p_T(i,k) & \text{if } i = j. \end{cases}$$

We claim that M_T is a finite, aperiodic, irreducible Markov chain. The condition on the neighborhood system implies that each state is reachable from every other state, so that M_T is irreducible. Since $p_T(i,i) > 0$ and P consists of non-negative entries, $p_T^{(t)}(i,i) > 0$ for all $t \geq 0$, so the period of each $i \in Q$ is 1; hence M_T is aperiodic. Thus by Theorem 2.17, M_T has unique stationary probabilities $(p_T^*(1), \ldots, p_T^*(n))$, also called the *equilibrium distribution*.

CLAIM $p_T^*(i) = \frac{e^{-E(i)/T}}{Z}$, where the partition function $Z = \sum_{j \in Q} e^{-E(j)/T}$.

PROOF Let $b_T(i)$ denote the Boltzmann probability $p_T^*(i) = \frac{e^{-E(i)/T}}{Z}$, for each $i \in Q$. By uniqueness of the stationary probabilities, it suffices to show that $(b_T(1), \ldots, b_T(n))$ is a left eigenvector of P_T with eigenvalue 1, i.e. that $b_T(j) = \sum_{i=1}^{n} b_T(i) p_T(i,j)$.

To this end, we first show that the Boltzmann probabilities satisfy the *detailed balance* equation

$$b_T(i) p_T(i,j) = b_T(j) p_T(j,i)$$

for all $i, j \in Q$. This clearly holds if $j \notin N_i$. Consider now the case that $E(j) > E(i)$. Then

$$\begin{aligned} b_T(i) p_T(i,j) &= \frac{e^{-E(i)/T}}{Z} \frac{e^{-[E(j)-E(i)]/T}}{|N_i|} \\ &= \frac{1}{|N_i| Z} e^{[E(i)-E(j)-E(i)]/T} \\ &= \frac{1}{|N_i| Z} e^{-E(j)/T} \\ &= \frac{e^{-E(j)/T}}{Z} \frac{1}{|N_i|} \\ &= \frac{e^{-E(j)/T}}{Z} \frac{1}{|N_j|} \\ &= b_T(j) p_T(j,i). \end{aligned}$$

Finally, the case that $E(i) \geq E(j)$ is proved analogously, by reversing the roles of i, j. This establishes detailed balance for the Boltzmann distribution.

Now we claim that the Boltzmann distribution is a left eigenvector of P_T with eigenvalue 1, in other words, $(b_T(1), \ldots, b_T(n)) \cdot P_T = (b_T(1), \ldots, b_T(n))$, or for each $1 \leq j \leq n$,

$$b_T(j) = \sum_{i=1}^{n} b_T(i) p_T(i,j).$$

By detailed balance,

$$\sum_{i=1}^{n} b_T(i) p_T(i,j) = \sum_{i=1}^{n} b_T(j) p_T(j,i) = b_T(j) \sum_{i=1}^{n} p_T(j,i) = b_T(j).$$

It follows by the uniqueness condition of Theorem 2.17 that M_T converges to the stationary Boltzmann distribution; i.e.

$$\lim_t p_T^{(t)}(i,j)\pi(i) = b_T(j) = \frac{e^{-E(j)/T}}{\sum_{k \in Q} e^{-E(k)/T}}$$

We have thus proved the following theorem.

THEOREM 2.20 (METROPOLIS *et al.* [MRR+53])
The above Markov chain M_T has equilibrium distribution

$$p_T^*(i) = \frac{e^{-E(i)/T}}{\sum_{j \in Q} e^{-E(j)/T}}$$

for $i \in Q$.

We have seen that for low temperature $T > 0$, if the energy has a unique global minimum at $i_0 \in Q$, then the Boltzmann probability $\frac{e^{-E(i_0)/T}}{Z}$ is large, so that with large probability, the Monte Carlo algorithm will converge to state i_0. What is the expected time until convergence? This follows from the previously cited Theorem 2.19 due to A. Sinclair, plus a small estimation for mean first passage time.

Annealing Schedule

Consider the version of simulated annealing given in Algorithm 2.2 with an exponentially slow annealing schedule. Let Q be a finite state space and consider the function $E : Q \to \mathbb{R}$, whose minimum we are interested in computing. Let $c \geq \Delta$ be arbitrary, where $\Delta = \max_{i \in Q}\{E(i)\} - \min_{i \in Q}\{E(i)\}$.

Algorithm 2.2 Simulated annealing – theoretical version

```
1   n = 1; T = c; i = initial
2   repeat {
3       choose random j ∈ N_i
4       if (E(j) ≤ E(i)) then
5           i = j
6       else
7           x = random(0,1)
8           if (x < e^{-[E(j)-E(i)]/T} then i = j
9       n = n+1; T = c/ln n
10  } until T ≈ 0
```

Let X_n denote the random variable whose value is the state (i.e. value of variable i in above pseudocode) in the nth pass through the repeat loop. Similarly let T_n denote the temperature in the nth pass through the repeat loop. With this notation, S. Geman and D. Geman proved convergence in probability of simulated annealing with the above exponentially slow annealing schedule.

THEOREM 2.21 (S. GEMAN AND D. GEMAN [GG84])
Let $c \geq \Delta$, $T_n \geq \frac{\Delta}{\ln n}$ for all $n \geq 1$, and $\lim_{n \to \infty} T_n = 0$. Then with π_0 as defined before Claim 2.8

$$\lim_{n \to \infty} Pr[X_n = i] = \pi_0(i)$$

for all $i \in Q$.

2.1.5 Markov Random Fields and Gibbs Sampler

Earlier, we mentioned that 56% of the genes of *M. jannaschii* are completely unfamiliar to biologists. How can completely new genes be detected in the genome? This will be explained in Chapter 5, but involves a Markov model to determine the likely *open reading frames* (ORF). In particular, a 4th-order Markov model was derived to predict coding regions of *M. jannaschii*, using the software GENEMARK of [BM93].

A kth-order Markov chain describes the value of X_t (at time t) in terms of a conditional probability distribution depending on the values of $X_{t-k}, \ldots, X_{t-1}$. One can imagine random variables $X_{i,j}$ for $1 \leq i, j \leq m$ placed on an $m \times m$ integer lattice, where the value of $X_{i,j}$ depends on the values of neighboring $X_{i',j'}$ (e.g. nearest neighbors, or neighbors within distance r, etc.). The generalization of Markov chains to this scenario is called a *Markov random field* (MRF) , defined as follows.

DEFINITION 2.22 (MARKOV RANDOM FIELD)
Let I be a finite set of N indices or sites. A neighborhood system[4] *$\mathcal{G}$ is a collection of subsets $G_i \subseteq I$, for each $i \in I = \{1, \ldots, N\}$, such that*

- $i \notin G_i$,
- $i \in G_j \iff j \in G_i$.

For indices $i, j \in I$, i is said to be a neighbor of j if $i \in G_j$. Let Q denote a finite state space, let $X = (X_i : i \in I)$ denote a sequence of random variables. Let $\mathcal{Q} = Q^I = \{(\omega_1, \ldots, \omega_N) \mid \omega_i \in Q, 1 \leq i \leq N\}$ be the space of all possible configurations of the family X, where $(\omega_1, \ldots, \omega_N)$ is abbreviated by ω, and the event $X_1 = \omega_1, \ldots, X_N = \omega_N$ by $X = \omega$. The family X is a Markov random field (MRF) if

- $Pr[X = \omega] > 0$, *for all $\omega \in \mathcal{Q}$,*
- $Pr[X_i = \omega_i | X_j = \omega_j, j \neq i] = Pr[X_i = \omega_i | X_j = \omega_j, j \in G_i]$, *for every $i \in I$ and $\omega \in \mathcal{Q}$ (this last condition is the local* Markov property*).*

Note that if R is the binary relation defined by $R(i,j) \iff i \in G_j$, then above neighborhood system requirement states simply that R is irreflexive and symmetric.

DEFINITION 2.23 (GIBBS DISTRIBUTION WRT $I, \mathcal{G}$)
A set $C \subseteq I$ of indices is called a clique if every two distinct elements of C are neighbors (thus $C \subseteq G_i \cup \{i\}$, for every $i \in I$). Let $\mathcal{C}$ denote the set of all cliques on I with neighborhood system $\mathcal{G}$. A Gibbs distribution relative to $I, \mathcal{G}$ is a probability distribution g_T on $\mathcal{Q}$ defined by

$$g_T(\omega) = \frac{e^{-U(\omega)/T}}{Z},$$

[4] *Warning*. This neighborhood system has nothing in common with that previously defined for MCMC.

where the partition function Z satisfies

$$Z = \sum_{\omega \in \mathcal{Q}} e^{-U(\omega)/T}.$$

The potential energy function U is defined by

$$U(\omega) = \sum_{C \in \mathcal{C}} V_C(\omega).$$

where $V_C(\omega)$ is required to depend only on the coordinates x_s of ω belonging to the clique C. Denoting the restriction of $\alpha \in \mathcal{Q}$ to clique $C \subseteq I$ by $\alpha|_C$, this last condition translates into

$$\alpha|_C = \beta|_C \Rightarrow V_C(\alpha) = V_C(\beta).$$

For $\omega \in \mathcal{Q}$, $i \in I$ and $q \in Q$, define $\omega(i,q) \in \mathcal{Q}$ by

$$\omega(i,q) = \begin{cases} q & \text{if } i = j, \\ \omega_j & \text{if } i \neq j. \end{cases}$$

If g_T is a Gibbs distribution wrt $I, \mathcal{G}$, then by definition of conditional probability,

$$Pr[X_i = \omega_i | X_j = \omega_j, j \neq i] = \frac{g_T(\omega)}{\sum_{q \in Q} g_T(\omega(i,q))}$$

for all $i \in I, \omega \in \mathcal{Q}$. We will soon show that if $g_T(\omega) > 0$ for all $\omega \in \mathcal{Q}$, then X is a Markov random field, since the condition

$$Pr[X_i = \omega_i | X_j = \omega_j, j \neq i] = Pr[X_i = \omega_i | X_j = \omega_j, j \in G_i]$$

is then satisfied.

In image processing [GG84], one can imagine the set $I = \{(i,j) \mid 1 \leq i,j \leq m\}$ (the set of pixel locations in an $m \times m$ grid), the set Q of colors, or gray-scale values, the set $\mathcal{Q}$ of all possible images, η an initial distribution on the set of images, and g_T a stationary distribution on the set of images, where the local energy value $V_C(\alpha)$ of an image α involves an averaging effect on colors of pixels within the clique C, and the global energy value $U(\alpha)$ is the sum of all local contributions. In this example, for $d > 0$, we can define the neighborhood system $\mathcal{G}$ consisting of

$$G_{i,j}^d = \{(u,v) \in I \mid (u-i)^2 + (v-j)^2 \leq d\}.$$

For $d = 2$, we have 1-cliques such as $\{i\}$, 2-cliques such as $\{(i,j),(i+1,j)\}$ and $\{(i,j),(i,j+1)\}$, 3-cliques such as $\{(i,j),(i+1,j),(i,j+1)\}$, and 4-cliques such as $\{(i,j),(i+1,j),(i,j+1),(i+1,j+1)\}$.

A surprising fact is the relation between Markov random fields and the Gibbs distribution, given in the following theorem. One direction of this theorem is not difficult to prove, yields insight into the nature of this connection, and more importantly is necessary for an efficient Gibbs sampler algorithm, and so will be given here. For a full discussion, proof and references, see [KS80].

THEOREM 2.24
Let $\mathcal{G}$ be a neighborhood system on I. Then X is a Markov random field if and only if $Pr[X = \omega]$ is a Gibbs distribution with respect to $I, \mathcal{G}$.

Let $M = (I, Q, \pi, g_T)$ be a stochastic process, where I is a finite set of *indices* or *sites*, Q is a finite set of *states*, $\mathcal{Q} = Q^I = \{\alpha \mid \alpha \text{ maps } I \text{ into } Q\}$, π is an initial distribution on $\mathcal{Q}$, and g_T is the Gibbs distribution on $\mathcal{Q}$, defined by

$$g_T(\alpha) = \frac{e^{-U(\alpha)/T}}{Z},$$

where the local energy function $V_C : \mathcal{Q} \to \mathbb{R}$ for $C \in \mathcal{C}$ is such that if $\alpha, \beta \in \mathcal{Q}$, then $\alpha|_C = \beta|_C \Rightarrow V_C(\alpha) = V_C(\beta)$ and $U : \mathcal{Q} \to \mathbb{R}$ is defined by $U(\alpha) = \sum_{C \in \mathcal{C}} V_C(\alpha)$.

If $i \in I$ and $\omega \in \mathcal{Q}$, then

$$g_T(\omega, i) = \frac{e^{-U(\omega,i)/T}}{Z(i)},$$

where $U(\omega, i) : (\mathcal{Q} \times I) \to \mathbb{R}$ is defined by $U(\omega, i) = \sum_{i \in C} V_C(\omega)$, the sum being taken over all cliques $C \in \mathcal{C}$ that contain the index i, and where $Z(i) = \sum_{q \in Q} e^{-U(\omega(i,q),i)/T}$.

LEMMA 2.25
$Pr[\omega_i = q_i|\omega_j = q_j, i \neq j] = Pr[\omega_i = q_i|\omega_j = q_j, j \in G_i]$.

PROOF We first compute $Pr[\omega_i = q_i|\omega_j = q_j, i \neq j]$. By the definition of conditional probability,

$$\begin{aligned} Pr[\omega_i = q_i|\omega_j = q_j, i \neq j] &= \frac{Pr[\omega]}{\sum_{q \in Q} Pr[\omega(i,q)]} \\ &= \frac{g_T(\omega)}{\sum_{q \in Q} g_T(\omega(i,q))} \\ &= \frac{e^{-U(\omega)/T}/Z}{\sum_{q \in Q} e^{-U(\omega(i,q))/T}/Z} \\ &= \frac{e^{-U(\omega)/T}}{\sum_{q \in Q} e^{-U(\omega(i,q))/T}}. \end{aligned}$$

Next we compute $Pr[\omega_i = q_i|\omega_j = q_j, j \in G_i]$. To this end, define

$$\mathcal{Q}(\omega, i) = \{\alpha \in \mathcal{Q} \mid \alpha(j) = \omega(j) \text{ for all } j \in G_i \cup \{i\}\}$$

and

$$\widetilde{\mathcal{Q}}(\omega, i) = \{\alpha \in \mathcal{Q} \mid \alpha(j) = \omega(j) \text{ for all } j \in G_i\}.$$

Thus

$$\widetilde{\mathcal{Q}}(\omega, i) = \mathcal{Q}(\omega, i) \cup \{\alpha(i,q) \mid \alpha \in \mathcal{Q}(\omega, i), q \in Q\} = \{\alpha(i,q) \mid \alpha \in \mathcal{Q}(\omega, i), q \in Q\}.$$

By the definition of conditional probability,

$$\begin{aligned} Pr[\omega_i = q_i | \omega_j = q_j, j \in G_i] &= \frac{Pr[\omega_j = q_j, j \in G_i \cup \{i\}]}{Pr[\omega_j = q_j, j \in G_i]} \\ &= \frac{\sum_{\alpha \in \mathcal{Q}(\omega,i)} Pr[\alpha]}{\sum_{\alpha \in \widetilde{\mathcal{Q}}(\omega,i)} Pr[\alpha]} \\ &= \frac{\sum_{\alpha \in \mathcal{Q}(\omega,i)} \frac{1}{Z} \prod_C e^{-V_C(\alpha)/T}}{\sum_{\alpha \in \widetilde{\mathcal{Q}}(\omega,i)} \frac{1}{Z} \prod_C e^{-V_C(\alpha)/T}} \\ &= \frac{\sum_{\alpha \in \mathcal{Q}(\omega,i)} \left(\prod_{C:i \in C} e^{-V_C(\alpha)/T} \prod_{C:i \notin C} e^{-V_C(\alpha)/T} \right)}{\sum_{\alpha \in \widetilde{\mathcal{Q}}(\omega,i)} \left(\prod_{C:i \in C} e^{-V_C(\alpha)/T} \prod_{C:i \notin C} e^{-V_C(\alpha)/T} \right)}. \end{aligned}$$

Note that $i \in C$ implies that $C \in G_i \cup \{i\}$. For $\alpha \in \mathcal{Q}(\omega, i)$, we have $\alpha(j) = \omega(j)$ for all $j \in G_i \cup \{i\}$. Now if $i \in C \subseteq G_i \cup \{i\}$, then for each such α,

$$e^{-V_C(\alpha)/T} = e^{-V_C(\omega)/T}.$$

This allows us to factor out $e^{-V_C(\alpha)/T}$ from the numerator. Recall that $\widetilde{\mathcal{Q}}(\omega, i) = \mathcal{Q}(\omega, i) \cup \{\alpha(i, q) \mid \alpha \in \mathcal{Q}(\omega, i), q \in Q\}$. This, together with the previous remark and the trivial observation that $V_C(\alpha) = V_C(\alpha(i, q))$ for $i \notin C$, allows us to conclude that $Pr[\omega_i = q_i | \omega_j = q_j, j \in G_i]$ equals

$$\frac{\left(\prod_{C:i \in C} e^{-V_C(\omega)/T}\right) \left(\sum_{\alpha \in \mathcal{Q}(\omega,i)} \prod_{C:i \notin C} e^{-V_C(\alpha)/T}\right)}{\left(\sum_{q \in Q} \prod_{C:i \in C} e^{-V_C(\omega(i,q))/T}\right) \left(\sum_{\alpha \in \mathcal{Q}(\omega,i)} \prod_{C:i \notin C} e^{-V_C(\alpha)/T}\right)},$$

and thus

$$\begin{aligned} Pr[\omega_i = q_i | \omega_j = q_j, j \in G_i] &= \frac{e^{-\sum_{C:i \in C} V_C(\omega)/T}}{\sum_{q \in Q} e^{-\sum_{C:i \in C} V_C(\omega(i,q))/T}} \\ &= \frac{e^{-U(\omega,i)/T}}{Z(i)}. \end{aligned}$$

We can now complete the proof of the claim. Now

$$\begin{aligned} e^{-U(\omega)/T} &= \prod_{C:i \in C} e^{-V_C(\omega)/T} \prod_{C:i \notin C} e^{-V_C(\omega)/T} \\ &= e^{-U(\omega,i)/T} \prod_{C:i \notin C} e^{-V_C(\omega)/T}. \end{aligned}$$

For $q \in Q$, by replacing ω by $\omega(i, q)$ in the previous equation, we have

$$e^{-U(\omega(i,q))/T} = e^{-U(\omega(i,q),i)/T} \prod_{C:i \notin C} e^{-V_C(\omega(i,q))/T}.$$

Now for $i \notin C$, by the local property in the definition of V_C, $V_C(\omega(i,q)) = V_C(\omega)$. Thus

$$\begin{aligned}
Pr[\omega_i = q_i | \omega_j = q_j, i \neq j] &= \frac{e^{-U(\omega)/T}}{\sum_{q \in Q} e^{-U(\omega(i,q))/T}} \\
&= \frac{e^{-U(\omega,i)/T} \prod_{C: i \notin C} e^{-V_C(\omega)/T}}{\sum_{q \in Q} e^{-U(\omega(i,q),i)/T} \prod_{C: i \notin C} e^{-V_C(\omega)/T}} \\
&= \frac{e^{-U(\omega,i)/T}}{Z(i)} \\
&= Pr[\omega_i = q_i | \omega_j = q_j, j \in G_i].
\end{aligned}$$

■

It thus follows that if g_T is a Gibbs distribution, then X is a Markov random field, yielding a proof of the simpler direction in Theorem 2.24. Computationally, this means that

$$Pr[\omega_i = q_i | \omega_j = q_j, i \neq j] = \frac{e^{-U(\omega,i)/T}}{Z(i)},$$

which is something exploited in the Gibbs sampler algorithm, and in a later application to multiple sequence alignment.

In analogy to Theorem 2.21, S. Geman and D. Geman proved the convergence of a Markov random field to the Gibbs distribution. In the following, $g_T(\omega)$ is as before, and $g_0(\omega) = \lim_{T \to 0} g_T(\omega)$.

THEOREM 2.26 (S. GEMAN, D. GEMAN [GG84])
Let I be an index set of N elements, and let $X = (X_i : i \in I)$ be a Markov random field on I, and assume that the indices are visited in the order $n_0, n_1, \ldots$. Assume there exists M such that for all $t \geq 1$, $I \subseteq n_{t+1}, n_{t+2}, \ldots, n_{t+M}$. Then for every initial configuration $\eta \in \mathcal{Q}$ and configuration $\omega \in \mathcal{Q}$,

$$\lim_{t \to \infty} Pr[X(t) = \omega | X(0) = \eta] = g_T(\omega).$$

Moreover, if $T(t)$ is a decreasing sequence of temperatures satisfying

- $\lim_{t \to \infty} T(t) = 0$
- $T(t) \geq N\Delta / \ln t$ *for all* $t \geq t_0$, *for some temperature* $t_0 \geq 2$,

then

$$\lim_{t \to \infty} Pr[X(t) = \omega | X(0) = \eta] = g_0(\omega).$$

Given an arbitrary configuration $\omega \in \mathcal{Q}$, such as a digitized image with noise, Algorithm 2.3 (Gibbs sampler) converges to an image with noise removed.

In our previous notation, the expression in (4) is $\frac{e^{-U(\omega(i,q),i)/T}}{Z(i)}$. In practice, just as in Monte Carlo with simulated annealing, the annealing schedule goes from an initial high temperature to a final low temperature, where in each pass through the repeat loop, an assignment such as `T = 0.9 * T` is performed.

Algorithm 2.3 Gibbs sampler

```
n = 1; T = c;
repeat {
    choose random site i ∈ I
    set ω_i = q with probability
```

$$\frac{e^{-\sum_{C:i\in C} V_C(\omega(i,q))/T}}{\sum_{r\in Q} e^{-\sum_{C:i\in C} V_C(\omega(i,r))/T}}$$

```
    n = n + 1; T = c/ln n
} until T ≈ 0
```

2.1.6 Maximum Likelihood

After constructing a mathematical model for a biological process, one would like to determine how well the model predicted observed data. This is the case when constructing phylogenetic trees that relate various species and in constructing an appropriate Markov model to predict coding regions of the genome.

Suppose that the data sequence $\mathcal{O} = o_0, \ldots, o_{T-1}$ has been observed, and that M is a stochastic *model* that supposedly generates $\mathcal{O}$. The *likelihood* $L_{\mathcal{O}}(M)$ of the model with respect to the observation sequence $\mathcal{O}$ is defined by

$$L_{\mathcal{O}}(M) = Pr[\mathcal{O}|M].$$

When attempting to determine the best parameters for the model, a common statistical approach is to determine the *maximum likelihood*.[5] In simple cases, the function $L(M)$ can be expressed as a closed formula in one (or several) parameter(s), and a *local maximum* can be obtained by setting the derivative (partial derivatives) equal to 0. In the case of several parameters, a *saddlepoint* rather than a local maximum might be obtained. Here is a simple example to fix ideas.

Consider a coin tossing experiment with a possibly biased coin. The outcomes after n tosses are

$$x_1, \ldots, x_n,$$

where $x_i = 1$ if *heads* was obtained, otherwise $x_i = 0$. The *model* considered is a single coin tossing model with parameter p, indicating the bias of the coin: with probability p one tosses heads. The likelihood

$$L(p) = p^{x_1}(1-p)^{(1-x_1)} \cdots p^{x_n}(1-p)^{(1-x_n)} = p^{\sum x_i}(1-p)^{\sum(1-x_i)},$$

so the derivative $L'(p)$ is

$$\left(\sum x_i\right) p^{(\sum x_i)-1}[(1-p)^{\sum(1-x_i)}] - \left[\sum(1-x_i)\right](1-p)^{[\sum(1-x_i)]-1}[p^{\sum x_i}];$$

[5] *Warning*. There are examples where the likelihood function has no maximum.

so, setting the derivative equal to 0, we obtain

$$\begin{aligned}
\left(\sum_{i=1}^{n} x_i\right)(1-p) &= p\sum_{i=1}^{n}(1-x_i), \\
\sum_{i=1}^{n} x_i - p\sum_{i=1}^{n} x_i &= pn - p\sum_{i=1}^{n} x_i, \\
\sum_{i=1}^{n} x_i &= pn, \\
p &= \frac{\sum_{i=1}^{n} x_i}{n}.
\end{aligned}$$

Thus, the maximum-likelihood value of the parameter p is $\frac{\sum x_i}{n}$, which is as expected. Since the logarithmic function is monotonic increasing, $L(M)$ attains a maximum at the same values as $\log L_{\mathcal{O}}(M)$. Usually one determines the maximum of $\log L_{\mathcal{O}}(M)$, which, even in the above case, is substantially easier.

In more complicated situations, various optimization techniques can be applied, including gradient ascent, Newton–Raphson, Monte Carlo with simulated annealing, *expectation maximization* (EM) [DLR77, Wu83], etc.

2.2 Combinatorial Optimization

Suppose that Q is a finite set of states of a physical system, for which each state $i \in Q$ has an associated energy $E(i)$. We would like to determine the state for which energy is a minimum, i.e. to minimize the function $E : Q \to \mathbb{R}$. For instance, Q could be the set of self-avoiding walks along a 3-dimensional face-centered cubic lattice, representing possible backbone conformations of a protein, and for $i \in Q$, $E(i)$ could represent the free energy of conformation i. Such *combinatorial optimization* problems involving exponentially large state spaces can often be shown to be NP-hard. Monte Carlo, simulated annealing, genetic algorithms, etc. are useful heuristic algorithms for attacking such problems, and will be considered in this section.

2.2.1 Lagrange Multipliers

Suppose we want to determine a local maximum or minimum of the function

$$f(x_1, \ldots, x_n)$$

subject to $m < n$ additional restrictions $\phi_1(x_1, \ldots, x_n) = 0, \ldots, \phi_m(x_1, \ldots, x_n) = 0$. Extend the function f to F by defining $F(x_1, \ldots, x_n, \lambda_1, \ldots, \lambda_m)$ to be

$$f(x_1, \ldots, x_n) + \lambda_1\phi_1(x_1, \ldots, x_n) + \cdots + \lambda_m\phi_m f(x_1, \ldots, x_n).$$

Lagrange's method[6] consists of determining stationary points of F; i.e. those $a = (a_1, \ldots, a_n)$ such that $\frac{\partial F}{\delta x_i}(a) = 0$ for $1 \leq i \leq n$, and $\frac{\partial F}{\delta \lambda_i}(a) = \phi_i(a) = 0$ for $1 \leq i \leq m$. Such points are either stationary, non-extremal (e.g. saddle points), or local maxima or local minima of f that additionally satisfy the given constraints.

2.2.2 *Gradient Descent*

In order to minimize a differentiable energy function E, possibly of many variables, the heuristic of gradient descent is to repeatedly set $p = p + \Delta p$, where the increment Δp is chosen to be a constant times the gradient. Since the update is always in the direction of greatest decrease in the energy function, this heuristic is an example of greedy algorithm.

For example, one can use gradient descent to determine the most likely bias p of a coin, where i many heads and $n - i$ many tails were observed. In that case, corresponding to the likelihood $L(p) = p^i(1-p)^{n-i}$, we define energy $E = -\ln L(p) = -i \ln p - (n-i)\ln(1-p)$. Define the increment $\Delta p = -k\frac{d}{dp}(E) = -k[i/p - (n-i)/(1-p)]$, where k is a constant. If one programs this example with $n = 20$, $i = 10$, initial value of p as 0.001, and constant $k = 0.00001$, then after 5350 steps, convergence to the correct probability of $p = 0.5$ occurs. However, for $k = 0.0001$, the algorithm diverges. A difficulty of this method, which we shall see again applied to hidden Markov models is that it appears difficult to find suitable parameters for correct convergence.

2.2.3 *Heuristics Related to Simulated Annealing*

Algorithm 2.4 Threshold Accepting (TA)

```
1  threshold θ = c > 0
2  i = initial
3  min = E(x)
4  repeat {
5     repeat {
6        choose random j ∈ N_i
7        if (E(j) - E(i) < θ) then
8             i = j
9     } until no change in energy
10    lower threshold θ > 0
11 } until convergence
12 return i and E(i)
```

In [Due92] G. Dueck introduced several variants of simulated annealing, and cited performance statistics of these algorithms against simulated annealing for the benchmark of Grötschel's 442-city version of the Euclidean traveling salesman problem. Unlike the situation for the Metropolis–Hastings algorithm (Markov chain Monte Carlo), whose

[6] Louis de Lagrange, French mathematician, 1736–1813.

convergence follows because of the existence of stationary probabilities of an underlying Markov chain, to our knowledge there is no theoretical underpinning for the heuristics of [Due92]. Nevertheless, Dueck's *threshold accepting* algorithm (TA, Algorithm 2.4) and *record-to-record Travel* algorithm (RRT, Algorithm 2.5) could prove useful in applications. Recall that simulated annealing makes a move from configuration i to j with probability $\min(1, e^{-\Delta E/T})$, where $\Delta E = E(j) - E(i)$. In contrast, the threshold accepting algorithm makes a move from configuration i to j if $\Delta E < \theta$. The record-to-record travel (RRT) algorithm is even more streamlined.

Algorithm 2.5 Record-To-Record Travel (RRT)

```
1  ε = c < 0
2  i = initial
3  min = E(i)
4  repeat {
5      choose random j ∈ N_i
6      if (E(j) < ε) then i = j
7      if (E(j) < min) then min = E(j)
8  } until convergence
9  return min
```

In [Gol93], N. Goldman applied the RRT algorithm to determine artificial genetic codes, whose fault tolerance is greater than the natural genetic code. Using a genetic algorithm, in a project in our group at Munich, A. Macri (unpublished) obtained substantially more fault-tolerant artificial codes than those of Goldman.

2.2.4 *Applications of Monte Carlo*

Optimality of the Genetic Code

Glancing at the block-structured form of Table 1.3 in Chapter 1, it is clear that the genetic code is *fault-tolerant*, in the sense that transcription errors in the third codon position frequently do not influence the amino acid expressed (*wobble*). Moreover, errors in the other codon positions often lead to amino acids having similar chemical properties. Several articles ([HH91, Gol93, Giu89] etc.) have studied the question of *optimality* of the genetic code with respect to fault tolerance in 1-site transcription errors.

In restricting attention only to block-respecting codes (i.e. permutations of the 20 amino acids while respecting the block structure of the code), Haig and Hurst [HH91] considered to what extent the natural code has been optimized with respect to fault tolerance concerning (a) polar requirement,[7] (b) hydropathy, (c) molecular volume, and (d) isoelectric point.

A *general genetic code* c is simply an onto map[8] $c : \{A, C, G, U\}^3 \to \{1, \ldots, 21\}$ from the 64 codons onto the 20 amino acids plus the stop signal. We do not include the 21st amino

---

[7] *Polar requirement*, as measured by C. Woese et al. [WDD$^+$66], is taken to be synonymous with hydrophobicity.

[8] An *onto map*, or surjection, is a map that, to every element of the range, maps some element of the domain.

acid *selenocysteine*. A code is *block-structured* if

- $c(xyz) = c(x'y'z')$ holds for any two codons xyz, $x'y'z'$ coding the same amino acid in the natural code, and
- $c(UAA) = c(UAG) = c(UGA) = \texttt{Stop}$.

In other words, a code is block-structured if it simply permutes the 20 amino acids, while retaining the block form of the table. Finally, a *shuffled-codon* code allows changes in the block form of the table, while retaining the assumption that 3 amino acids have 6 codons, 4 amino acids have 4 codons, 1 amino acid has 3 codons, the stop signal has 3 codons, 9 amino acids have 2 codons, and 2 have 1 codon. There are

$$20! = 2432902008176640000 > 2.43 \times 10^{18}$$

block-respecting codes, and

$$\frac{64!}{(2!)^{10}(3!)^2(4!)^4(6!)^3} > 10^{65}$$

many shuffled codon codes (there are 2 amino acids having only one codon (Met, Trp), and in this computation, the stop signal is counted as having 3 codons). In comparison, the number of general codes can be calculated as follows. There are $21! \cdot S(64, 21)$ many general codes, where $S(n, m)$ is a Stirling number of the second kind. Since $S(64, 21) = 2.95572845518811 \times 10^{64}$, it follows that there are more than 1.51×10^{84} general codes (thanks to R. Matthes for the computation).

Following [HH91], define the polar requirement *mean square difference* $MS(c)$ of code c to be

$$\sum_{xyz \in \{A,C,G,T\}^3} \left[\frac{(2.7) + (2.8) + (2.9)}{C} \right], \tag{2.6}$$

where C is the number of 1-site mutations from a non-stop codon to a non-stop codon,[9] and

$$\sum_{x' \in \{A,C,G,T\} - \{x\}} D(c(xyz), c(x'yz)), \tag{2.7}$$

$$\sum_{y' \in \{A,C,G,T\} - \{y\}} D(c(xyz), c(xy'z)), \tag{2.8}$$

$$\sum_{z' \in \{A,C,G,T\} - \{z\}} D(c(xyz), c(xyz')), \tag{2.9}$$

the sums being taken over non-stop codons. If X, Y denote amino acids, then $D(X, Y) = [W(X) - W(Y)]^2$, where $W(X)$, $W(Y)$ respectively denote the polar requirement values from Table 1.2 in Chapter 1, as determined by Woese *et al.* [WDD+66].

Similarly, define fault tolerance in the first position $MS_1(c)$ of code c by considering only term (2.7) in (2.6), and by replacing the denominator C by C_1, the number of mutations of

---

[9] Without the restriction concerning non-stop codons, C would be $64 \cdot 3 \cdot 3$, since there are 64 codons, and for each codon 3 sites, and for a fixed codon and site 3 remaining nucleotides for the mutation choice.

the first position from a non-stop codon to a non-stop codon. Similarly, define MS_2 and MS_3 for the second and third positions. Thus $MS(c)$ is a measure of the *fault tolerance* of genetic code c, while $MS_1(c)$, $MS_2(c)$, $MS_3(c)$ measure the fault tolerance of c with respect to 1-site mutations in the first, second, and third positions, respectively. For clarity, in Algorithm 2.6, we list a function, written in C, that computes $MS(\text{code})$.

Algorithm 2.6 C program to compute *MS*(code)

```
float MS(gcode code) {
  int i, j, k, m;
  int num = 12*64;
      /* potentially this many swaps between codon (no stop) */
  float sum = 0.0;

  for (i = 0; i < 4; i++)
    for (j = 0; j < 4; j++)
      for(k = 0; k < 4; k++) {
        if (code[i][j][k] != STOP)
          for (m = 0; m < 4; m++) {
            if (  ( i != m )  && (code[m][j][k] != STOP) )
                sum += D[code[i][j][k]][code[m][j][k]];
            else
                num--; /* subtract 1 from number of swaps */
            if (  ( j != m )  && (code[i][m][k] != STOP) )
                sum += D[code[i][j][k]][code[i][m][k]];
            else
                num--;
            if (  ( k != m )  && (code[i][j][m] != STOP) )
                sum += D[code[i][j][k]][code[i][j][m]];
            else
                num--;
          }
        else
          num -= 12;   /* disallow swap of stop codon   */
      }
  return sum/num;
}
```

By measuring the values MS_1, MS_2, MS_3, MS for the mean square change in an attribute's value (e.g. polar requirement, molecular volume, etc.) for all single-base substitutions in first, second, third, and resp. for all three codon positions for the natural and random block-respecting codes, Haig–Hurst concluded that 'single-base substitutions are strongly conservative with respect to changes in polar requirement and hydropathy in the first and third codon positions, but much less so in the second codon position.' Moreover, the polar requirement mean square difference MS for the natural code was determined to be 5.194, while only 2 out of 10 000 random codes were found to be more conservative with respect to polar requirement (MS values of 5.167 and 5.189). Polar requirement MS_1, MS_2, and MS_3 values for the natural code were determined to be 4.88, 10.56, and 0.14, which reflects the

Table 2.1 Freeland–Hurst genetic code.

	U	C	A	G	
	Ile	Ala	Gln	His	U
	Ile	Ala	Gln	His	C
U	Cys	Ala	Stop	Stop	A
	Cys	Ala	Stop	Gly	G
	Cys	Leu	Thr	Ser	U
	Cys	Leu	Thr	Ser	C
C	Cys	Leu	Phe	Ser	A
	Cys	Leu	Phe	Ser	G
	Trp	Pro	Asp	Ala	U
	Trp	Pro	Asp	Ala	C
A	Trp	Pro	Glu	Ser	A
	Val	Pro	Glu	Ser	G
	Tyr	Met	Asn	Arg	U
	Tyr	Met	Asn	Arg	C
G	Tyr	Met	Lys	Arg	A
	Tyr	Met	Lys	Arg	G

fact that error tolerance is highest for transcription errors in the third codon position (i.e. on average there is greatest conservation of polar requirement for single-base substitutions in the second codon position).

In a recent paper, Freeland and Hurst [FH98] sharpened the results of Haig–Hurst by studying the polar requirement mean square difference where *transversions* are less likely to occur than *transitions*.[10] In 1 million randomly generated codes, they found only one code more fault tolerant than the natural code (code displayed in Table 2.1).

In contrast to the block-respecting codes of [HH91, FH98], Goldman [Gol93] considered the more general *shuffled codon* codes, which maintain the same number of codons per amino acid as in the natural code, but do not require the block structure of the natural code table. Goldman computed the mean square difference MS over all single-base substitutions (for amino acid, non-stop codons) for artificial codes obtained by a heuristic for optimization called the *record-to-record travel* algorithm [Due92]. The most conservative code found by in [Gol93] had polar requirement value $MS = 4.005$, and for this code more uniformly spread MS_1, MS_2, and MS_3 values of 3.06, 3.67, and 5.28.

In [Giu89], Di Giulio estimated that the natural code has achieved 68% minimization of polarity distance, by comparing the natural code with random block-respecting codes. When considering single base changes in the codons, let $N_{i,j}$ be the number of times the ith amino acid changes into the jth amino acid, and X_i be the polarity index [WDD$^+$66] of the ith

[10] A transversion is a mutation from a purine to a pyrimidine, or vice versa, while a transition is a mutation from a purine to a purine, or from a pyrimidine to a pyrimidine. Transitions are more frequent than transversions.

Table 2.2 Di Giulio's genetic code.

	U	C	A	G	
U	4.8	x_5	x_8	x_{13}	U
	4.8	x_5	x_8	x_{13}	C
	x_1	x_5	Stop	Stop	A
	x_1	x_5	Stop	x_{14}	G
C	x_1	5.4	x_9	x_{15}	U
	x_1	5.4	x_9	x_{15}	C
	x_1	5.4	x_{10}	x_{15}	A
	x_1	5.4	x_{10}	x_{15}	G
A	x_2	x_6	12.5	x_5	U
	x_2	x_6	12.5	x_5	C
	x_2	x_6	13.0	x_{15}	A
	x_3	x_6	13.0	x_{15}	G
G	x_4	x_7	x_{11}	7.9	U
	x_4	x_7	x_{11}	7.9	C
	x_4	x_7	x_{12}	7.9	A
	x_4	x_7	x_{12}	7.9	G

amino acid. DiGiulio's *percent minimization* is defined by

$$\frac{MS_{\text{mean}} - MS_{\text{code}}}{MS_{\text{mean}} - MS_{\text{low}}}, \tag{2.10}$$

where

$$MS(c) = \frac{\sum_{i,j}(X_i - X_j)^2 N_{i,j}}{\sum_{i,j} N_{i,j}},$$

MS_{mean} is the average $MS(c)$ value, obtained by averaging over many random block-respecting codes c, and MS_{low} is an approximation of the lowest possible MS value obtained using the method of Lagrange multipliers to solve a constrained minimization problem. Specifically, define the function

$$G(x_1, \ldots, x_{15}, \lambda) = MC(c) + \lambda\Phi$$

where c is the code in Table 2.2, and $\Phi = \sum_{i=1}^{15} x_i - 104.8$. DiGiulio selects the largest and smallest polar requirement values, along with median values, and places these values on the diagonal, since (with a small exception) no 1-site mutation from a diagonal element can mutate diagonal elements to other diagonal elements.[11] This produces diagonal values of 13.0, 12.5, The real-valued variables $x_1, \ldots, x_{15}$ represent polar requirement values (not required to lie in Table 1.2 in Chapter 1), which, together with the previously mentioned values add

[11] Without setting certain polar requirement values in the code, the minimum MS is trivially obtained by assigning each amino acid the polar requirement (hydrophobicity) value $148.4/20$.

up to 148.4, the total polar requirement in Table 1.2. Following Lagrange's method, set the partial derivatives of G equal to 0

$$\frac{\partial G}{\partial x_1} = 0, \quad \ldots, \quad \frac{\partial G}{\partial x_{15}} = 0, \quad \frac{\partial G}{\partial \lambda} = 0 \tag{2.11}$$

and then solve for $x_1, \ldots, x_{15}, \lambda$, to obtain a local extremum for the constrained optimization problem. Taking MS_{low} to be the mean square difference for the resulting solution to system (2.11), and applying (2.10), DiGiulio estimates that the natural genetic code has been optimized 68%.

2.2.5 *Genetic Algorithms*

Drawing on an analogy with sexual evolution (crossover and pointwise mutation), John Holland introduced the notion of *genetic algorithm* (GA), subsequently generalized in evolutionary programs. Our interest in GAs is in that of a combinatorial optimization tool for exploration of large search spaces. The general form is given in Algorithm 2.7.

Algorithm 2.7 Genetic algorithm

```
t = 0
initialize population P(t)
compute F(t)
best = argmax { F(x) | x ∈ P(t) }
repeat {
     t++
     amplify fit individuals
     crossover
     pointwise mutation
     compute F(t) for new population P(t)
     max = argmax { F(x) | x ∈ P(t) }
     if ( f(max) > f(best) ) best = max
}
until convergence  // cook until done
```

We now give a few further details to refine the previous pseudocode. The population size m is constant throughout the algorithm, where at the beginning of the program, the population is initialized to consist of random bit strings of length n. Suppose that at time t, the population $P(t)$ consists of bit strings (also called chromosomes) $x_1, \ldots, x_m$. Each bit string x_i has an associated fitness $f(x_i)$, where our goal is to determine where f attains its maximum. The fitness $F(t)$ of the entire population is just the sum of the individual fitness values; i.e. $F(t) = \sum_{i=1}^{m} f(x_i)$. For instance, in an application to protein folding on lattice models given in a later chapter, $f(x_i)$ will be the number of hydrophobic–hydrophobic contacts in a conformation on a 2-dimensional lattice, corresponding to a possible conformation of a given amino acid sequence.

In the amplification step, the intent is to produce a temporary population $P'(t)$ where the expected number of occurrences of x_i in $P'(t)$ is mp_i, and p_i is the relative fitness, given by

Algorithm 2.8 Roulette wheel

```
q_0 = 0
for i = 1 to n
    q_i = q_{i-1} + p_i
for i = 1 to m {
    z = random real in (0,1)
    find least i_0 such that z ≤ q_{i_0}
    output x_{i_0}
}
```

$p_i = \frac{f(x_i)}{F(t)}$. To achieve this, use cumulative probabilities and the *roulette wheel* technique, as given in Algorithm 2.8.

Suppose the individuals x_i of the population $P'(t)$ are now given in array A. We shuffle the contents of A by using a temporary array B, and then recopy B into A. This is described in Algorithm 2.9.

We can now pair A[0] with A[1], and A[2] with A[3], etc. and perform crossover. Crossover is performed for a given pair consisting of mother chromosome $a_1, \ldots, a_n$ and father chromosome $b_1, \ldots, b_n$, by randomly choosing a crossover position $1 \leq i < n$, and forming the new pair consisting of $a_1, \ldots, a_i, b_{i+1}, \ldots, b_n$ and $b_1, \ldots, b_i, a_{i+1}, \ldots, a_n$. Pointwise mutation is performed by changing each bit of each chromosome (toggling the bit in the case of bit strings) with some small probability. This completes our description of genetic algorithms. For more, consult the very readable references [Mic96, Mit98].

2.3 Entropy and Applications to Molecular Biology

In the latter part of the 19th century, Boltzmann introduced the concept of *entropy* as a measure of disorder in a closed container containing an ideal gas. In this section, following E.T. Jaynes [Jay57], we will show how Shannon's information theoretic notion of entropy can be used to derive Boltzmann's notion of thermodynamic entropy.

Algorithm 2.9 Shuffle

```
j=0; k=m;
for i = 1 to m {
    x = random integer in { 0,...,k-1 }
    B[j] = A[x]; j=j+1;
    swap( A[k-1], A[x] );
    k=k-1;
}
for i = 1 to m A[i] = B[i];
```

2.3.1 Information Theoretic Entropy

Let Σ be an alphabet of size N, and consider a word $w = w_1 \cdots w_n$ consisting of n letters from Σ, for instance $\Sigma = \{A, C, G, T\}$. What is the information content of w? Suppose that we have the partial word $w_1 \cdots w_i$ and have just ascertained w_{i+1}. How much extra information does w_{i+1} give us? Shannon's information theoretic concept of entropy was developed to answer these questions.

Suppose that we receive a symbol from an alphabet of size N, where letters are sent with equiprobability (i.e. uniform distribution). If a symbol is transmitted as a binary signal, then we must wait until all $\log_2 N$ bits of the symbol are sent before we know with certainty which symbol has been transmitted. Thus the *information* or *information theoretic entropy* contained in a symbol from an alphabet of size N is $\log_2 N$, under the uniform probability distribution. Assume that $p_1 = Pr[A_1], \ldots, p_N = Pr[A_N]$ are the probabilities of outputting characters A_i in a message, where $\sum_{i=1}^{N} p_i = 1$. Suppose that n, the length of a random message M (or length of nucleotide sequence) is large, and let $n_i = np_i$ be the expected number of occurrences of A_i. Then the message M belongs with high probability to a set of size given by the multinomial coefficients

$$N_n = \frac{N!}{n_1! \cdots n_N!}$$

representing the number of ways of partitioning N into a collection of sets of sizes $n_1, \ldots, n_N$. The average information should then equal

$$I = \frac{\log_2 N_n}{n}.$$

Stirling's formula from Theorem 2.1 yields that

$$N_n \sim \frac{\sqrt{2\pi n} n^n e^{-n}}{\sqrt{2\pi n_1} n_1^{n_1} e^{-n_1} \cdots \sqrt{2\pi n_N} n_N^{n_N} e^{-n_N}},$$

so that

$$\begin{aligned}
\ln N_n &\sim n \ln n - np_1 \ln(np_1) - \cdots - np_N \ln(np_N) \\
&= n \ln n - (np_1 + \cdots + np_N) \ln n - np_1 \ln p_1 - \ldots - np_N \ln p_N \\
&= -n \sum_{i=1}^{N} p_i \ln p_i.
\end{aligned}$$

Since ln and $\log_2$ are related by a constant, it follows that

$$\log_2 N_n \sim -n \sum_{i=1}^{N} p_i \log_2 p_i,$$

thus motivating the definition of *entropy* $H(p_1, \ldots, p_n)$ in Shannon's formula:

$$H(p_1, \ldots, p_n) = I = \frac{\log_2 N_n}{n} = -\sum_{i=1}^{N} p_i \log_2 p_i. \tag{2.12}$$

Shannon's entropy, sometimes called *information*, is usually defined using logarithms to the base 2. However in various settings, we may consider natural logarithms or logarithms to another base – this modifies the value only by a multiplicative constant. By convention $0 \cdot \log_2(0) = 0$, so for any probability distribution $p_1, \ldots, p_n$ which concentrates on i_0 in the sense that $p_i = 1$ if $i = i_0$, otherwise $p_i = 0$, we have that $H(p_1, \ldots, p_n) = 0$.

2.3.2 *Shannon Implies Boltzmann*

In this section, following [Jay57], we show how Shannon's entropy function gives rise to the Boltzmann probability distribution, and can be used to derive Boltzmann's law. Thus information theoretic entropy appears to be a more primitive notion than that of energy and gas kinetics.

In [Khi57] Khinchin proved that information theoretic entropy

$$H(p_1, \ldots, p_n) = -\sum_{i=1}^{n} p_i \ln p_i \tag{2.13}$$

achieves a unique maximum for the uniform distribution;[12] i.e. $p_i = \frac{1}{n}$ for $1 \leq i \leq n$. This is intuitively clear, since under equiprobability of sending a character in an alphabet Σ of size n, the information (or entropy) of a character is $\log_2 n$, i.e. until all binary bits in the transmitted message are sent, we do not know the proper character transmitted.

For small values of n, Khinchin's result is easy to derive using (partial) differentiation. Suppose $0 \leq p, q \leq 1$ satisfy $p + q = 1$. Using (2.13), define the function $F(p) = H(p, 1-p) = -[p \ln p + (1-p) \ln(1-p)]$. Then

$$\frac{dF}{dp} = -[1 + \ln p - 1 - \ln(1-p)] = -[\ln p - \ln(1-p)]$$

Thus $\frac{dF}{dp} = 0$ iff $\ln p = \ln(1-p)$ iff $p = 1 - p$, so that $p = \frac{1}{2} = q$.

The case for probabilities p, q, r summing to 1 is left as an exercise. The general case of Khinchin's result can be proved using the method of Lagrange multipliers. Define

$$\begin{aligned} h(p_1, \ldots, p_n, \lambda) &= H(p_1, \ldots, p_n) + \lambda \left(\sum_{i=1}^{n} p_i - 1 \right) \\ &= -\sum_{i=1}^{n} p_i \ln p_i + \lambda \left(\sum_{i=1}^{n} p_i - 1 \right). \end{aligned}$$

Setting the partial derivatives to 0, we obtain

$$\frac{\partial h}{p_i} = -(1 + \ln p_i) + \lambda = 0,$$

so that $p_i = e^{\lambda - 1}$ for $1 \leq i \leq n$. Additionally, we have the requirement that $\sum_{i=1}^{n} p_i = 1$ and so $1 = ne^{\lambda - 1}$; hence $\lambda - 1 = \ln(1/n)$. From this expression, we have that $p_i = e^{\ln(1/n)} = 1/n$ for $1 \leq i \leq n$, thus proving that the uniform distribution yields a stationary point for

---

[12] Laplace's *principle of insufficient reason* states that if we have no knowledge to the contrary, then we should assume that events are equiprobable. Thus Khinchin's result states that Laplace's principle is justified by maximal entropy.

the entropy function. Additional scrutiny shows that it is the unique maximum of the entropy function.

In [Jay57], Jaynes considered computing the maximum of the entropy function, under the additional requirement that the average energy

$$\langle E(x)\rangle = \sum_{i=1}^{n} p_i E_i(x)$$

is known. Surprisingly, this approach leads immediately to Boltzmann's probability distribution, in place of the uniform distribution in Khinchin's result. Let us start with some definitions.

Assume that variable x can take on values in the discrete space $\Omega = \{x_1, \ldots, x_n\}$, where

$$p_i = Pr[x = x_i].$$

Suppose the probabilities p_i are not known, but only the expected values

$$\langle E(x)\rangle = \sum_{i=1}^{n} p_i E(x_i) \tag{2.14}$$

of a function E. From this point on, let us fix x and write E_i in place of $E(x_i)$. As the p_i are probabilities summing to 1,

$$\sum_{i=1}^{n} p_i = 1. \tag{2.15}$$

The Shannon entropy function H measures the *entropy* or *uncertainty* of the probability distribution $p_1, \ldots, p_n$, and is defined by

$$H(p_1, \ldots, p_n) = -\sum_{i=1}^{n} p_i \ln p_i. \tag{2.16}$$

In the absence of other criteria, the most likely probability distribution satisfying equations (2.14) and (2.15) is that having maximum entropy (2.16). To maximize (2.16) subject to the constraints (2.14) and (2.15), we use the method of Lagrange multipliers. Let $a = a_1, \ldots, a_n$ be a local maximum of H, and α, β be such that

$$\frac{\partial H}{\partial p_j}(a) = \alpha \frac{\partial}{\partial p_j}\left(\sum_{i=1}^{n} p_i E_i - \langle E\rangle\right)(a) + \beta \frac{\partial}{\partial p_j}\left(\sum_{i=1}^{n} p_i - 1\right)(a) \tag{2.17}$$

for $j = 1, \ldots, n$. Now $\langle E\rangle$ and $E_1, \ldots, E_n$ are constant with respect to the variables $p_1, \ldots, p_n$, so

$$\begin{aligned} \frac{\partial}{\partial p_j}\left(\sum_{i=1}^{n} p_i E_i - \langle E\rangle\right) &= E_j, \\ \frac{\partial}{\partial p_j}\left(\sum_{i=1}^{n} p_i - 1\right) &= 1; \end{aligned}$$

hence

$$\frac{\partial H}{\partial p_j} = \alpha E_j + \beta. \tag{2.18}$$

From equation (2.16),

$$\frac{\partial H}{\partial p_j} = -(1 + \ln p_j); \tag{2.19}$$

hence from equations (2.18) and (2.19), we have

$$\ln p_j = -\alpha E_j - (\beta + 1),$$

so that

$$p_j = e^{-\alpha E_j - (\beta+1)}.$$

Write the previous equation as

$$p_j = e^{-\lambda - \mu E_j}. \tag{2.20}$$

Substitute (2.20) into (2.14) and (2.15) to solve for λ, μ. Thus from (2.14),

$$1 = \sum_{i=1}^{n} p_i = \sum_{i=1}^{n} e^{-\lambda - \mu E_i},$$

so

$$e^{\lambda} = \sum_{i=1}^{n} e^{-\mu E_i}. \tag{2.21}$$

From (2.15),

$$\langle E \rangle = \sum_{i=1}^{n} p_i E_i = \frac{\sum_{i=1}^{n} e^{-\mu E_i} E_i}{e^{\lambda}} = \frac{\sum_{i=1}^{n} e^{-\mu E_i} E_i}{\sum_{i=1}^{n} e^{-\mu E_i}}.$$

Let

$$Z(\mu) = \sum_{i=1}^{n} e^{-\mu E_i} \tag{2.22}$$

be the *partition function*. Differentiation of $Z(\mu)$ with respect to μ shows that

$$\langle E \rangle = -\frac{\partial}{\partial \mu} \ln Z(\mu), \tag{2.23}$$

and by taking logarithms of (2.21),

$$\lambda = \ln Z(\mu). \tag{2.24}$$

From (2.20)–(2.22), we have the maximal entropy probability distribution

$$p_j = \frac{e^{-\mu E_j}}{Z(\mu)}.$$

By (2.20), (2.15), and (2.14) the distribution has entropy

$$\begin{aligned} H_{\max} &= -\sum_{i=1}^{n} p_i \ln p_i \\ &= -\sum_{i=1}^{n} p_i(-\lambda - \mu E_i) \\ &= \lambda \sum_{i=1}^{n} p_i + \mu \sum_{i=1}^{n} p_i E_i \\ &= \lambda + \mu \langle E \rangle. \end{aligned}$$

From the kinetic theory of gases,

$$p_j = \frac{e^{-E_j/kT}}{\sum_{i=1}^{n} e^{-E_i/kT}}, \tag{2.25}$$

where T is the absolute temperature in degrees Kelvin and K is Boltzmann's constant. As outlined in [Jay57], this approach, as applied to statistical mechanics, where averages of the energy levels E_i of a system are known, yields $\lambda_1 = \frac{1}{kT}$, so that

$$G = U - TS = -kT \ln Z, \tag{2.26}$$

$$\tag{2.27}$$

and, as in equation (2.23),

$$S = -\frac{\partial G}{\partial T} = -\frac{1}{kT} \sum_{i=1}^{n} p_i \ln p_i. \tag{2.28}$$

Here, T is absolute temperature, S is thermodynamic entropy, U is *internal energy* (in a molecule, this is *enthalpy*, i.e. energy from ionic, hydrogen and covalent bonds, etc.). and G is (Gibbs) free energy.

2.3.3 Simple Statistical Genomic Analysis

We have just seen Khinchin's result that the maximum entropy is achieved with the uniform distribution; i.e. maximum entropy is

$$H_1^{\max} = -\sum_{i=1}^{n} \frac{1}{n} \log_2 \frac{1}{n} = \log_2 n,$$

or approximately the number of bits in the binary representation of n. In the case of nucleotides A, C, G, T, the maximum entropy is then $-\log \frac{1}{4} = 2$.

When analyzing a nucleotide sequence, one can measure the frequencies $p(A), p(C), p(G), p(T)$ of A, C, G, T, and define the nucleotide entropy

$$H_1 = -p(A)\log p(A) - p(C)\log p(C) - p(G)\log p(G) - p(T)\log p(T).$$

Following [Gat72], the *divergence* D_1 from equiprobability is defined as

$$D_1 = H_1^{\max} - H_1.$$

A computation using TIGR's sequence data for the *M. jannaschii* genome yields frequencies

$$p(G) = 0.157, \quad p(A) = 0.344, \quad p(T) = 0.343, \quad p(C) = 0.155$$

and so the mononucleotide entropy H_1 is 1.89653 bits.[13] Since maximal entropy $H_1^{\max}$ is 2, it follows that D_1 is 0.103473. A similar computation yields the frequencies

$$p(G) = 0.190, \quad p(A) = 0.310, \quad p(T) = 0.308, \quad p(C) = 0.192$$

for *Haemophilus influenzae* (6824 main), with corresponding mononucleotide entropy H_1 of 1.9591 bits, and divergence $D_1 = 2 - 1.9591 = 0.0409$ bits.

One can compute the dinucleotide entropy, i.e. with respect to all subwords of length 2 (and more generally for any fixed length). Define

$$H_2 = -p(AA)\log p(AA) - p(AC)\log p(AC) - p(AG)\log p(AG) - p(AT)\log p(AT) - \cdots.$$

A computation using TIGR's genomic data yields dinucleotide frequency data for *M. jannaschii* given in the following table:

	G	A	T	C
G	0.034	0.057	0.039	0.027
A	0.060	0.134	0.111	0.039
T	0.055	0.098	0.134	0.056
C	0.008	0.055	0.059	0.033

A computation then gives $H_2 = 3.760$ bits for *M. jannaschii*. The following table consists of dinucleotide frequencies for *M. jannaschii* assuming the independence of occurrence of different nucleotides, i.e. $p(AA) = p(A)p(A)$, $p(AC) = p(A)p(C)$, etc:

	G	A	T	C
G	0.024649	0.054008	0.053851	0.024335
A	0.054008	0.118336	0.117992	0.05332
T	0.053851	0.117992	0.117649	0.053165
C	0.024335	0.05332	0.053165	0.024025

A computation then yields $H_2^{\text{ind}} = 3.787$ bits, so that $D_2 = H_2^{\text{ind}} - H_2 = 3.787 - 3.760 = 0.027$ bits.

Sometimes a genomic entropy plot is made, where the nucleotide or dinucleotide entropy of the contents of a window of the genome is plotted as a function of the starting position of the window. See Figure 2.1 for an example.

When presented with raw data consisting of millions of nucleotides, an immediate consideration is to perform a frequency count of nucleotides, dinucleotides, etc. Moreover, in coding regions the G,C content may be higher than A,T content, perhaps because of the

---

[13] Data from *M. jannaschii* 1070 main, 1069 ECL, 1063 ECS. The computed frequencies of Watson–Crick base pairs are not equal because of uncertain sequencing data such as N, R, Y, etc.

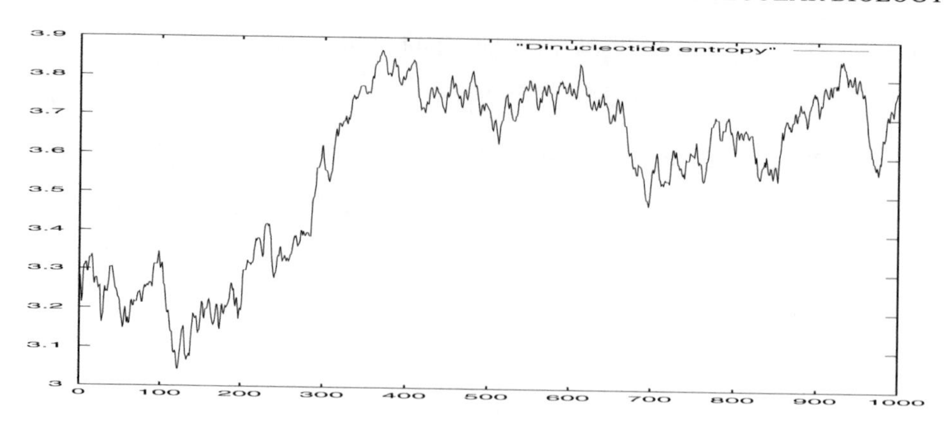

Figure 2.1 Dinucleotide entropy plot of *M. jannaschii* genome fragment.

additional stability provided by three rather than two hydrogen bonds between G and C. For instance, *Haemophilus influenza* has 1.83 Mb with G + C content of 38%; *Mycoplasma genitalia* has 580 kb with G + C content of 32%; *Mycoplasma genitalia* has 816 kb with G + C content of 40%; *Methanococcus jannaschi* has 1.665 Mb with G + C content of 31%. Following Karlin [Kar97a], define the following *odds ratio* measures for dinucleotides and trinucleotides, where x, y, z are nucleotide letters A, C, G, T:

$$\begin{aligned} \rho_{x,y} &= \frac{f_{x,y}}{f_x f_y}, \\ \beta_{x,y,z} &= \frac{f_{x,y,z} f_y}{f_{x,y} f_{y,x}}, \\ \gamma_{x,y,z} &= \frac{f_{x,y,z} f_x f_y f_z}{f_{x,y} f_{y,z} f_{x,z}}. \end{aligned}$$

For a random nucleotide sequence s of length n, Karlin observed that the *relative abundance* $\rho_{x,y}$ is approximately $\frac{c}{\sqrt{n}}$, where c is a constant, and using simulations has shown that for $n = 10^5$, $\rho_{x,y}$ lies between 0.92 and 1.08. From Karlin's observations reported in [Kar97a], the dinucleotide TA has almost universally low relative abundance $\rho_{T,A}$, while the relative abundance of GC is persistently high in gamma enterobacteria, etc. Exactly what such measures indicate is largely unknown, though it is speculated that low TA dinucleotide relative abundance is because of the low thermodynamic stability of TA hydrogen bonds, and the fact that TA is part of certain regulatory sequences such as the TATA box, and the terminator signal AATAAA in higher eukaryotes. Relative abundances for *M. jannaschii*, as computed from TIGR's genomic data, are given in Table 2.3.

Following [Kar97a], for nucleotide sequences $\mathbf{a} = a_1 \cdots a_n$ and $\mathbf{b} = b_1 \cdots b_m$, define

$$\delta^*(\mathbf{a}, \mathbf{b}) = \frac{1}{16} \sum_{i \in \{A,C,G,T\}} \sum_{j \in \{A,C,G,T\}} |\rho_{i,j}(\mathbf{a}) - \rho_{i,j}(\mathbf{b})|$$

Thus $\delta^*(\mathbf{a}, \mathbf{b})$ measures the average dinucleotide relative abundance between two nucleotide sequences. Noting that species appear to have an approximately constant *genomic signature* when measured over 50 kbp (50 kilobasepairs) or larger, Karlin observes that δ^* could be used to construct phylogeny trees for very unrelated species.

Table 2.3 Relative abundance data for *M. jannaschii*.

	G	A	T	C
G	1.389	1.051	0.714	1.123
A	1.103	1.131	0.944	0.728
T	1.027	0.830	1.137	1.047
C	0.320	1.033	1.109	1.375

2.3.4 Genomic Segmentation Algorithm

In [RRBGO98], a segmentation algorithm is introduced, which computes a partition of an input DNA sample, yielding segments or regions of the genome where the purine/pyrimidine entropy is homogeneous. This is done by iteratively computing segmentation points for which a maximum, statistically significant *Jensen–Shannon divergence* from the DNA background entropy is obtained. In this section, we explain the notion of divergence, and present pseudocode for the algorithm developed in [RRBGO98].

Specifically, given a DNA sequence of length n, consider the sequence $W = w_1 \cdots w_n \in \{R, Y\}^*$ of corresponding purines (R) and pyrimidines (Y).[14] The sequence W can be broken into two segments U, V, where $U = w_1 \cdots w_m$ and $V = w_{m+1} \cdots w_n$ for some $1 \leq m < n$. Suppose that W is a sequence of length n, consisting of r purines. Define the entropy $H(W) = -\frac{r}{n}\log_2(\frac{r}{n}) - \frac{n-r}{n}\log_2(\frac{n-r}{n})$, and similarly for U, V.

The *Jensen–Shannon divergence* $JS_2(U, V)$ is defined by

$$JS_2(U, V) = H(W) - \frac{m}{n}H(U) - \frac{n-m}{n}H(V)$$

where W is the concatenation of U and V, written $W = UV$, and $m = |U|, n - m = |V|$.

To gain intuition for this notion, let us consider several simple examples. Clearly the entropy of any sequence composed only of purines is 0. Thus if $W = \texttt{RR...R}$, then $JS_2(U, V) = 0$ for all segmentations of the form $W = UV$. Consider now the example of a sequence $W = \texttt{RYRYRY ... RY}$ of alternating purines and pyrimidines of even length n. Clearly $H(W) = 1$. For any segmentation of the form $W = UV$, where U, V are both of even length, $H(U) = 1 = H(V)$, and hence $JS_2(U, V) = 0$. Suppose now that U is a sequence of purines of length m, and V a sequence of pyrimidines of length k, and that $W = UV$. Then $H(U) = 0 = H(V)$ and $JS_2(U, V) = H(W) = -\frac{m}{m+k}\log_2(\frac{m}{m+k}) - \frac{k}{m+k}\log_2(\frac{k}{m+k}) > 0$. Moreover any other segmentation of W into different U', V' will have smaller Jensen–Shannon divergence. Consult Figure 2.2 for a graph of divergence as a function of segmentation point for a fragment of the *M. jannaschii* genome.

The *statistical significance* of the segmentation $W = UV$, where $|W| = n$, $|U| = m$, $|V| = n - m$, is defined as the probability that a random sequence, having the same base composition and length, when split into a first segment of length m and a second segment of length $n - m$, has Jensen–Shannon divergence at most that of $JS_2(U, V)$. For short sequences, this can be computed exactly using the hypergeometric distribution, where $h(n, r; m, k)$ is the probability of drawing k red balls in a sample of size m, given that there are r red balls in

[14] In [RRBGO98] it is reported that better results were obtained for the alphabet R,Y (purine, pyrimidine), than for S (strong, i.e. cytosine, guanine) and W (weak, i.e. adenine, thymine).

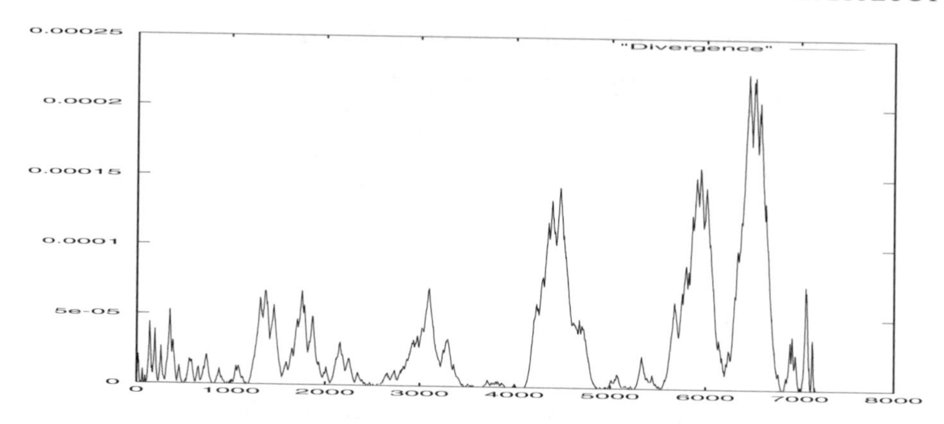

Figure 2.2 Jensen–Shannon divergence of 717112 base fragment of *M. jannaschii* genome as a function of segmentation point (x-axis in units of 100 bases).

a collection of n balls. For longer sequences, a more easily calculable approximation to the hypergeometric distribution can be used [BGRRO96]. Algorithm 2.10 computes, given the segmentation $W = UV$, the statistical significance s, where $0 \leq s \leq 1$.

If W is broken into contiguous segments $W = U_1 \cdots U_m$, with $|U_i| = \ell_i$, and $\sum \ell_i = n$, then

$$
\begin{aligned}
JS_m(U_1, \ldots, U_m) &= H(W) - \sum_{i=1}^{m} \frac{\ell_i}{n} H(U_i) \\
&= \sum_{i=1}^{m} \frac{\ell_i}{n} [H(W) - H(U_i)]
\end{aligned}
$$

represents the weighted sum of histogram entropy divergences of the domains $U_1, \ldots, U_m$ from the average, a measure of the number and compositional bias of the domains in the segmentation. Finally define $JS^*_m(s)$ to be the largest value $JS_m(U_1, \ldots, U_m)$ over all possible segmentations of W into m contiguous segments $U_1 \cdots U_m$, provided the statistical significance of $U_1 \cdots U_m$ is at least s. Note that $JS^*_m(s)$ monotonically increases as s decreases, since for $s' < s$, more partitions are considered. Let $JS^*(s)$ be the maximum of $JS^*_m(s)$ over all $2 \leq m \leq n$.

In [BGRRO96], a heuristic algorithm for approximating $JS^*(s)$ is given. Namely iteratively split a segment R into subsegments S, T that maximize $JS_2(S,T)$, provided that the statistical significance of the subsegmentation S, T is at least s. (This avoids the trivial segmentation where each segment consists of a single nucleotide.) To formalize this, let `statSig(i,j,m)` represent the statistical significance, as computed by Algorithm 2.10 with $U = w_i \cdots w_m$, $V = w_{m+1} \cdots w_j$, $|U| = m - i + 1$, $|V| = j - m$. Note that the main loop in this algorithm is of the form `for k = 0 to m`, except that we rule out non-positive arguments of the logarithm function. This leads to the form `for k =` $\max(0, r + m - n)$ `to` $\min(m, r)$. Now we have Algorithm 2.11 for the heuristic to approximate $JS^*(s)$.

In [RRBGO98], it is claimed without proof that the computation of $JS^*(s)$ is an NP-complete problem. It would be interesting to provide a proof of this assertion as well as an analysis of how well the heuristic `segment(i,j,s)` performs; i.e. for what values of ϵ

Algorithm 2.10 StatSignificance(U,V)

$s = 0$

$n = ||W||$

$m = ||U||$

r = number of purines in W

r_u = number of purines in U

// $n - r$ = number of pyrimidines in W

// $r - r_u$ = number of purines in V

$H(W) = -\frac{r}{n}\log_2(\frac{r}{n}) - \frac{n-r}{n}\log_2(\frac{n-r}{n})$

for k = $\max(0, r + m - n)$ to $\min(m, r)$ {

let U^* have k purines and m-k pyrimidines

let V^* have r-k purines and n-m-(r-k) pyrimidines

// $m - r_u$ = number of pyrimidines in U

// $n - m - (r - r_u)$ = number of pyrimidines in V

$H(U^*) = -\frac{k}{m}\log_2(\frac{k}{m}) - \frac{m-k}{m}\log_2(\frac{m-k}{m})$

$H(V^*) = -\frac{r-k}{n-m}\log_2(\frac{r-k}{n-m}) - \frac{n-m-(r-k)}{n-m}\log_2(\frac{n-m-(r-k)}{n-m})$

if $(H(W) - \frac{k}{n}H(U^*) - \frac{n-k}{n}H(V^*) \leq JS_2(U,V))$

$s = s + h(n, r; m, k)$

}

return s

Algorithm 2.11 segment(i,j,s)

```
void segment( int i, int j, double s) {
   max=0; splitPoint=undefined;
   for k=1 to j-1{
      U = w_1 ··· w_k; V = w_{k+1} ··· w_j;
      if statSig(U,V) > max then {
         max = statSig(U,V); splitPoint = k;
      }
      if max > s then {
         k = splitPoint; segment(i,k,s);
         output k; segment(k+1,j,s);
      }
   }
}
```

does the above algorithm output a segmentation whose divergence is within factor ϵ of the maximum $JS^*(s)$?

2.4 Exercises

Probability

1. It is estimated that the nucleotide substitution rate λ per site per year for nuclear DNA of higher primates is 1.3×10^{-9}. Histone H4 consists of 105 amino acids, and hence 315 nucleotides. Assuming that nucleotide substitutions occur uniformly across the 315 sites of histone H4, what is the least number of nucleotide substitutions, where with probability at least 0.5 two substitutions have occurred at the same site? Using λ, estimate the amount of time that has elapsed for this to occur.
   HINT Determine the least n such that $\prod_{i=0}^{n-1} \frac{315-i}{315} < 0.5$ and compute the time n/λ.
2. Before sequencing was widespread, A. Kornberg[15] pioneered a laboratory technique to measure the relative frequencies $p(A|A)$, etc. in a nucleotide sequence, where $p(X|Y)$ is the conditional probability that nucleotide X follows nucleotide Y. Suppose that one has all the 16 relative frequencies. Show how to compute the base frequencies $p(A), p(C), p(G), p(T)$.
   HINT

$$\begin{aligned}
p(A)p(A|A) + p(C)p(A|C) + p(G)p(A|G) + p(T)p(A|T) &= p(A), \\
p(C)p(C|C) + p(C)p(C|C) + p(G)p(C|G) + p(T)p(C|T) &= p(C), \\
p(G)p(G|G) + p(C)p(G|C) + p(G)p(G|G) + p(T)p(G|T) &= p(G), \\
p(T)p(T|T) + p(C)p(T|C) + p(G)p(T|G) + p(T)p(T|T) &= p(T).
\end{aligned}$$

   This yields a linear system of 4 equalities with 4 unknowns.
3. Write a program to compute powers P^n of the matrices given below. State whether each matrix is stochastic, doubly-stochastic, or substochastic, and whether the corresponding Markov chains are irreducible and aperiodic. For the latter, compute the stationary probabilities.

(a)

$$P = \begin{pmatrix} 0 & 1 & 0 \\ 1 & 0 & 0 \\ 0 & 0 & 1 \end{pmatrix}.$$

(b)

$$P = \begin{pmatrix} 0.5 & 0.5 & 0 \\ 0 & 0.5 & 0.5 \\ 0.5 & 0 & 0.5 \end{pmatrix}.$$

[15] Nobel Prize for discovery of DNA polymerase.

(c)

$$P = \begin{pmatrix} 0.5 & 0.0 & 0.5 \\ 0.25 & 0.25 & 0.5 \\ 0.25 & 0.75 & 0.0 \end{pmatrix}.$$

(d)

$$P = \begin{pmatrix} 0.5 & 0.0 & 0.5 \\ 0.9 & 0.0 & 0.1 \\ 0.25 & 0.75 & 0.0 \end{pmatrix}.$$

(e)

$$P = \begin{pmatrix} 0.5 & 0.0 & 0.25 \\ 0.25 & 0.25 & 0.25 \\ 0.25 & 0.75 & 0.0 \end{pmatrix}.$$

4. Determine the maximum likelihood of the coin-flipping example from Section 2.1.6,

$$L(p) = p^{x_1}(1-p)^{(1-x_1)} \cdots p^{x_n}(1-p)^{(1-x_n)} = p^{\sum x_i}(1-p)^{\sum(1-x_i)},$$

by computing the maximum of $\log L(M)$.
5. A car drives through n stoplights. For each stoplight, the probability of a red light is 0.6, while the probability of a green light is 0.4 (there is no yellow light). Compute the probability $Pr[X = k]$, where $X = k$ is the event that the car stops for the first time in going through the kth stoplight. Here $X = 0$ is the event that the car stops at no stoplight. Compute the expectation $E[X]$, i.e. the average number of stoplights at which a car first stops. What is the name of this probability distribution?

Combinatorial Optimization

6. Consider the function

$$f(x) = x^4 + 2x^3 - x^2 - x + 2 \tag{2.29}$$

on the closed interval $[-2, 2]$. Implement a Monte Carlo program with simulated annealing in order to determine the minimum of f in this interval.
7. Determine the maximum of the function in the previous exercise by implementing a genetic algorithm.
HINT Divide the interval $[-2, 2]$ into M subintervals, where $M = 2^n$ (for 2-place decimal accuracy, let $M = 512$, which is larger than 4×10^2). Your program should maintain a population of size $m \ll M$ of bit strings of length n, corresponding to appropriate subintervals. Fitness of the bit string $x = x_{n-1}, \ldots, x_0 \in P(t)$ is $f(-2 + 4\frac{\sum_{i=0}^{n-1} x_i \cdot 2^i}{2^n})$, where f is given in equation (2.29).
8. Using gradient descent, find the minimum of f, where f is given in equation (2.29).
9. Certain virus genomes actually code for different proteins in overlapping regions by shifting the reading frame; this fantastic optimization of code allows for very short genomes. Devise a method of testing whether the genetic code has been optimized in part to allow overlapping reading frames, for the current proteins in the PDB.

10. In a similar manner to the simulated annealing algorithm described in the text, implement a genetic algorithm to find optimized (artificial) genetic codes.
11. Define $O(n, m)$ to be the number of onto maps from $\{1, \ldots, n\}$ onto $\{1, \ldots, m\}$. Give a recurrence relation for $O(n, m)$ and write a program to compute $O(n, m)$. Using this, determine the number of general genetic codes.
    HINT. Note that $O(n+1, m+1) = (m+1)O(n, m) + (m+1)O(n, m+1)$, where the first term arises by mapping the last element $n+1$ to one of the $m+1$ elements of the range, and then mapping the remaining n elements of the domain in an onto manner to the remaining m elements of the range.
12. S. Kauffman [Kau70] introduced the *random boolean cellular automaton* as a possible model for the regulation of gene expression (how genes are turned off and on).
    A *boolean cellular automaton* B is given by $(G, \vec{f}, \vec{x})$, where
    - $G = (V, E)$ is a directed graph, $V = \{1, \ldots, n\}$, and $E \subseteq V \times V$ (so that loops but no multiple directed edges are allowed);
    - $\vec{f} = (f_1, \ldots, f_n)$, where for each $1 \leq i \leq n$, the fan-in of vertex i is $m(i)$ and $f_i : \{0,1\}^{m(i)} \to \{0,1\}$;
    - $\vec{x} = (x_1, \ldots, x_n) \in \{0,1\}^n$.

    Suppose that vertex i of fan-in $m(i)$ has in-edges from vertices $v_{i,1} < \cdots < v_{i,m(i)}$. Then for state $\vec{y} = (y_1, \ldots, y_n) \in \{0,1\}^n$, define

    $$B(\vec{y}) = (f_1(v_{1,1}, \ldots, v_{1,m(1)}), \ldots, f_n(v_{n,1}, \ldots, v_{n,m(n)})).$$

    The state of $B = (G, \vec{f}, \vec{x})$ at time 0 is the *initial state* $\vec{x}$. At time 1, the state of B is $B(\vec{x})$, and generally at time t the state of B is $B^{(t)}(\vec{x}) = B(B(\cdots B(\vec{x}) \cdots))$ where there are t occurrences of B.

    Boolean cellular automata were simulated on computer by S. Kauffman, who reported a surprising stability manifested by random automata. The first formal proofs of certain stable behavior were worked out by Luczak and Cohen [LC91], and a refutation of one of Kauffman's claims was given by J. Lynch [Lyn93, Lyn95]. Write a simulation program for Kaufmann's boolean cellular automata. Your program should be able to read in a text file description of a boolean cellular automaton, and then simulate it, as well as be capable of generating and simulating random boolean cellular automata. With this, test Kaufmann's conjecture that the cycle time is $O(\sqrt{n})$ for a random boolean cellular automaton (a conjecture disproved by J. Lynch [Lyn93, Lyn95] by difficult probabilistic analysis). Compare your results with the impressive random boolean cellular automaton simulation program of A. Wuensche at the Santa Fe Institute (see the web page of A. Wuensche at `http://www.santafe.edu/`).
13. Certain organisms, such as yeast, can exist in both haploid and diploid form. Genes can be masked when in diploid form; in particular, when one gene is not functional, a diploid organism can survive because of the functional allele. Devise a program to simulate an organism that under environmental stress (lack of nutrients, high salt or pH, etc.) will switch from diploid to haploid state, thus expressing all genes.
14. Consider the following intriguing idea, first proposed by Pudlák and Pudlák [PP97]. Assume that there are genes that control whether another gene is dominant or recessive. Devise a simulation program to determine whether the survival chances

of a population are increased by the ability, under environmental stress, to switch a dominant gene to a recessive gene, or vice versa.

Entropy

15. Write a program to compute the frequencies of nucleotides and of dinucleotides from an input file. Using the data for *M. jannaschii*, answer the following questions:

    (a) What is the G + C content of *M. jannaschii*? How does this G + C content compare with that of *Haemophilus influenzae* and of *Mycoplasma genitalium* (see [FAW+95, FGW+95])?
    (b) How many Ys occur in the chromosome of *M. jannaschii*? How about Ns? What percentage of the genome was not uniquely determined (i.e. codes different from A,C,G,T)?
    (c) TA is a dinucleotide often appearing in regulatory segments of the genome (e.g. the TATA box). What is the *relative abundance* $\rho_{TA} = \frac{f_{TA}}{f_T f_A}$ of TA in *M. jannaschii*? Is it less than the relative abundance of other dinucleotides? How does the TA relative abundance in *M. jannaschii* compare with that in *Haemophilus influenzae* and *Mycoplasma genitalium*?

16. Roughly 10% of the codons in the genetic code (6 out of the 61 codons that code for amino acids) code the amino acid leucine. Write a program to determine the frequency of leucines[16] in a fragment of the protein database. Is the frequency you compute roughly 10%? Try the same for other amino acids.
17. What is the average ratio of hydrophobic to hydrophilic residues in proteins from the protein database? What about for specific classes of proteins (globular, kinases, etc.)? Use a modification of the *polar requirement* of C. Woese *et al.* [WDD+66] as a measure of hydrophobicity.
18. With the genomic data of *M. jannaschii*, compute H_1, $H_1^{\max}$, D_1 and H_2, H_2^{ind}, D_2.
19. Following R.Verin [CV99], implement as follows a linear time technique for the possible identification of genomic regions having repetitions (either many short repetitions, or a few large repetitions). If Σ is a finite alphabet (for instance $\Sigma = \{A, C, G, T\}$) then Σ^* denotes the set of all finite words in this alphabet. If u, w are words in alphabet Σ, then u is a *subword* u of w if there exists a prefix x and suffix y, for which $xuy = w$. Denote the set of subwords of word w by $S(w)$, and denote the subword count of a current window by wc .

    Let $a_1 \cdots a_N \in \{A, C, G, T\}^*$ be genomic data, and fix the size k of a *window*. Count the number of distinct subwords in every window of the genome. A window consisting of

    $$AAA \cdots A$$

    has $k + 1$ many non-empty subwords – namely A, AA, ... ,A^k together with the empty word *lambda*. On the other hand there are at most $\binom{k}{2}$ many subwords in a window of size k. Verin's heuristic is that areas where a repetition is found have linear wc values (i.e. low), while in other regions, the value is nonlinear (i.e. high).

---

[16] That is, the average of the ratio *number of leucines in a protein* over the *number of residues in a protein*.

To compare two windows, define

$$\mathcal{S}(w) = S(w) \cup S(\widetilde{w}),$$

where $(\widetilde{w})$ is the reverse complement of w. *Jaccard's index* is defined by

$$ind_J(u,v) = \frac{|\mathcal{S}(u) \cap \mathcal{S}(v)|}{|\mathcal{S}(u) \cup \mathcal{S}(v)|}.$$

Clearly $ind_J(u,u) = 1$, and if u and v do not share many subwords, then $ind_J(u,v)$ is a small value near 0. Implement these heuristics and investigate the genome of *E. coli*.

20. By definition if p, q are probabilities whose sum is 1, then

$$H(p,q) = -p \ln p - (1-p) \ln(1-p).$$

Letting $F(p)$ denote the right-hand side of this equation, graph F as a function of p on the interval $[0,1]$.

21. Generalize the previous exercise to probability distribution p, q, r, where $p+q+r = 1$. Thus

$$H(p,q,r) = -p \ln p - q \ln q - (1-p-q) \ln(1-p-q).$$

Letting $F(p,q)$ denote the right-hand side of this equation, graph F as a function of p, q on $[0,1] \times [0,1]$.

22. Discretize some common continuous probability distributions (such as the normal, exponential, Poisson, and Boltzmann distributions), and compute the entropies.

23. Suppose that $0 \leq p, q, r \leq 1$ satisfy $p + q + r = 1$. Prove that the entropy $H(p,q,r) = -(p \ln p + q \ln q + r \ln r)$ attains a unique maximum when $p = q = r = \frac{1}{3}$.

HINT Define the function

$$G(p,q) = H(p,q,1-p-q) = -[p \ln p + q \ln q + (1-p-q) \ln(1-p-q)]$$

and set G's partial derivatives equal to 0. Thus

$$\begin{aligned} \frac{\partial G}{\partial p} &= -[1 + \ln p - 1 - \ln(1-p-q)] \\ &= -[\ln p - \ln(1-p-q)], \end{aligned}$$

so $\frac{\partial G}{\partial p} = 0$ iff $\ln p = \ln(1-p-q)$, and similarly $\frac{\partial G}{\partial q} = 0$ iff $\ln q = \ln(1-p-q)$. Thus the point (p,q) is a stationary point of G iff $p = 1-p-q$ and $q = 1-p-q$, where one readily computes that $p = q = r = \frac{1}{3}$.

2.5 Appendix: Modification of Bezout's Lemma

In this section, we prove an apparently new technical Lemma 2.16, used in our proof that finite, aperiodic, irreducible Markov chains have an equilibrium distribution. We begin with a classic number theoretic result, known as Bezout's Lemma.

LEMMA 2.27 (BEZOUT)
Suppose that $x_1, \ldots, x_m$ are positive integers, and that $d = gcd(x_1, \ldots, x_m)$. Then there exist $a_1, \ldots, a_m \in \mathbb{Z}$ such that $d = a_1x_1 + \cdots + a_mx_m$.

PROOF Define d to be the smallest positive element in the ideal $I = \{c_1x_1 + \cdots + c_mx_m \mid c_1, \ldots, c_m \in \mathbb{Z}\}$, and define $J = \{cd \mid c \in \mathbb{Z}\}$. We claim that $I = J$ and that $d = gcd(x_1, \ldots, x_m)$.

Clearly $J \subseteq I$. We claim that $I \subseteq J$. If not, then let e be the smallest positive element in $I - J$. Since I is closed under multiplication by elements in $\mathbb{Z}$ and by addition of elements in I, $e \bmod d \in I$ and is smaller than d, contradicting the choice of d. Thus $I = J$.

Since every element of I is a multiple of d, it follows that d divides each of $x_1, \ldots, x_m$. Now suppose that e divides each of $x_1, \ldots, x_m$, so that there exist $q_1, \ldots, q_m$, such that $x_i = eq_i$ for $1 \leq i \leq m$. Then since $d \in I$, there exist $a_1, \ldots, a_m$ such that

$$\begin{aligned} d &= a_1x_1 + \cdots + a_mx_m \\ &= a_1q_1e + \cdots + a_mq_me, \end{aligned}$$

and hence e divides d. Thus d is the greatest common divisor of $x_1, \ldots, x_m$. ■

An alternate, somewhat longer proof of Bezout's Lemma uses the Euclidean `gcd` algorithm to compute the a_i. It follows from Bezout's Lemma that if $x_1, \ldots, x_m$ are relatively prime, then for every n there exist $a_1, \ldots, a_m \in \mathbb{Z}$ such that $n = a_1x_1 + \cdots + a_mx_m$. We plan to show that for n sufficiently large, the a_i can be chosen to be non-negative.

LEMMA 2.28
Let x, y be positive integers and $d = gcd(x, y)$. For all integers n, if $nd \geq xy$, then there exist $r, s \in \mathbb{N}$ such that $nd = rx + sy$.

PROOF Let $nd \geq xy$. Bezout's Lemma implies the existence of $r', s' \in \mathbb{Z}$ such that $d = r'x + s'y$, and hence $nd = nr'x + ns'y$.

CASE 1: $nr', ns' \geq 0$

In this case, the statement of the current lemma is satisfied with $r = nr'$ and $s = ns'$.

CASE 2 Without loss of generality, assume that $nd = ax - by$, for $a, b \geq 1$.

Let

$$\begin{aligned} a &= a'y + r_1, \quad \text{where } 0 \leq r_1 < y, \\ b &= b'x + r_2, \quad \text{where } 0 \leq r_2 < x. \end{aligned}$$

so that

$$\begin{aligned} nd &= (a'y + r_1)x - (b'x + r_2)y \\ &= xy(a' - b') + r_1 x - r_2 y. \end{aligned}$$

Now if $a' - b' \leq 0$, then $nd \leq xy(a' - b') + r_1 x \leq r_1 x < xy$, contradicting the hypothesis that $nd \geq xy$. Thus $a' > b'$ and so

$$\begin{aligned} nd &= xy(a' - b' - 1) + xy + r_1 x - r_2 y \\ &= xy(a' - b' - 1) + y(x - r_2) + r_1 x \\ &= x[y(a' - b' - 1) + r_1] + y(x - r_2). \end{aligned}$$

Letting $r = y(a' - b' - 1) + r_1$ and $s = x - r_2$, we have $nd = rx + sy$, where $r, s \geq 0$. ■

LEMMA 2.29
Let $d = gcd(x_1, \ldots, x_m)$. For all n, if $nd \geq x_1 \cdots x_m$, then there exist non-negative $a_1, \ldots, a_m$ such that $nd = a_1 x_1 + \cdots + a_m x_m$.

PROOF The general inductive proof is a notational variant of the case, where $m = 3$, which we prove here. Suppose that $d = gcd(x, y, z)$ and let n satisfy $nd \geq xyz$. By Bezout's Lemma, there exist $r, s, t \in \mathbb{Z}$ such that $nd = rx + sy + tz$. Let $e = gcd(y, z)$, so that $d = gcd(x, e)$.

If any of x, y, z equals 1, then the lemma easily follows. Suppose, for instance, that $x = 1$. By setting $a = 1$, $b = 0$, $c = 0$, we have $ax + by + cz = 1$. Thus without loss of generality, assume that $x, y, z \geq 2$. By Lemma 2.28, there exist $a, b \geq 0$ such that $nd = ax + be$. Let

$$\begin{aligned} a &= a'e + r_1, \quad \text{where } 0 \leq r_1 < e, \\ b &= b'x + r_2, \quad \text{where } 0 \leq r_2 < x, \end{aligned}$$

and so

$$\begin{aligned} nd &= (a'e + r_1)x + (b'x + r_2)e \\ &= xr_1 + e(a'x + b'x + r_2). \end{aligned}$$

CLAIM $e(a'x + b'x + r_2) \geq yz$.

PROOF If not, then $nd < r_1 x + yz < ex + yz \leq yx + yz = y(x + z) \leq xyz$, where we have used the facts that $e = gcd(y, z) \leq \min(y, z) \leq y$ and that $x, z \geq 2$. Since we assumed that $nd \geq xyz$, this contradiction establishes the claim.

Thus $e(a'x + b'x + r_2) \geq yz$, and by Lemma 2.28, there exist $s, t \geq 0$ such that $e(a'x + b'x + r_2) = sy + tz$. Setting $r = r_1$, it follows that $nd = rx + sy + tz$. Since the general inductive case for $m > 3$ is analogous, this establishes the lemma. ■

Specializing the previous lemma to the case where $gcd(x_1, \ldots, x_m) = 1$, we have the following.

COROLLARY 2.30
If $x_1, \ldots, x_m \geq 1$ are relatively prime, then for all $n \geq x_1 \cdots x_m$, there exist $a_1, \ldots, a_m \geq 0$ such that $n = a_1 x_1 + \cdots + a_m x_m$.

We can now prove the *Positive Transition Matrix* Lemma 2.16, which is restated for convenience.

LEMMA 2.31 (POSITIVE TRANSITION MATRIX)
If $M = (Q, \pi, P)$ is a finite, aperiodic, irreducible Markov chain, then there exists $N \geq 0$ such that P^N is strictly positive.

PROOF Suppose that $|Q| = n$, and fix state $i \in Q$. By aperiodicity, the period of i is 1; i.e. $gcd(\{t \geq 1 \mid p_{i,i}^{(t)} > 0\}) = 1$. It follows that there exist m and $t_1, \ldots, t_m$ such that $p_{i,i}^{(t_1)} > 0, \ldots, p_{i,i}^{(t_m)} > 0$ and $gcd(t_1, \ldots, t_m) = 1$. Define $N_i = t_1 \cdots t_m$. By Corollary 2.30, for $N \geq N_i$, there exist $a_1, \ldots, a_m \geq 0$ such that $N = a_1 t_1 + \cdots + a_m t_m$. Since P consists of non-negative entries, if $p_{i,i}^{(t)} > 0$ and $p_{i,i}^{(u)} > 0$, then then $p_{i,i}^{(kt)} > 0$ for all multiples kt, and $p_{i,i}^{(t+u)} > 0$. Thus for all $N \geq N_i$, $p_{i,i}^{(N)} > 0$.

The previous analysis holds for each fixed $i \in Q$. It follows that for any $N \geq \max(N_1, \ldots, N_n)$, $p_{i,i}^{(N)} > 0$. Now clearly if $p_{i,i}^{(N)} > 0$ and $p_{i,j}^{(M)} > 0$, then $p_{i,j}^{(N+M)} > 0$. By finiteness of Q and the definition of irreducibility, it thus follows that there exists N such that $p_{i,j}^{(N)} > 0$ for all $i, j \in Q$ (in fact the proof establishes that $p_{i,j}^{(N')} > 0$ for all $i, j \in Q$ and $N' \geq N$). This completes the proof of the lemma. ■

Acknowledgments and References

An elegant, concise introduction to probability theory is Rozanov [Roz77]. Feller [Fel68b, Fel68a] is an excellent text on probability, while Karlin–Taylor [KT75] is a classic introduction to stochastic processes and Kemeny–Snell [KS60] treats finite Markov chains. Our motivation for the definition of Shannon's entropy follows that of [Roz77]. Our presentation of the derivation of the exponential distribution from consideration of interarrival times comes from S. Garland [Gar86]. Our definition of Markov random field and statements of the results of Geman–Geman come from [GG84]. Michalewicz [Mic96] presents an introduction to genetic algorithms. Our direct proof of Lemma 2.16 using the apparently new Lemma 2.29 seems to be new, and to the best of our knowledge does not appear in the literature. L. Gatlin [Gat72] gives early applications of entropy to biological systems; in particular, we have followed her applications in computing the entropy of single nucleotides and dinucleotides in bacteria.

3

Sequence Alignment

> We have seen that the members of the same class, independently of their habits of life, resemble each other in the general plan of their organization. This resemblance is often expressed by the term 'unity of type', or by saying that the several parts and organs in the different species of the class are homologous. (C. Darwin, *Origin of Species*, 1859 [Dar58])

As already mentioned, *M. jannaschii* is a methane-generating archaebacterium living at depths over 2.5 km on the ocean floor, near deep-sea thermal vents. From sequence comparisons of *M. jannaschii* with prokaryotes and eucharyotes, Bult *et al.* [BWO$^+$96] firmly established the validity of Woese's theory that *Archea* is a third life domain, along with the other two domains *Prokarya* and *Eukarya*. *M. jannaschii* has 1738 genes, and by performing sequence alignment with other genomes, it appears that archaebacteria share an evolutionary heritage with eukaryotes for transcription, translation, and DNA replication, yet share certain other features with prokaryotes, since for instance archaebacteria have no nuclear membrane.

In this chapter, we will cover dynamic programming sequence alignment algorithms, and some of their applications. A good example of a recent commercial application of sequence alignment and homology testing was given by Richard Roberts[1] in [Rob97]. Restriction enzymes appear to be certain bacteria's defense system against phages. Type II restriction enzymes (such as the well-known *ecoRI* endonuclease) recognize a small fragment of double-stranded DNA and cut both strands, either with a blunt cut, or leaving sticky ends. For instance, *ecoRI* recognizes the hexanucleotide palindrome $GAATTC$, and cuts, leaving sticky ends as indicated below:

$$\begin{array}{l} \cdots G|AATTC\cdots \\ \cdots CTTAA|G\cdots \end{array} \longrightarrow \begin{array}{c} |AATTC\cdots \\ \quad\;\; |G\cdots \\ \text{and} \\ \cdots G| \\ \cdots CTTAA| \end{array}$$

[1] Nobel Prize for the discovery of introns.

In order to protect itself from the action of its own restriction enzymes, a bacterium methylates[2] restriction enzyme recognition sites of its own genome. At the current time,[3] there are 3154 discovered restriction enzymes, some 5604 publications, and out of 16 (resp. 64) possible tetranucleotide (resp. hexanucleotide) palindromes (such as AAGG (resp. GAATTC)), all but 2 (resp. 2) have been shown to be recognition sites for certain restriction enzymes. Though there appears to be *no* recognizable sequence homology between different restriction enzyme genes, there *is* sequence homology between different recognition site methylation genes. Using this principle, the group of R. Roberts has successfully developed the REBASE [RM99] software to locate likely candidates for potential restriction enzymes within genomic databases.

We should like to define the distance between two sequences $a_1 \cdots a_n$ and $b_1 \cdots b_m$, allowing inexact matches, in producing an alignment. Of course, the usual distance measures (Euclidean distance, Hamming distance, etc.) have little bearing on this problem, since the sequences have possibly different lengths (though there is work in mathematical evolution theory, where Hamming distance is used).

For instance, given an initial sequence ACGTACGT of length 8, after 9540 generations, the sequence ACACGGTCCTAATAATGGCC was generated, assuming probability of deletion 0.0001, probability of insertion 0.001, probability of transitional substitution 0.00008, and probability of transversional substitution 0.00002. On a different program run with the same parameters, after 9540 generations, ACGTACGT evolved into the rather different sequence CAGGAAGATCTTAGTTC. The sequences ACACGGTCCTAATAATGGCC and CAGGAAGATCTTAGTTC are related in that both evolved from the same ancestor sequence ACGTACGT. What is their *true*, or historically accurate alignment? This question makes sense, because in the simulation program that generated the mutated sequences, the type and position of *edit* operation (delete, insertion, substitution) were taken into account. The true alignments are

```
--ACG-T-A---CG-T----
ACACGGTCCTAATAATGGCC
```

and

```
---AC-GTA-C--G-T--
CAG-GAAGATCTTAGTTC
```

so, by superposition, we obtain

```
-ACAC-GGTCCTAAT--AATGGCC
CAG-GAA-G-AT--CTTAGTTC--
```

In contrast, Gotoh's algorithm with mismatch penalty of 3 and gap penalty function $g(k) = 2 + 2k$ for length k gap (discussed in this chapter), yields a completely different alignment:

```
ACACG--GTCCTAATAATGGCC
-CAGGAAGATCT--TAGTT--C
```

There is recent work to determine the choice of mismatch and gap parameter, using in part the significance of an alignment score (significance is determined by how many random sequences having the same base composition have a better alignment score).

---

[2] To methylate means to add a CH_3 group to a nucleotide base.

[3] November 1, 1999. Thanks to Dr R.J. Roberts for information (personal correspondence).

3.1 Motivating Example

In this section, we present a motivating example to give a more concrete understanding of some of the problems involved in sequence comparison. One of the purposes of sequence alignment is to detect *homologous proteins*. Two proteins (or protein domains) are *homologous* if they have evolved from a common predecessor. Usually, this implies that the two proteins have a similar structure and function. Homology is a very important concept in biology that is used as a fine-grained criterion to group together different proteins (with known structures). There are currently two principal databases for the hierarchical structural classification of proteins – namely CATH [OMJ+97] and **scop** [MBHC95]. CATH classifies proteins according to **C**lass (where the different protein classes are *mainly Alpha*, *mainly Beta*, *Alpha and Beta*, and proteins having *few secondary structures*), **A**rchitecture (which is the description of the gross overall arrangement of the secondary structure elements), **T**opology (which takes into account the overall shape as well as the connectivity of the secondary structure elements, using structural comparison algorithms), and **H**omology (where proteins are grouped together if they are believed to have a common ancestor). In deciding whether two proteins are homologous, sequence alignment is *one* of the techniques used. **scop** (**S**tructural **C**lassifcation **o**f **P**roteins) has a similar hierarchical structure, where proteins are grouped according to fold (major structural similarity), superfamily (probable common evolutionary origin) and family (clear evolutionary relationship).

As an example of homologous proteins, we consider the family of reverse transcriptases. Reverse transcriptases are used by retroviruses to replicate their own genetic information, which is stored in RNA. This RNA is translated back into DNA (for duplication) using reverse transcriptase. As an example, we take reverse transcriptase of Moloney murine leukemia virus, and the reverse transcriptase of HIV Type 1 virus. The structures of the homologous domains

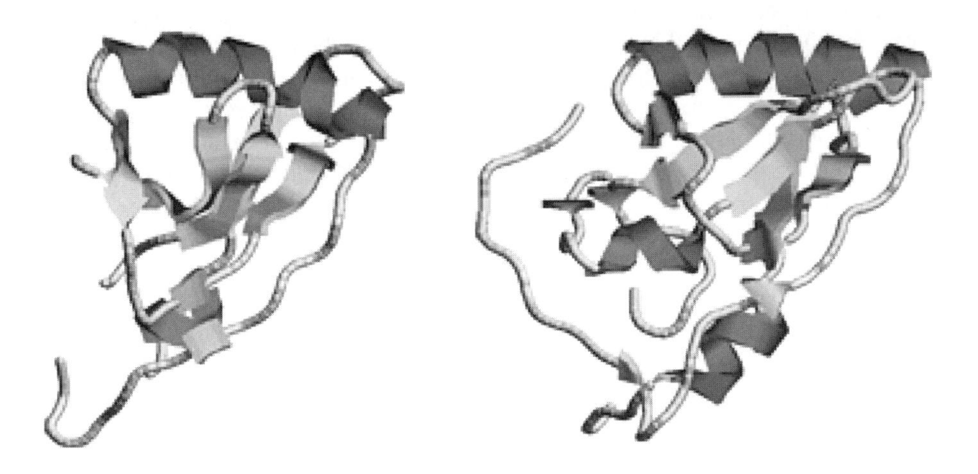

Figure 3.1 Two homologous proteins of the reverse transcriptase family. The first protein is domain 1 of reverse transcriptase of the Moloney murine leukemia virus (PDB code 1MML [GJO+95]), and the second is domain 1 of the A chain of the HIV Type 1 reverse transcriptase (PDB code 1RTH [REG+95]). For both proteins, we show the backbone structure and indicate the location of the secondary structure elements (α helices and β strands). The two structures show a high degree of similarity in the arrangement of the secondary structure elements (especially the long helix and the four long β strands forming a β sheet).

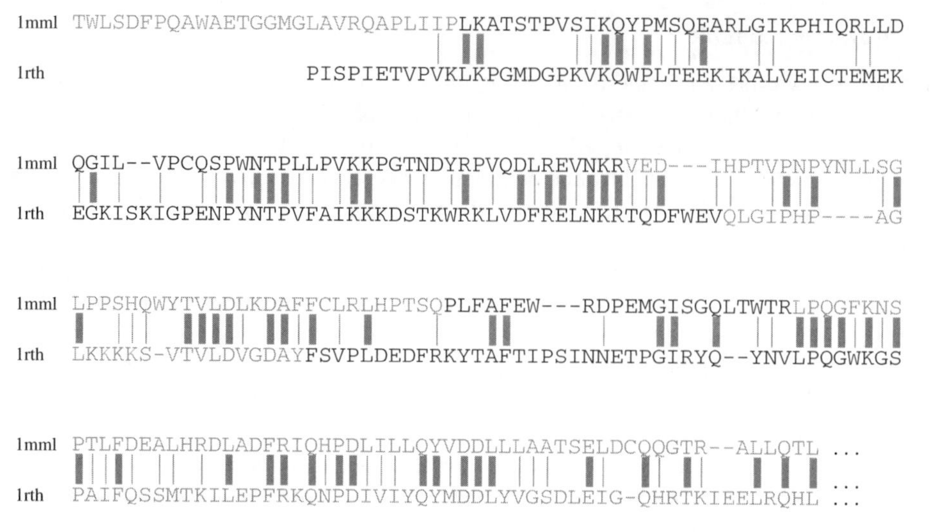

Figure 3.2 Part of an alignment of Mmlv reverse transcriptase (PDB code 1MML) and HIV-1 reverse transcriptase (PDB code 1RTH). The alignments were produced using the FASTA package [PL88] with the Blosum50 substitution matrix [HH92]. The sequence of the first domains of both chains are printed in black, while all other amino acids are printed in gray. Thicker gray bars indicate an exact matching of amino acids, whereas the thinner gray bars indicate amino acids that are similar according to the Blossum50 similarity matrix.

are shown in Figure 3.1. They are classified as homologous by both the CATH database and the SCOP database.

As already mentioned, sequence alignment is one of the steps used to determine whether two proteins are homologous. Figure 3.2 shows the sequence alignment for the amino acid sequences of both protein domains. There are several problems involved in finding a good sequence alignment. First, the question is whether to align a complete protein sequence or only part of it (since it may be the case that only a part of the protein has been conserved). This is the difference between global and local sequence alignment. Second, major parts of the alignment are substitutions of amino acids by other amino acids, which have 'similar properties'. These substitutions are shown using thin gray bars (instead of thicker ones, which are used for exact matches). Now the question is how to weight the different substitutions (the weights are in fact measures of the similarity between amino acids). This will be discussed in the next section, which treats scoring matrices.

3.2 Scoring Matrices

Alignments are used to reveal homologous proteins, or regions of proteins that are conserved in different proteins. Since evolution is a stochastic process, it is clear that there cannot be one single correct alignment. Instead, there are different possible alignments, and we have to choose the 'correct' ones (although we never know for sure which alignment is correct). To do this, one associates a *score* with every alignment, where the alignments with the highest score are the ones to be selected.

Before we can choose an appropriate scoring function, we have first to fix the underlying evolutionary model. Usually, we have an evolutionary model where sequences can be mutated using insertions, deletions or substitutions. Of course, this is a simplification, and there are approaches to incorporate other kinds of mutations as well. An example is tandem repeats, which will be discussed in this chapter.

If we take insertion/deletions and substitution as possible evolutionary events, a scoring function can easily be based on scores for the different operations (since any alignment can be decomposed in a unique way in deletions/insertions and substitutions). Under the assumption that the different operations occurred independently, the complete score can be taken to be the sum of scores associated with the different operations.

We will show how to generate a scoring function for substitutions of amino acids. The reason is that there is a well-established theory for generating scores for substitutions. In this case, the scores for the different substitutions are stored in a *substitution score matrix*, which contains for every pair (A, B) of amino acids an entry s_{AB} (the score for aligning A with B). Clearly, the selection of an appropriate substitution score matrix is crucial for achieving good alignments.

There are different manners in which a substitution score matrix can be derived. The first is an ad hoc approach – i.e. a biologist can always set up a score matrix that produces good alignments. The second approach is to derive a score matrix from physical/chemical properties. The idea behind this approach is that an amino acid is more likely to be substituted by another if their properties are similar. The third and most often used approach is a statistical one. To be more precise, let s and s' be two amino acid sequences of length n, whose alignment score we would like to give, assuming that only substitutions (no insertions or deletions) are used in transforming s into s'. Using reliable amino acid substitution matrices allows one later to compute optimal *local* sequence alignments between two proteins of possibly different length; i.e. to find subsequences of the same length in the two proteins that are conserved.

Score Matrices Based on a Statistical Model

The score for aligning s with s' is generated by comparing two different hypothesis. The first hypothesis (the null hypothesis) is that s and s' *do not stem* from a common predecessor, i.e., that we have an alignment by chance. This hypothesis assumes an underlying random model R that generates two sequences s and s' randomly. Assuming a probability q_A for producing an amino acid A in the model R (where we set q_A to be the relative frequency of the amino acid A in proteins), the probability $P(s, s'|R)$ for the null hypothesis is given by

$$P(s, s'|R) = \prod_{1 \leq l \leq n} q_{s_l} \prod_{1 \leq l \leq n} q_{s'_l} = \prod_{1 \leq l \leq n} q_{s_l} q_{s'_l}.$$

The second hypothesis (or the homologous hypothesis) is that s and s' *stem* from the same predecessor. The underlying model E is an evolutionary model. One assumes that there is an unknown ancestor sequence r of length n, from which s and s' are generated by random substitution of amino acids. Furthermore, we assume in the model E that p_{AB} is the probability that the amino acids A and B are aligned and have hence been derived from an ancestor amino acid C. How to determine these probabilities will be explained later. Given the probabilities p_{AB}, the probability $P(s, s'|E)$ for the homologous hypothesis is given by

$$P(s, s'|E) = \prod_{1 \leq l \leq n} p_{s_l s'_l}.$$

In principle, we want to determine what the chance (or what the odds are) is that the alignment of s and s' reflects the fact that we have aligned conserved regions. Hence, a reasonable score for the alignment of s with s' is generated by comparing the probability of the homologous hypothesis with the null hypothesis. This is called the *odds ratio*:

$$\frac{P(s, s'|E)}{P(s, s'|R)} = \frac{\prod_{1 \leq l \leq n} p_{s_l s'_l}}{\prod_{1 \leq l \leq n} q_{s_l} q_{s'_l}} = \prod_{1 \leq l \leq n} \frac{p_{s_l s'_l}}{q_{s_l} q_{s'_l}}.$$

Therefore, we can generate an appropriate score for the alignment of s and s' from scores $\frac{p_{s_l s'_l}}{q_{s_l} q_{s'_l}}$ for the amino acid pairings (s_l, s'_l) (with $1 \leq l \leq n$) in the alignment of s and s'. To achieve a scoring function that is additive instead of multiplicative, one defines the entries of the substitution score matrix by *log odds ratios*:

$$s_{AB} = \log \frac{p_{AB}}{q_A q_B}. \tag{3.1}$$

In fact, it has been shown that any score matrix consists of log odds ratios under reasonable assumptions (see [Alt91]).

PAM and Amino Acid Pair Probabilities

The next problem is to determine an appropriate evolutionary model E. To be more precise, the model E is a model $E((p_{AB})_{AB})$ that is parameterized in the pair probabilities p_{AB}, and we want to find these probabilities. In a statistical approach, these probabilities have to be estimated from data. This can be done if we have two sequences s and s' for which we *know* that they are homologous.[4] Estimating p_{AB} from the known alignment of s with s' means that we want to find parameters p_{AB} that maximize

$$P(E((p_{AB})_{AB})|s, s').$$

To do this, we apply a maximum-likelihood approach, i.e., we compute parameters p_{AB} that maximize

$$P(s, s'|E((p_{AB})_{AB})).$$

A simple calculation using Lagrange multipliers (see the appendix to this chapter, Section 3.9) shows that the maximum is achieved if we set

$$p_{AB} = \frac{n_{AB}(s, s')}{n},$$

where $n_{AB}(s, s')$ is the number of times the amino acids A and B are aligned in one column in the alignment of s and s', and n is the length of s (and s'). Thus $\frac{n_{AB}(s,s')}{n}$ is the relative frequency of a pair (A, B) in the alignment of s and s'.

This approach was used for the first time by Dayhoff, Schwartz, and Orcutt [DSO78] to generate the so-called PAM matrices, perhaps the most widely used substitution matrices.

---

[4] Of course, this approach can be generalized to several homologous sequences.

To construct the PAM matrix, Dayhoff *et al.* were faced with the problem that the only way to find sequences s and s' believed to be homologous is to apply (local) sequence alignment. Hence, to find good substitution matrices for sequence alignment, one needs to apply sequence alignment, which sounds like a vicious circle. To overcome this problem, Dayhoff *et al.* considered only very closely related sequences (i.e., sequences that differ in at most 15% of their amino acids). The resulting alignment is thus very likely to reflect homology.

This approach yields a substitution score matrix (often simply called substitution matrix), which is valid only for sequences that are as closely related as the sequences used in generating the score matrix. On the other hand, one would like to align sequences that are more distantly related as well. To this end, Dayhoff *et al.* introduced an additional parameter t, which models the evolutionary time scale; i.e. they defined pair probabilities

$$p_{AB}(t) = P(A, B|t)$$

that depend on time t, rather than p_{AB} in our earlier discussion. How can one generate probabilities $P(A, B|t)$ for different values of t? Let

$$P(B|A, t)$$

denote the probability that amino acid A is substituted by B within evolutionary distance t. Assuming that time directionality of evolution can be ignored, we have

$$P(A, B|t) = P(B|A, t)P(A|t).$$

Assuming that the distribution of amino acids does not change during evolution, we have that $P(A|t) = P(A) = q_A$, hence $P(A, B|t) = P(B|A, t)q_A$, which implies that

$$P(B|A, t) = \frac{P(A, B|t)}{q_A}. \tag{3.2}$$

By equation (3.2), $P(B|A, t)$ can be estimated from the relative frequency of the pair (A, B) in the known alignment of two sequences s and s' with distance t, and from the relative frequency q_A of the amino acid A.

The advantage of this approach is that from the matrix

$$M = (P(B|A, t))_{AB}$$

of substitution probabilities for all pairs of amino acids for a specific evolutionary distance t, we can generate the matrix

$$M' = (P(B|A, kt))_{AB}$$

of substitution probabilities for any $k \in \mathbb{N}$, simply by setting

$$M' = M^k.$$

This allows the extrapolation of substitution probabilities over a longer time scale. From the substitution probabilities, a score matrix can be generated by equations (3.2) and (3.1).

How can the values $P(B|A, t)$ be adjusted to allow for arbitrary distances t? For this purpose, Dayhoff *et al.* introduced the evolutionary time scale PAM, which is an acronym for '**p**oint **a**ccepted **m**utation'. Two sequences s and s' have an evolutionary distance of 1 PAM if s was converted into s' by a *series* of accepted substitutions with an average of 1 accepted substitution per 100 amino acids. Since substitution can occur at the same amino acid site several times, a distance of n PAM between two sequences s and s' does not necessarily imply that exactly $n\%$ of the amino acid positions of s and s' differ (models for substitution of amino acids and nucleotides will be discussed in Chapter 4). For this reason, it is possible to consider sequences whose PAM distance is greater than 100 (the PAM-250 substitution matrix is widely used). A substitution matrix

$$M = (P(B|A,t))_{AB}$$

is defined to be 1 PAM if the expected number of substitutions in a 'typical' protein using the substitution probabilities $P(A,B|t)$ is 1%. By the previous discussion, an n PAM matrix N can be generated from a 1 PAM matrix M by computing the nth power of M; i.e. $N = M^n$.

We can hardly expect to find homologous sequences s and s' having exactly 1 PAM distance. Dayhoff *et al.* therefore scaled the values $P(B|A,t)$ for $A \neq B$ by a value λ, such that the resulting substitution probabilities imply an expected substitution frequency of 1% (one amino acid among every 100 amino acids). More precisely, in the estimation of $P(B|A,t)$ from a known alignment of sequences s and s', Dayhoff *et al.* set

$$P(B|A,t) = \lambda \frac{n_{AB}(s,s')}{q_A}$$

for $B \neq A$, and defined

$$P(A|A,t) = 1 - \sum_{B \neq A} P(B|A,t),$$

where λ was chosen so that the resulting substitution matrix is 1 PAM.

The PAM matrices have been used with much success, their only problem being that since the entries in the 1 PAM matrix were estimated from very closely related sequences, it is therefore difficult to extrapolate to more distantly related proteins. For this reason, Henikoff and Henikoff [HH92] used highly conserved regions in multiple sequence alignments of several distantly related proteins, rather than pairwise alignment, in generating their BLOSUM matrices. Experimental results have shown that the BLOSUM matrices are better suited for aligning more distantly related proteins.

3.3 Global Pairwise Sequence Alignment

3.3.1 Distance Methods

There are two different problems involved when comparing two similar sequences. The first one is to calculate the distance between the two different sequences (i.e., are the sequences similar or not?). The second problem is to align the sequences in order to find conserved regions. Hence, we will give two different definitions for comparing sequences, namely *edit distance* and *alignment* (and *alignment distance*), and we will show under which conditions these two definitions yield the same result.

Let Σ be a finite alphabet. A word $a \in \Sigma^*$ is simply a sequence $a_1 \dots a_n$, where $n \in \mathbb{N}$ and $a_i \in \Sigma$, for $1 \leq i \leq n$. The length of word a is denoted $|a|$. The empty word is denoted λ. With Σ^+ we denote the set $\Sigma^* \backslash \lambda$. The gap symbol is $-$.

We first define the simplest notion of edit distance for two sequences a, b. The edit distance is interpreted as the number of evolutionary operations to transform a string a into b using mutation, insertion and deletion of one symbol as the only evolutionary operations. The edit distance, unlike the approach of Waterman et al. [WSB76], has no special treatment for gaps; i.e. a size k gap is treated as k many single gaps (either insertions or deletions). Biologically, this is not realistic, so we retain the term edit distance for the approach of Needleman–Wunsch [NW70].

DEFINITION 3.1 (EDIT OPERATION)
Given an alphabet Σ with $- \notin \Sigma$, an edit operation *is a pair*

$$(x, y) \in (\Sigma \cup \{-\}) \times (\Sigma \cup \{-\}).$$

We say that an edit operation (x, y) is a

- substitution *if $x, y \in \Sigma$ with $x \neq y$,*
- insertion *if $x = -$ and $y \in \Sigma$,*
- deletion *if $x \in \Sigma$ and $y = -$.*

Given $a, b \in (\Sigma \cup \{-\})^*$ and an edit operation (x, y) such that $x \neq y$, we write

$$a \rightarrow_{(x,y)} b$$

if b can be obtained from a by replacing one occurrence of x by y (if $x, y \in \Sigma$), or by deleting one occurrence of x (if $y = -$), or by inserting one occurrence of y (if $x = -$). If $S = s_1 \ldots s_r$ is a sequence of edit operations, then we write

$$a \Rightarrow_S b$$

if $a = a^{(0)} \rightarrow_{s_1} a^{(1)} \rightarrow_{s_2} \cdots \rightarrow_{s_r} a^{(r)} = b$ for words $a^{(0)} \ldots a^{(r)}$.

We will use *indels* as short for insertions or deletions. Note that the edit operations can be applied at every position of the string. *Context-sensitive* edit operations (i.e., edit operations that for instance produce lethal mutations) are computationally more powerful than the edit operations defined above [BC97].

In evolution, some types of mutation happen more often than others. This is simulated in the definition of an edit distance by weighing the edit operations. Let $w(x, y)$ be a *cost function* assigning weights to the edit operations.

DEFINITION 3.2 (EDIT DISTANCE)
Given a cost function $w : (\Sigma \cup \{-\}) \times (\Sigma \cup \{-\}) \rightarrow \mathbb{R}$ and two words $a, b \in \Sigma^*$, the cost of a sequence $S = s_1 \ldots s_r$ of edit operations is defined as $\sum_{i=1}^{r} w(s_i)$. The edit distance *of a, b is defined as*

$$d_w(a, b) \quad = \quad \min\{w(S) \mid a \Rightarrow_S b\}.$$

Usually, when w is implicit, we write $D(a, b)$ rather than $d_w(a, b)$.

One must be careful which kind of cost function can be used. For example, one must avoid the cost of the sequence $(A, C)(C, T)$ being smaller than of the substitution (A, T) itself. This and other properties are combined in the notion of a metric.

DEFINITION 3.3 (METRIC)
A metric d must satisfy the following:

1. $d(x, y) = 0$ *iff* $x = y$.
2. $d(x, y) = d(y, x)$ *(symmetry).*
3. $d(x, z) \leq d(x, y) + d(y, z)$ *(triangle inequality).*

PROPOSITION 3.4
If w is a metric, then d_w is also a metric.

For the calculation of edit distance, the straightforward definition of edit distance is not very useful. For this reason, we define the notions of alignment and alignment distance explicitly. We show that the definitions of alignment distance and edit distance agree in the case that the cost function is a metric.

DEFINITION 3.5 (ALIGNMENT AND ALIGNMENT DISTANCE)
Let Σ be an alphabet with $- \notin \Sigma$. For every $u \in (\Sigma \cup \{-\})^*$ we define $u|_\Sigma$ to be the restriction of u to Σ (by deleting all occurrences of $-$ in u). An alignment *is a pair $(a^\diamond, b^\diamond)$ with $a^\diamond, b^\diamond \in (\Sigma \cup \{-\})^*$ such that*

$$|a^\diamond| = |b^\diamond|$$

and there is no position i such that

$$a_i^\diamond = - = b_i^\diamond.$$

An alignment $(a^\diamond, b^\diamond)$ is an alignment of (a, b) *with $a, b \in \Sigma^*$ if*

1. $a^\diamond|_\Sigma = a$, *and*
2. $b^\diamond|_\Sigma = b$.

Given a cost function w, we define the cost of an alignment by

$$w(a^\diamond, b^\diamond) = \sum_{i=1}^{|a^\diamond|} w(a_i^\diamond, b_i^\diamond).$$

The alignment distance *of a, b is*

$$d_w^a(a, b) \quad = \quad \min\{w(a^\diamond, b^\diamond) \mid (a^\diamond, b^\diamond) \text{ alignment of } (a, b)\}.$$

Again we write $D(a, b)$ if w is clear from the context. The alignment $(a^\diamond, b^\diamond)$ is optimal *if $d_w^a(a, b) = w(a^\diamond, b^\diamond)$.*

PROPOSITION 3.6 (ADDITIVITY OF ALIGNMENTS)
Let $(a^\diamond, b^\diamond)$ and $(c^\diamond, d^\diamond)$ be two alignments. Then $(a^\diamond c^\diamond, b^\diamond d^\diamond)$ is also an alignment with

$$w(a^\diamond c^\diamond, b^\diamond d^\diamond) = w(a^\diamond, b^\diamond) + w(c^\diamond, d^\diamond).$$

PROPOSITION 3.7
Let w be a metric cost function, and $a, b \in \Sigma^*$. Then for every alignment $(a^\diamond, b^\diamond)$ of (a, b) there is a sequence S of edit operations such that $a \Rightarrow_S b$ and

$$w(S) = w(a^\diamond, b^\diamond).$$

For every sequence S such that $a \Rightarrow_S b$, there is an alignment $(a^\diamond, b^\diamond)$ of (a, b) such that

$$w(a^\diamond, b^\diamond) \leq w(S).$$

PROOF For every alignment $(a^\diamond, b^\diamond)$, the sequence $S = s_1 \ldots s_{|a^\diamond|}$ with

$$s_i \quad = \quad (a_i^\diamond, b_i^\diamond)$$

is a sequence of edit operations with $a \Rightarrow_S b$ and $w(a^\diamond, b^\diamond) = w(s)$.

The other direction is proven by induction on sequence length. Let S be a sequence of edit operations with $|S| = n + 1$. Then $S = S's$ and there is a $c \in \Sigma^*$ such that

$$a \Rightarrow_{S'} c \rightarrow_s b.$$

We prove the existence of an alignment $(a^\diamond, b^\diamond)$ for (a, b) with $w(a^\diamond, b^\diamond) \leq w(S)$ for the case that s is a substitution. The other cases can be proven analogously.

By induction hypothesis, we know that there is an alignment $(a^\diamond, c^\diamond)$ such that $w(a^\diamond, c^\diamond) \leq w(S')$. Note that we can use an arbitrary alignment with this condition. Let i be the position changed by $c \rightarrow_s b$ (i.e., i is the unique position with $c_i \neq b_i$), and let j be the corresponding position in $c^\diamond$ (i.e., the position of the ith Σ-letter in $c^\diamond$). We define $(a^\diamond, b^\diamond)$ with

$$b^\diamond = c_1^\diamond \ldots c_{j-1}^\diamond b_i c_{j+1}^\diamond \ldots c_{|c^\diamond|}^\diamond,$$

which is an alignment of (a, b). Then

$$\begin{aligned} w(a^\diamond, b^\diamond) &= w(a^\diamond, c^\diamond) - w(a_j^\diamond, c_j^\diamond) + w(a_j^\diamond, b_j^\diamond) \\ &\overset{w \text{ metric}}{\leq} w(a^\diamond, c^\diamond) + w(c_j^\diamond, b_j^\diamond) \\ &= w(a^\diamond, c^\diamond) + w(s) \\ &\overset{\text{Ind. Hyp.}}{\leq} w(S') + w(s) \quad = \quad w(S). \end{aligned}$$

■

Consider as an example the alphabet $\Sigma = \{A, C, G, T\}$ and the two words $a = ACCGGTA$ and $b = AGGCTG$. Then one possible alignment is

$$\begin{aligned} a^\diamond &= ACCGG{-}TA \\ b^\diamond &= A{-}{-}GGCTG \end{aligned}$$

A corresponding sequence S of edit operations with $a \Rightarrow_S b$ and $w(S) = w(a^\diamond, b^\diamond)$ is

$$S = (C, -)(C, -)(-, C)(A, G).$$

REMARK 3.8 Note that Proposition 3.7 is not true if w is not a metric.

THEOREM 3.9 (NEEDLEMAN–WUNSCH EDIT DISTANCE)
Let $w : (\Sigma \cup \{-\}) \times (\Sigma \cup \{-\}) \rightarrow \mathbb{R}$ be a metric cost function for the given alphabet Σ. Let $a, b \in \Sigma^*$ with $|a| = n$ and $|b| = m$. Define the matrix $(D_{i,j})$ with $0 \leq i \leq |a|$ and

$0 \leq j \leq |b|$ by

$$
\begin{aligned}
D_{0,0} &= 0, \\
D_{0,j} &= \sum_{k=1}^{j} w(-, b_k), \\
D_{i,0} &= \sum_{k=1}^{i} w(a_k, -), \\
\forall i, j > 0 : D_{i,j} &= \min \left\{ \begin{array}{l} D_{i,j-1} + w(-, b_j), \\ D_{i-1,j-1} + w(a_i, b_j), \\ D_{i-1,j} + w(a_i, -) \end{array} \right\}.
\end{aligned}
\tag{3.3}
$$

Then $D_{i,j}$ is the minimum global sequence alignment distance between sequences a and b, i.e.

$$D_{i,j} = D(a_1 \ldots a_i, b_1 \ldots b_j).$$

This theorem furnishes a dynamic programming algorithm (i.e. fills in the distance matrix) with runtime $O(nm) = O(n^2)$ if $n \geq m$.

PROOF By induction on pairs (i, j), where $(k, l) < (i, j)$ if and only if $k < i$, or $k = i$ and $l < j$ (lexicographic ordering). So assume as induction hypothesis that we have proven the theorem for all $(k, l) < (i, j)$.

We have to show that for all alignments $(u^\diamond, v^\diamond)$ of $a_1 \ldots a_i, b_1 \ldots b_j$,

$$D_{i,j} \leq D(u^\diamond, v^\diamond),$$

and that there is an alignment $(u^\diamond, v^\diamond)$ such that

$$D(u^\diamond, v^\diamond) \leq D_{i,j}.$$

For the first direction consider an arbitrary alignment $(u^\diamond, v^\diamond)$ of $a_1 \ldots a_i$ and $b_1 \ldots b_j$. Let $r = |u^\diamond| = |v^\diamond|$. Then by Proposition 3.6 we have that

$$D(u_1^\diamond \ldots u_r^\diamond, v_1^\diamond \ldots v_r^\diamond) = D(u_1^\diamond \ldots u_{r-1}^\diamond, v_1^\diamond \ldots v_{r-1}^\diamond) + D(u_r^\diamond, v_r^\diamond). \tag{3.4}$$

Since $u_r^\diamond = v_r^\diamond = -$ is excluded by the definition of an alignment, we have the three cases

$$
\begin{array}{lll}
u_r^\diamond = a_i & \quad u_r^\diamond = a_i & \quad u_r^\diamond = - \\
v_r^\diamond = - & \quad v_r^\diamond = b_j & \quad v_r^\diamond = b_j.
\end{array}
$$

Applying the induction hypothesis, we get for the different cases

$$
\begin{array}{rll}
& D_{i,j-1} & \leq D(u_1^\diamond \ldots u_{r-1}^\diamond, v_1^\diamond \ldots v_{r-1}^\diamond), \\
\text{resp.} & D_{i-1,j-1} & \leq D(u_1^\diamond \ldots u_{r-1}^\diamond, v_1^\diamond \ldots v_{r-1}^\diamond), \\
\text{resp.} & D_{i-1,j} & \leq D(u_1^\diamond \ldots u_{r-1}^\diamond, v_1^\diamond \ldots v_{r-1}^\diamond).
\end{array}
\tag{3.5}
$$

With (3.4) and the recursion equation (3.3), this yields

$$D_{i,j} \leq D(a_1 \ldots a_i, b_1 \ldots b_j).$$

For the other direction, the case distinction shows how to produce an optimal alignment of $(a_1 \ldots a_i, b_1 \ldots b_j)$ given optimal alignments of $(a_1 \ldots a_i, b_1 \ldots b_{j-1})$, $(a_1 \ldots a_{i-1}, b_1 \ldots b_{j-1})$ or $(a_1 \ldots a_{i-1}, b_1 \ldots b_j)$.

■

With a small modification, we can also easily find all optimal alignments (i.e., alignments whose distance is minimal). This is achieved via an additional matrix called the trace matrix. In principle, at every matrix cell we just store pointers that indicate which cells have been used in calculating the actual cell. The *trace matrix* $(\mathrm{tr}_{i,j})$ for two words $a, b \in \Sigma^*$ is defined to be the matrix of elements $\mathrm{tr}_{i,j} \subseteq \{\leftarrow, \uparrow, \nwarrow\}$ with

$$\begin{aligned}
\mathrm{tr}_{0,0} &= \emptyset, \\
\mathrm{tr}_{0,j} &= \{\leftarrow\}, \\
\mathrm{tr}_{i,0} &= \{\uparrow\}, \\
\forall i, j > 0 : \nwarrow \in \mathrm{tr}_{i,j} &\Leftrightarrow D_{i,j} = D_{i-1,j-1} + w(a_i, b_j), \\
\uparrow \in \mathrm{tr}_{i,j} &\Leftrightarrow D_{i,j} = D_{i-1,j} + w(a_i, -), \\
\leftarrow \in \mathrm{tr}_{i,j} &\Leftrightarrow D_{i,j} = D_{i,j-1} + w(-, b_j).
\end{aligned}$$

A *traceback* t is an element of $\{\leftarrow, \uparrow, \nwarrow\}^*$ that is a path in $(\mathrm{tr}_{i,j})$ starting from the lower right corner of the trace matrix and following the pointer in $\mathrm{tr}_{i,j}$ until we reach the upper left corner. The path is written left to right, i.e., the right end corresponds to $\mathrm{tr}_{|a|,|b|}$ and the left end to $\mathrm{tr}_{0,0}$. The alignment $(a^\diamond, b^\diamond)$ associated with t is generated as follows. Define $a^\diamond \in (\Sigma \cup \{-\})^*$ from t by substituting a_i for the ith occurrence of $\nwarrow$ or $\uparrow$, and by replacing all occurrences of $\leftarrow$ with $-$. Similarly, we get $b^\diamond$ by substituting b_j for the jth occurrence of $\leftarrow$ or $\nwarrow$, and by replacing all occurrences of $\uparrow$ with $-$.

EXAMPLE 3.10 Consider the words $a = AT$ and $b = AAGT$ and a cost function where $\forall x \neq y : w(x, y) = 1$ and $w(x, x) = 0$. Then the distance matrix $(D_{i,j})$ for a, b is

		A	A	G	T
	0	1	2	3	4
A	1	0	1	2	3
T	2	1	1	2	2

and the corresponding trace matrix is

		A	A	G	T
	$\emptyset$	$\{\leftarrow\}$	$\{\leftarrow\}$	$\{\leftarrow\}$	$\{\leftarrow\}$
A	$\{\uparrow\}$	$\{\nwarrow\}$	$\{\leftarrow, \nwarrow\}$	$\{\leftarrow\}$	$\{\leftarrow\}$
T	$\{\uparrow\}$	$\{\uparrow\}$	$\{\leftarrow\}$	$\{\leftarrow, \nwarrow\}$	$\{\nwarrow\}$

One possible traceback, indicated by black arrows, is

$$\nwarrow \leftarrow \leftarrow \nwarrow .$$

The corresponding alignment is

$$\begin{aligned}
a^\diamond &= A{-}{-}T\,, \\
b^\diamond &= AAGT\,.
\end{aligned}$$

The other possible traceback, indicated by black and dark gray arrows, is

$$\leftarrow \nwarrow \leftarrow \nwarrow,$$

which produces the alignment

$$\begin{aligned} a^\diamond &= -A-T\,, \\ b^\diamond &= AAGT\,. \end{aligned}$$

Figure 3.3 shows the distance matrix together with the two tracebacks, which is the more conventional way of displaying tracebacks. A more complex example can be found in Section 3.3.2 (see Figure 3.6).

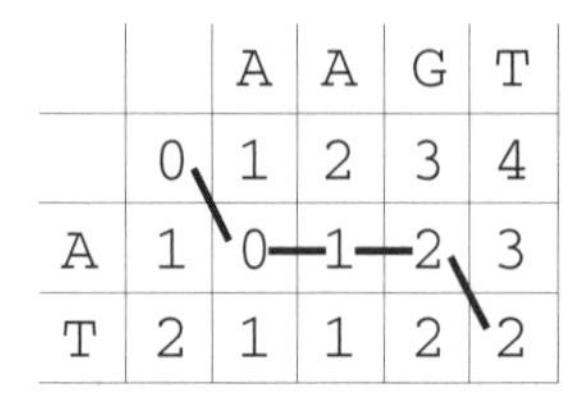
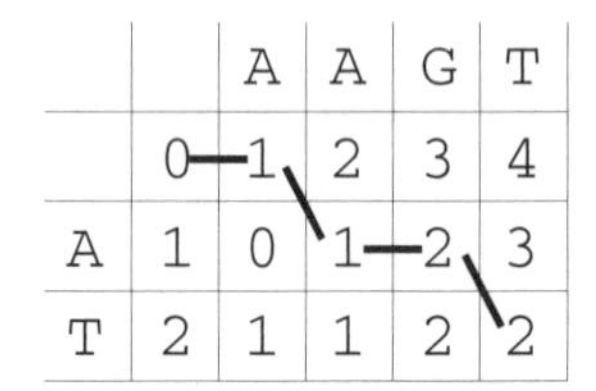

Figure 3.3 Distance matrix and traceback.

By the definition of distance, the insertion (resp. deletion) of a gap of length n has exactly the same cost as the insertion of n gaps of length 1. This is often biologically unrealistic. So we refine the definition of alignment distance as given in Definition 3.5 as follows. Given a word $u \in (\Sigma \cup \{-\})^*$, we say that u contains a *gap* of length k at position i if $u_i \ldots u_{i+k-1} \in \{-\}^*$, and there is no other subword of u extending $u_i \ldots u_{i+k-1}$ that is composed uniquely of $-$s. We define $\Delta_k(u)$ to be the number of different gaps of u that have length k.

DEFINITION 3.11 (ALIGNMENT WITH GAP PENALTY)
A gap penalty *is a function $g(k) : \mathbb{N} \to \mathbb{R}$ that is subadditive, i.e.,*

$$\forall k, l : g(k+l) \leq g(k) + g(l).$$

A gap penalty is called affine *if there are $\alpha, \beta \in \mathbb{R}$ such that*

$$g(k) = \alpha + k\beta.$$

Given a cost function w on $\Sigma \times \Sigma$, the cost of an alignment $(a^\diamond, b^\diamond)$ of length $r = |a^\diamond| = |b^\diamond|$ is defined as

$$w(a^\diamond, b^\diamond) = \sum_{\substack{1 \leq i \leq r \\ a_i^\diamond \neq -,\ b_i^\diamond \neq -}} w(a_i^\diamond, b_i^\diamond) + \sum_{1 \leq k \leq r} \Delta_k(a^\diamond) g(k) + \sum_{1 \leq k \leq r} \Delta_k(b^\diamond) g(k).$$

$d_w^a(a, b)$ is defined as usual by

$$d_w^a(x, y) \quad = \quad \min\{w(a^\diamond, b^\diamond) \mid (a^\diamond, b^\diamond) \text{ alignment of } (a, b)\}.$$

Again we write $D(a, b)$ if w is clear from the context.

We have to redefine the notion of additivity for alignment with gap penalties. Then we present an $O(n^3)$ algorithm for arbitrary gap penalties, and finally an $O(n^2)$ optimization for affine gap penalties.

PROPOSITION 3.12 (ADDITIVITY WITH GAP PENALTY)
Given two alignments $(a^\diamond, b^\diamond)$ with length $r = |a^\diamond| = |b^\diamond|$ and $(c^\diamond, d^\diamond)$. If

$$a_r^\diamond \neq - \vee c_1^\diamond \neq -$$

and

$$b_r^\diamond \neq - \vee d_1^\diamond \neq -,$$

then $(a^\diamond c^\diamond, b^\diamond d^\diamond)$ is also an alignment with

$$w(a^\diamond c^\diamond, b^\diamond d^\diamond) = w(a^\diamond, b^\diamond) + w(c^\diamond, d^\diamond).$$

THEOREM 3.13 (WATERMAN, SMITH AND BEYER [WSB76])
Let $g : \mathbb{N} \to \mathbb{R}$ be a gap penalty and w be cost function on $\Sigma \times \Sigma$. Let $a = a_1 \dots a_n$ and $b = b_1 \dots b_m$ be two words in Σ^*. We define $(D_{i,j})$ with $1 \leq i \leq n$ and $1 \leq j \leq m$ by

$$\begin{aligned} D_{0,0} &= 0, \\ D_{0,j} &= g(j), \\ D_{i,0} &= g(i), \\ D_{i,j} &= \min \left\{ \begin{array}{l} \min\limits_{1 \leq k \leq j} \{D_{i,j-k} + g(k)\}, \\ D_{i-1,j-1} + w(a_i, b_j), \\ \min\limits_{1 \leq k \leq i} \{D_{i-k,j} + g(k)\}\} \end{array} \right\}. \end{aligned}$$

Then $D_{i,j} = D(a_1 \dots a_i, b_1 \dots b_j)$.

PROOF This is similar to the proof for Theorem 3.9, using the case distinction

$$\begin{array}{lll} u^\diamond = \cdots \sigma\, a_{i-k+1} \cdots a_i & u^\diamond = \cdots a_i & u^\diamond = \cdots a_i \quad - \quad \cdots - \\ v^\diamond = \cdots b_j \quad - \quad \cdots - & v^\diamond = \cdots b_j & v^\diamond = \cdots \sigma' b_{j-k+1} \cdots b_j \end{array}$$

with $\sigma \in \{a_{i-k}, -\}$ and $\sigma' \in \{b_{j-k}, -\}$, together with the modified Additivity Proposition 3.12. ■

This theorem furnishes a dynamic programming algorithm with runtime $O(nm \cdot (n+m)) = O(n^3)$ if $n \geq m$. In the case of affine gap penalty, one can even apply an algorithm that has an $O(n^2)$ runtime. The main idea is to use three matrices instead of a single one. In the two additional matrices $(P_{i,j})$ and $(Q_{i,j})$, one stores the minimal distance for all alignments ending with a gap. That is, we define $P_{i,j}$ (resp. $Q_{i,j}$) to be the minimal distance for all alignments $(u^\diamond, v^\diamond)$ for $(a_1 \dots a_i, b_1 \dots b_j)$ that are of the form

$$\begin{array}{l} u^\diamond = \cdots a_i \\ v^\diamond = \cdots -. \end{array} \qquad \left(\text{resp.} \quad \begin{array}{l} u^\diamond = \cdots - \\ v^\diamond = \cdots b_j \end{array} \right).$$

Since we have affine gap penalties, it is easy to define the recursion equation for $P_{i,j}$ and $Q_{i,j}$. When considering $P_{i,j}$, we have two possible cases for the optimal alignment $(u^\diamond, v^\diamond)$ (where optimal means optimal under the condition that a_i is aligned with a gap):

1. $(u^\diamond, v^\diamond)$ extends an existing gap, i.e., $(u^\diamond, v^\diamond)$ is of the form

$$\begin{array}{rcl} u^\diamond & = & \cdots a_{i-1} a_i, \\ v^\diamond & = & \cdots - \quad -. \end{array}$$

Then we have to add only the gap extension cost β, i.e., we get $P_{i,j} = P_{i-1,j} + \beta$.

2. $(u^\diamond, v^\diamond)$ is creates a new gap, i.e.,

$$\begin{array}{rcl} u^\diamond & = & \cdots - a_i \\ v^\diamond & = & \cdots b_j - \end{array} \quad \text{or} \quad \begin{array}{rcl} u^\diamond & = & \cdots a_{i-1} a_i \\ v^\diamond & = & \cdots b_j \quad -. \end{array}$$

Then we have to add the cost for a gap of length 1, i.e., $P_{i,j} = P_{i-1,j} + g(1)$.

This is basic idea applied in the following theorem.

THEOREM 3.14 (GOTOH [GOT82])
Let $g(k) = \alpha + k\beta$ be an affine gap penalty, and let $w : \Sigma \times \Sigma \to \mathbb{R}$ be a cost function and $a, b \in \Sigma^*$. Define the matrices $(D_{i,j})$, $(P_{i,j})$, and $(Q_{i,j})$ recursively by

$$\begin{array}{rcl} D_{0,0} & = & 0, \\ D_{0,j} & = & g(j), \\ D_{i,0} & = & g(i), \\ D_{i,j} & = & \min \left\{ \begin{array}{l} D_{i-1,j-1} + w(a_i, b_j) \\ P_{i,j} \\ Q_{i,j} \end{array} \right\}, \end{array}$$

with $i, j \geq 1$, where for $1 \leq i \leq |a|, 1 \leq j \leq |b|$,

$$\begin{array}{rcl} P_{0,j} & = & \infty, \\ P_{i,j} & = & \min \left\{ \begin{array}{l} D_{i-1,j} + g(1) \\ P_{i-1,j} + \beta \end{array} \right\} \end{array}$$

and

$$\begin{array}{rcl} Q_{i,0} & = & \infty, \\ Q_{i,j} & = & \min \left\{ \begin{array}{l} D_{i,j-1} + g(1) \\ Q_{i,j-1} + \beta \end{array} \right\}. \end{array}$$

Then

$$D_{i,j} = D(a_1 \ldots a_i, b_1 \ldots b_j). \tag{3.6}$$

PROOF This is by induction over pairs (i, j). As an additional claim, we show that

$$P_{i,j} = \min \left\{ D(u^\diamond, v^\diamond) \middle| \begin{array}{l} (u^\diamond, v^\diamond) \text{ alignment of } (a_1 \ldots a_i, b_1 \ldots b_j), \\ \text{with } u^\diamond_{|u^\diamond|} = a_i \text{ and } v^\diamond_{|v^\diamond|} = - \end{array} \right\} \tag{3.7}$$

and

$$Q_{i,j} = \min \left\{ D(u^\diamond, v^\diamond) \middle| \begin{array}{l} (u^\diamond, v^\diamond) \text{ alignment of } (a_1 \ldots a_i, b_1 \ldots b_j), \\ \text{with } u^\diamond_{|u^\diamond|} = - \text{ and } v^\diamond_{|v^\diamond|} = b_j \end{array} \right\}. \tag{3.8}$$

Assume we have proven the claims for all $(k,l) < (i,j)$. We first prove that

$$P_{i,j} \quad \leq \quad \min\left\{ D(u^\diamond, v^\diamond) \left| \begin{array}{l} (u^\diamond, v^\diamond) \text{ alignment of } (a_1 \ldots a_i, b_1 \ldots b_j), \\ \text{with } u^\diamond_{|u^\diamond|} = a_i \text{ and } v^\diamond_{|v^\diamond|} = - \end{array} \right. \right\}.$$

Let $(u^\diamond, v^\diamond)$ with $r = |u^\diamond| = |v^\diamond|$ be an alignment of $(a_1 \ldots a_i, b_1 \ldots b_j)$ such that

$$\begin{aligned} u^\diamond_r &= a_i, \\ v^\diamond_r &= -. \end{aligned} \tag{3.9}$$

If $i = 1$, then there is only one possible alignment given condition (3.9), namely

$$\begin{array}{l} - \ldots - a_i, \\ b_1 \ldots b_j -. \end{array}$$

Hence,

$$\begin{aligned} D(u^\diamond, v^\diamond) &= g(j) + g(1) \\ &= D_{0,j} + g(1) \\ &= \min \left\{ \begin{array}{l} D_{0,j} + g(1) \\ \infty \end{array} \right\} = P_{1,j}. \end{aligned}$$

Otherwise, we know that $(u'^\diamond, v'^\diamond)$ with $u'^\diamond = u^\diamond_1 \ldots u^\diamond_{r-1}$ and $v'^\diamond = v^\diamond_1 \ldots v^\diamond_{r-1}$ is an alignment of $(a_1 \ldots a_{i-1}, b_1 \ldots b_j)$. If $v^\diamond_{r-1} \neq -$, then we get

$$\begin{aligned} D(u^\diamond, v^\diamond) &\overset{\text{Prop. 3.12}}{=} D(u'^\diamond, v'^\diamond) + D(a_i, -) \\ &= D(u'^\diamond, v'^\diamond) + g(1) \\ &\overset{\text{Ind. Hyp.}}{\geq} D_{i-1,j} + g(1). \end{aligned}$$

If $v^\diamond_{r-1} = -$, then $D(u'^\diamond, v'^\diamond) \geq P_{i-1,j}$ by induction hypothesis. Let $k > 0$ be the length of the final gap of $v'^\diamond$. Then

$$\begin{aligned} D(u^\diamond, v^\diamond) &= D(u'^\diamond, v'^\diamond) - g(k) + g(k+1) \\ &= D(u'^\diamond, v'^\diamond) - \alpha - k\beta + \alpha + (k+1)\beta \\ &= D(u'^\diamond, v'^\diamond) + \beta, \\ &\overset{\text{Ind. Hyp.}}{\geq} P_{i-1,j} + \beta. \end{aligned}$$

It remains to be shown that there is an alignment $(u^\diamond, v^\diamond)$ satisfying (3.9) such that

$$D(u^\diamond, v^\diamond) \leq P_{i,j}.$$

If $P_{i,j} = P_{i-1,j} + \beta$, then take an alignment $(u'^\diamond, v'^\diamond)$ of $a_1 \ldots a_{i-1}, b_1 \ldots b_j$ that is optimal under the condition that

$$\begin{aligned} u'^\diamond &= a_{i-1}, \\ v'^\diamond &= -. \end{aligned}$$

By induction hypothesis, this implies $D(u'^{\diamond}, v'^{\diamond}) = P_{i-1,j}$. Hence,

$$\begin{aligned} D(u^{\diamond}, v^{\diamond}) &= D(u'^{\diamond} a_i, v'^{\diamond} -) \\ &= D(u'^{\diamond}, v'^{\diamond}) + \beta \\ &= P_{i-1,j} + \beta \\ &= P_{i,j}. \end{aligned}$$

If $P_{i,j} = D_{i-1,j} + g(1)$, then take an optimal alignment $(u'^{\diamond}, v'^{\diamond})$ of $a_1 \dots a_{i-1}, b_1 \dots b_j$. Then

$$D(u'^{\diamond}, v'^{\diamond}) = D_{i-1,j}$$

by induction hypothesis, and hence

$$D(u'^{\diamond} a_i, v'^{\diamond} -) \leq D_{i-1,j} + g(1).$$

An analogous argument can be given for (3.8). The proof for the main claim (3.6) is given by the same case distinction as in Theorem 3.9. ■

Again we can achieve an optimal alignment using trace matrices and tracebacks. Since we are using different matrices, we must use a different pointer set P, namely

$$\mathcal{P} = \{{}^{D}\nwarrow, {}^{Q}\bullet, {}^{P}\bullet, {}^{D}\leftarrow, {}^{Q}\leftarrow, {}^{D}\uparrow, {}^{P}\uparrow\}.$$

Thus, a traceback pointer points to a matrix *and* a cell in this matrix. Instead of taking one trace matrix, we define three trace matrices (tr^D), (tr^P), and (tr^Q), where

$$\begin{aligned} \mathrm{tr}^D_{0,0} &= \emptyset, \\ \mathrm{tr}^D_{0,j} &= \{{}^{D}\leftarrow\}, \\ \mathrm{tr}^D_{i,0} &= \{{}^{D}\uparrow\}, \end{aligned}$$

and

$$\begin{aligned} \forall i,j > 0 : {}^{D}\nwarrow \in \mathrm{tr}^D_{i,j} &\Leftrightarrow D_{i,j} = D_{i-1,j-1} + w(a_i, b_j), \\ {}^{Q}\bullet \in \mathrm{tr}^D_{i,j} &\Leftrightarrow D_{i,j} = Q_{i,j}, \\ {}^{P}\bullet \in \mathrm{tr}^D_{i,j} &\Leftrightarrow D_{i,j} = P_{i,j}; \\ \forall i,j > 0 : {}^{P}\uparrow \in \mathrm{tr}^P_{i,j} &\Leftrightarrow P_{i,j} = P_{i-1,j} + \beta, \\ {}^{D}\uparrow \in \mathrm{tr}^P_{i,j} &\Leftrightarrow P_{i,j} = D_{i-1,j} + g(1); \\ \forall i,j > 0 : {}^{Q}\leftarrow \in \mathrm{tr}^Q_{i,j} &\Leftrightarrow Q_{i,j} = P_{i,j-1} + \beta, \\ {}^{D}\leftarrow \in \mathrm{tr}^Q_{i,j} &\Leftrightarrow Q_{i,j} = D_{i,j-1} + g(1). \end{aligned}$$

A traceback is defined using intuitive functions on pairs of integers for the pointers $\{{}^{D}\nwarrow, {}^{Q}\bullet, {}^{P}\bullet, {}^{D}\leftarrow, {}^{Q}\leftarrow, {}^{D}\uparrow, {}^{P}\uparrow\}$, where we use triples (M, i, j) to name matrix elements:

$$\begin{aligned} &\forall i \geq 0, j \geq 0 : \quad {}^{M}\bullet(i,j) = (M, i, j) && \text{for } M \in \{Q, P\}, \\ &\forall i \geq 0, j > 0 : \quad {}^{M}\leftarrow(i,j) = (M, i, j-1) && \text{for } M \in \{D, Q\}, \\ &\forall i > 0, j \geq 0 : \quad {}^{M}\uparrow(i,j) = (M, i-1, j) && \text{for } M \in \{D, P\}, \\ &\forall i > 0, j > 0 : \quad {}^{D}\nwarrow(i,j) = (D, i-1, j-1). \end{aligned}$$

An element $t \in \{{}^{D}\nwarrow, {}^{Q}\bullet, {}^{P}\bullet, {}^{D}\leftarrow, {}^{Q}\leftarrow, {}^{D}\uparrow, {}^{P}\uparrow\}^*$ is a *traceback for* (a, b) if there is a sequence of matrix cells

$$(M_0, i_0, j_0) \ldots (M_{|t|}, i_{|t|}, j_{|t|})$$

such that

1. $(M_{|t|}, i_{|t|}, j_{|t|}) = (D, |a|, |b|)$ and $(M_0, i_0, j_0) = (D, 0, 0)$;
2. for all $1 \leq k \leq |t|$:

$$t_k \in \mathrm{tr}_{i_k, j_k}^{M_k} \text{ and } t_k(i_k, j_k) = (M_{k-1}, i_{k-1}, j_{k-1}).$$

The corresponding alignment is defined analogously to the case of alignment without gap penalty.

3.3.2 *Alignment with Tandem Duplication*

In the previous alignment algorithms, we considered edit operations of the forms insertion, deletion and substitution. Benson [Ben97] considered an additional operation, tandem duplication, which generates tandem repeats. The corresponding evolutionary model is is called the DSI model (**d**uplications, **s**ubstitutions and **i**ndels). The SI model (**s**ubstitutions and **i**ndels) is the underlying evolutionary model of standard alignment algorithms.

A tandem duplication is an operation that replaces a subsequence w in a sequence a by w^k. An example of the application of tandem duplication is

```
ACGAG|CCGTAGAA|TACCG
          ↓
ACGAG|CCGTAGAA|CCGTAGAA|CCGTAGAA|CCGTAGAA|TACCG.
```

Here, 3 copies of CCGTAGAA have been inserted. Benson considered the problem of tandem repeats under two assumptions:

1. Tandem duplication occurs before the other types of operations.
2. There is no removal of copies of a tandem repeat region.

Both assumptions together imply that one could can generate the alignment using the usual alignment algorithm if one knew the regions that have been duplicated.

Given the above assumptions, Benson [Ben97] showed that one can extend Gotoh's algorithm for affine gap penalty to incorporate also duplications. Let a and b be two sequences that should be aligned under the DSI model (with affine gap penalties). Let g be an affine gap penalty, $d : \mathbb{N} \to \mathbb{R}$ be an affine duplication cost (i.e., $d(l) = \gamma + \delta l$, where γ is the duplication initiation cost and δ is the duplication extension cost) and $w : \Sigma \to \mathbb{R}$ be a cost function on Σ. Then we define matrices $(D_{i,j})$, $(P_{i,j})$ and $(Q_{i,j})$ as follows. $(P_{i,j})$ and $(Q_{i,j})$ are defined as in Theorem 3.14, and $(D_{i,j})$ is defined by

$$\begin{aligned}
D_{0,0} &= 0, \\
D_{0,j} &= g(j), \\
D_{i,0} &= g(i), \\
D_{i,j} &= \min \left\{ \begin{array}{l} D_{i-1,j-1} + w(a_i, b_j) \\ P_{i,j} \\ Q_{i,j} \\ Dup[i,j] \end{array} \right\},
\end{aligned}$$

where $Dup[i,j]$ is defined as follows. For a word w, let $w_{r...s}$ be the subsequence $w_r \dots w_s$ of w. Then

$$Dup[i,j] = \min_{r,s} \begin{cases} D_{i-r,j-s} + \min_{k \in \mathbb{N}}\{d(k) + D\big(a_{i-r+1\dots i}, (b_{(j-s+1)\dots j})^k\big)\}, \\ D_{i-r,j-s} + \min_{k \in \mathbb{N}}\{d(k) + D\big((a_{i-r+1\dots i})^k, b_{j-s+1\dots j}\big)\}. \end{cases}$$

Here, the notation $(b_{(j-k+1)\dots j})^l$ is a shortcut for the word consisting of l copies of $b_{(j-k+1)\dots j}$, i.e. the word

$$\underbrace{b_{(j-k+1)\dots j} \dots b_{(j-k+1)\dots j}}_{l \text{ copies}},$$

and similarly for $(a_{i-h+1\dots i})^l$. Thus, the remaining problem is to calculate

$$\min_{k \in \mathbb{N}}\{d(k) + D(a, b^k)\}.$$

We consider only the problem of aligning a word a with an unknown number k of copies of a word b, ignoring the duplication cost. The *alignment distance with tandem repeat* $D^{\text{tandem}}(a,b)$ is defined by

$$D^{\text{tandem}}(a,b) = \min\{D(a, b^k) \mid k \in \mathbb{N}\}.$$

As we will see, the algorithm for calculating $D^{\text{tandem}}(a,b)$ can easily be extended to incorporate the duplication cost, too. An *alignment with tandem repeats* is defined to be a pair $(a^\diamond, b_1{}^\diamond \dots b_k{}^\diamond)$, where $(a^\diamond, b_1{}^\diamond \dots b_k{}^\diamond)$ is an alignment as usual with

- $a^\diamond|_\Sigma = a$, and
- for all $1 \le l \le k : b_l{}^\diamond|_\Sigma = b$.

In the following, we consider only an alignment distance D with an affine gap penalty $g(l) = \alpha + l\beta$ and a cost function w. The principle idea is to consider all possible alignment matrices $(D^k_{i,j})$ for the usual alignment of a with b^k for $k \in \mathbb{N}$. Each matrix is then divided into k submatrices of size $|a| \times |b|$, and all these sub-matrices are overlayed onto a single matrix using minimization as shown in Figure 3.4.

This general idea is formulated as follows. Let $(D^k_{i,j})$, $(P^k_{i,j})$, and $(Q^k_{i,j})$ be the matrices for alignment distance between a and b^k as defined by Gotoh's algorithm. The *overlay matrices* $(D_{i,j})$, $(P_{i,j})$, and $(Q_{i,j})$ of size $|a|+1 \times |b|+1$ are defined by

$$\begin{aligned} &D_{0,0} = 0, \quad D_{0,j} = g(j), \quad D_{i,0} = g(i), \\ &Q_{i,0} = \infty, \quad P_{0,j} = \infty, \end{aligned}$$

and

$$\forall i>0, j>0: \begin{aligned} D_{i,j} &= \min\{D^k_{i,l|b|+j} \mid k,l \in \mathbb{N} \wedge l < k\}, \\ P_{i,j} &= \min\{P^k_{i,l|b|+j} \mid k,l \in \mathbb{N} \wedge l < k\}, \\ Q_{i,j} &= \min\{Q^k_{i,l|b|+j} \mid k,l \in \mathbb{N} \wedge l < k\}. \end{aligned} \tag{3.10}$$

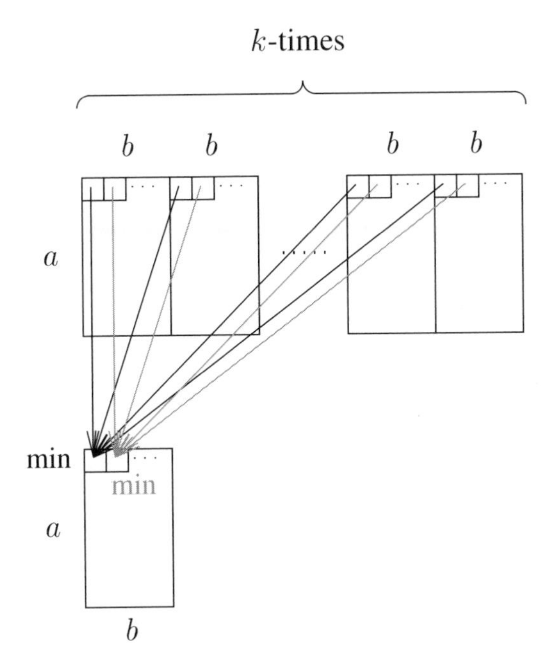

Figure 3.4 Overlay matrix for tandem repeats

PROPOSITION 3.15
For all $1 \leq i \leq |a|, 1 \leq j \leq |b|$,

$$\exists k \in \mathbb{N} : D_{i,j} = D(a_1 \dots a_i, b^k b_1 \dots b_i).$$

PROOF This follows directly from the definition of $D_{i,j}$. ■

Now we want to find a recursive definition for $D_{i,j}$ that can be used for a dynamic programming algorithm similar to Gotoh's algorithm. The main idea is again to simulate the Gotoh's algorithm on k copies of b using a single copy of b, except that we have a special treatment in the case that the dynamic programming algorithm hits the boundary. This is called the wraparound step (see Figure 3.5). The following two lemmas give a recurrence equation for the matrices (D_{ij}), (Q_{ij}) and (P_{ij}) for matrix elements that do not depend on matrix elements beyond a boundary in Lemma 3.16, and for the matrix elements that depend on matrix elements beyond a word boundary in Lemma 3.17.

LEMMA 3.16
Let $(D_{i,j})$, $(Q_{i,j})$, *and* $(P_{i,j})$ *be defined as in (3.10). Then*

$$\forall i > 0, j > 1: \quad D_{i,j} = \min \left\{ \begin{array}{l} D_{i-1,j-1} + w(a_i, b_j) \\ P_{i,j} \\ Q_{i,j} \end{array} \right\}, \tag{3.11}$$

$$\forall i > 0, j > 0: \quad P_{i,j} = \min \left\{ \begin{array}{l} D_{i-1,j} + g(1) \\ P_{i-1,j} + \beta \end{array} \right\}, \tag{3.12}$$

$$\forall i > 0, j > 1: \quad Q_{i,j} = \min \left\{ \begin{array}{l} D_{i,j-1} + g(1) \\ Q_{i,j-1} + \beta \end{array} \right\}. \tag{3.13}$$

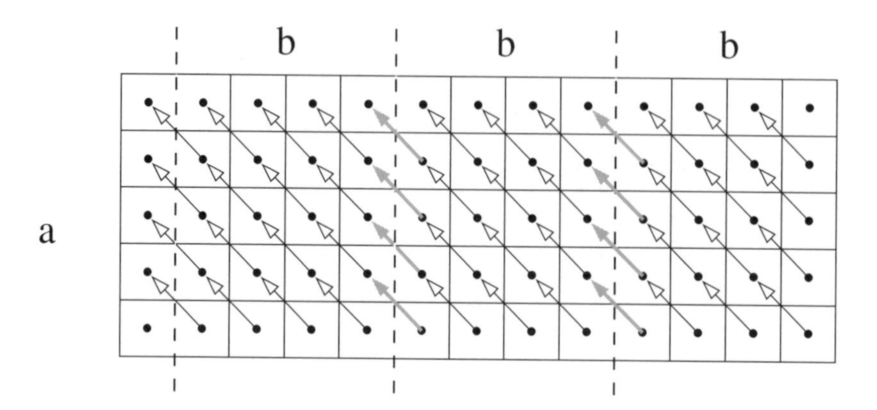

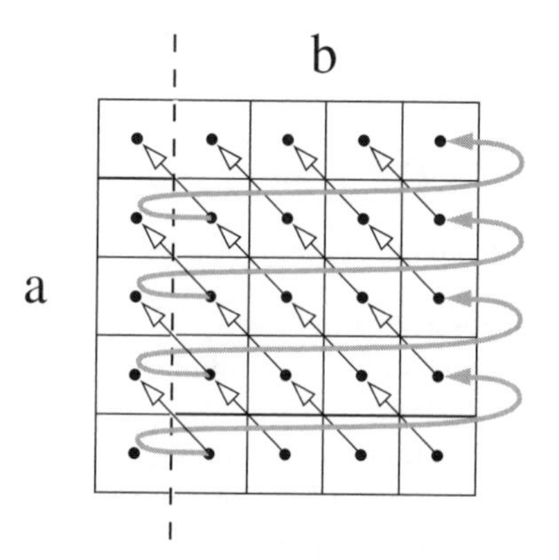

Figure 3.5 Wraparound step. The first picture shows which matrix elements of (D_{ij}) are evaluated in Gotoh's algorithm to calculate a specific matrix element D_{kl} for the alignment a with three copies of b. Note that we have ignored the (P_{ij}) and (Q_{ij}) matrices. The second picture shows which matrix elements are evaluated in the wraparound dynamic programming. All arrows in the first picture that cut a boundary between two successive copies of b (indicated in gray) can be thought to be superimposed onto one wraparound step (indicated in gray) in the second picture.

PROOF By Theorem 3.14, we get

$$
\begin{aligned}
D_{i,j} &= \min_{k,l<k} \{D^k_{i,lb+j}\} \\
&= \min_{k,l<k} \left\{ \min \left\{ \begin{array}{l} D^k_{i-1,l|b|+j-1} + w(a_i, b_j) \\ P^k_{i,l|b|+j} \\ Q^k_{i,l|b|+j} \end{array} \right\} \right\} \\
&= \min \left\{ \begin{array}{l} \min_{k,l<k} \{D^k_{i-1,l|b|+j-1}\} + w(a_i, b_j) \\ \min_{k,l<k} \{P^k_{i,l|b|+j}\} \\ \min_{k,l<k} \{Q^k_{i,l|b|+j}\} \end{array} \right\} \\
&\overset{(j-1>0)}{=} \min \left\{ \begin{array}{l} D_{i-1,j-1} + w(a_i, b_j) \\ P_{i,j} \\ Q_{i,j} \end{array} \right\}.
\end{aligned}
$$

The proofs for the other matrices are analogous. ■

Now we have to consider the special case $D_{i,1}$ and $Q_{i,1}$. For example, we know that in the calculation of $D_{i,1}$, the left or diagonal step either marks the beginning of the alignment (i.e., takes into account the cell $D_{i,0}$ or $D_{i-1,0}$), or it is the start of a new tandem repeat. In the latter, the costs for the previous alignment are stored in $D_{i,|b|}$ or $D_{i-1,|b|}$.

LEMMA 3.17
Let $(D_{i,j})$, $(Q_{i,j})$ and $(P_{i,j})$ be defined as in (3.10). Then

$$
\forall i > 0: \quad D_{i,1} = \min \left\{ \begin{array}{ll} D_{i-1,0} + w(a_i, b_j) & \\ D_{i-1,|b|} + w(a_i, b_j) & \text{(wraparound)} \\ P_{i,1} & \\ Q_{i,1} & \end{array} \right\}, \tag{3.14}
$$

$$
\forall i > 0: \quad Q_{i,1} = \min \left\{ \begin{array}{ll} D_{i,0} + g(1) & \\ D_{i,|b|} + g(1) & \text{(wraparound)} \\ Q_{i,0} + \beta & \\ Q_{i,|b|} + \beta & \text{(wraparound)} \end{array} \right\}. \tag{3.15}
$$

PROOF By Theorem 3.14, we get again

$$\begin{aligned}
D_{i,1} &= \min_{k,l<k}\{D^k_{i,lb+1}\} \\
&= \min_{k,l<k}\left\{\min\left\{\begin{array}{l} D^k_{i-1,l|b|}+w(a_i,b_j) \\ P^k_{i,l|b|+1} \\ Q^k_{i,l|b|+1}\end{array}\right\}\right\} \\
&= \min\left\{\begin{array}{l} \min\limits_{k,l<k}\{D^k_{i-1,l|b|}\}+w(a_i,b_j) \\ \min\limits_{k,l<k}\{P^k_{i,l|b|+1}\} \\ \min\limits_{k,l<k}\{Q^k_{i,l|b|+1}\}\end{array}\right\} \\
&= \min\left\{\begin{array}{l} \min\limits_{k,l<k}\{D^k_{i-1,l|b|}\}+w(a_i,b_j) \\ P_{i,1} \\ Q_{i,1}\end{array}\right\}.
\end{aligned}$$

The problem here is that there is no definition for $\min\limits_{k,l<k}\{D^k_{i-1,l|b|}\}$ in (3.10). But we can write the following:

$$\begin{aligned}
\min_{k,l<k}\{D^k_{i-1,l|b|}\}+w(a_i,b_j) &= \min\left\{\begin{array}{l} \min\limits_{k}\{D^k_{i-1,0}\} \\ \min\limits_{k,1\le l<k}\{D^k_{i-1,l|b|}\}\end{array}\right\}+w(a_i,b_j) \\
&= \min\left\{\begin{array}{l} \min\limits_{k}\{D^k_{i-1,0}\} \\ \min\limits_{k,0\le l<k-1}\{D^k_{i-1,l|b|+|b|}\}\end{array}\right\}+w(a_i,b_j).
\end{aligned}$$

By Theorem 3.14, we have $D^k_{i-1,0}=g(i-1)=D_{i-1,0}$, and therefore

$$\min_k\{D^k_{i-1,0}\}=D_{i-1,0}.$$

Furthermore, we know that given $k\le k'\in\mathbb{N}$, then for all $l<k$ and $j\le|b|$ we have

$$D^k_{i-1,l|b|+j}=D^{k'}_{i-1,l|b|+j}.$$

This implies

$$\min_{k,0\le l<k-1}\{D^k_{i-1,l|b|+|b|}\}=\min_{k-1,0\le l<k-1}\{D^{k-1}_{i-1,l|b|+|b|}\}=D_{i-1,|b|}.$$

Hence,

$$\min_{k,l<k}\{D^k_{i-1,l|b|}\}+w(a_i,b_j) = \min\left\{\begin{array}{l} \min\limits_{k}\{D^k_{i-1,0}\} \\ D_{i-1,|b|}\end{array}\right\}+w(a_i,b_j),$$

which proves the claim. The proof for the $(Q_{i,j})$ matrix is analogous. ■

The last two lemmas give a recursion equation that is satisfied by $((D_{i,j}),(P_{i,j}),(Q_{i,j}))$ defined by (3.10). The remaining part is to show that every triple $((D_{i,j}),(P_{i,j}),(Q_{i,j}))$ satisfying the recursion equation also satisfies (3.10).

LEMMA 3.18
Let a, b be given such that $|b| > 1$. Furthermore, let the gap penalty be $g(k) = \alpha + \beta k$ such that $\alpha \geq 0$ and $\beta > 0$. Then every triple $((D_{i,j}), (P_{i,j}), (Q_{i,j}))$ of matrices that satisfies (3.11)–(3.15) also satisfies (3.10).

PROOF Let $(D_{i,j})$, $(P_{i,j})$, and $(Q_{i,j})$ satisfy (3.11)–(3.15). We have to show that for all $j > 0$,

$$M_{i,j} = \min\{M^k_{i,l|b|+j} \mid k, l \in \mathbb{N} \wedge l < k\}$$

for $M \in \{D, P, Q\}$. Thus, we have to show that for all $j > 0$,

$$\forall k, l \in \mathbb{N} : M_{i,j} \leq M^k_{i,l|b|+j} \tag{3.16}$$

and

$$\exists k, l \in \mathbb{N} : M_{i,j} \geq M^k_{i,l|b|+j}. \tag{3.17}$$

- For the first direction, assume that (3.16) does not hold. Let (k^*, l^*, i^*, j^*) with $1 \leq j^* \leq |b|$ be a minimal quadruple (under the lexicographic ordering) such that (3.16) is not satisfied, i.e., such that

$$M_{i^*,j^*} > M^{k^*}_{i^*,l^*|b|+j^*}$$

for some $M \in \{D, P, Q\}$. Note that $l^* < k^*$.

We derive a contradiction for $M = Q$. The contradiction for the other matrices can be proven analogously. Since the matrices $(D^k_{i,j})$, $(P^k_{i,j})$, and $(Q^k_{i,j})$ satisfy Gotoh's recursion equations (Theorem 3.14), we get

$$Q^{k^*}_{i^*,l^*|b|+j^*} = \min\left\{\begin{array}{l} D^{k^*}_{i^*,l^*|b|+j^*-1} + g(1) \\ Q^{k^*}_{i^*,l^*|b|+j^*-1} + \beta. \end{array}\right\}.$$

We have two cases:

1. $j = 1$. If $l = 0$, then

$$D^{k^*}_{i^*,0\cdot|b|+0} = D_{i^*,0} \qquad \text{and} \qquad Q^{k^*}_{i^*,0\cdot|b|+0} = Q_{i^*,0}$$

by definition of the matrices. If $l > 1$, then we get again by minimality of (k^*, l^*, i^*, j^*) that

$$D^{k^*}_{i^*,l^*|b|+1-1} \overset{l\geq 1}{=} D^{k^*}_{i^*,(l^*-1)|b|+|b|} \geq D_{i^*,|b|}$$

and

$$Q^{k^*}_{i^*,l^*|b|+1-1} \overset{l\geq 1}{=} Q^{k^*}_{i^*,(l^*-1)|b|+|b|} \geq Q_{i^*,|b|}.$$

Hence,

$$\begin{aligned} Q^{k^*}_{i^*,l^*|b|+j^*} &\geq \min\left\{\begin{array}{l} D_{i^*,0} + g(1) \\ D_{i^*,|b|} + g(1) \\ Q_{i^*,0} + \beta \\ Q_{i^*,|b|} + \beta \end{array}\right\} \\ &= Q_{i^*,1}, \qquad \text{(by (3.15))} \end{aligned}$$

which is a contradiction.

2. $j > 1$. This is similar to the above case.

- For the other direction, assume that (3.17) is not satisfied. Let $(i^\bullet, j^\bullet)$ be the minimal quadruple such that

$$\forall k \in \mathbb{N}, l < k : M_{i^\bullet, j^\bullet} < M^k_{i^\bullet, l|b|+j^\bullet}. \tag{3.18}$$

If $M = P$, then by the recursion equation (3.12) we get

$$P_{i^\bullet, j^\bullet} = D_{i^\bullet-1, j^\bullet} + g(1) \qquad (\text{resp. } P_{i^\bullet, j^\bullet} = P_{i^\bullet-1, j^\bullet} + \beta).$$

By the minimality of $(i^\bullet, j^\bullet)$, we get that there are k, l such that

$$D_{i^\bullet-1, j^\bullet} \geq D^k_{i^\bullet-1, l|b|+j^\bullet} \qquad (\text{resp. } P_{i^\bullet-1, j^\bullet} \geq P^k_{i^\bullet-1, l|b|+j^\bullet}).$$

Since $(D^k_{i,j})$ (resp. $(P^k_{i,j})$) satisfy Gotoh's recursion equation, this gives us $P_{i^\bullet, j^\bullet} \geq P^k_{i^\bullet, l|b|+j^\bullet}$ in both cases, and therefore a contradiction.

A similar argument can be applied if $M = D$ or $M = Q$, and $j^\bullet \neq 1$.

The remaining case is $M = D$ or $M = Q$, and $j^\bullet = 1$. The main idea is to apply a similar argument by generating a traceback in the row $i^\bullet$ starting at $M_{i^\bullet, j^\bullet}$. The cells are connected via the recursion equation. If the traceback went back to the previous row, then the same traceback could be done in some $\left((D^k_{i,j}), (P^k_{i,j}), (Q^k_{i,j})\right)$, which would yield a direct contradiction to (3.18). If this is not the case, then we will show that the recursion equations are not satisfied.

For the formal definition, let $(M(1), j(1)) = (M, j^\bullet)$, and let the tuple

$$(M(r+1), j(r+1))$$

be generated from $(M(r), j(r))$ as follows:

$$(M(r+1), j(r+1)) = \begin{cases} & (P, j(r)) & \text{if } M(r) = D \text{ and } D_{i^\bullet, j(r)} = P_{i^\bullet, j(r)} \\ \text{else} & (Q, j(r)) & \text{if } M(r) = D \text{ and } D_{i^\bullet, j(r)} = Q_{i^\bullet, j(r)} \\ \text{else} & (D, j(r)-1) & \text{if } M(r) = Q \\ & & \text{and } Q_{i^\bullet, j(r)} = D_{i^\bullet, j(r)-1} + g(1) \\ \text{else} & (Q, j(r)-1) & M(r) = Q \\ & & \text{and } Q_{i^\bullet, j(r)} = Q_{i^\bullet, j(r)-1} + \beta \\ \text{else} & (D, |b|) & \text{if } j(r) = 1, M(r) = Q \\ & & \text{and } Q_{i^\bullet, 1} = D_{i^\bullet, |b|} + g(1) \\ \text{else} & (Q, |b|) & \text{if } j(r) = 1, M(r) = Q \\ & & \text{and } Q_{i^\bullet, 1} = Q_{i^\bullet, |b|} + \beta \\ \text{else} & \textit{undefined}. & \end{cases}$$

We have the following cases:

1. The traceback is finite. Let r be the maximal number such that $(M(r), i(r))$ is defined, and let $M = M(r)$. By definition of the tuples, there must be a matrix M' and a constant $c \in \mathbb{R}$ such that either $j(r) \neq 1$ and

$$M_{i^\bullet, j(r)} = M'_{i^\bullet-1, j(r)} + c \quad \text{or} \quad M_{i^\bullet, j(r)} = M'_{i^\bullet-1, j(r)-1} + c,$$

or $j(r) = 1$ and

$$M_{i^\bullet,1} = M'_{i^\bullet-1,|b|} + c.$$

In any case, we know by the minimality of $(i^\bullet, j^\bullet)$ that there are k, l such that

$$\begin{aligned} & M_{i^\bullet-1,j(r)} \geq M^k_{i^\bullet-1,l|b|+j(r)}, \\ \text{resp.} \quad & M_{i^\bullet-1,j(r)-1} \geq M^k_{i^\bullet-1,l|b|+j(r)-1}, \\ \text{resp.} \quad & M_{i^\bullet-1,|b|} \geq M^k_{i^\bullet-1,(l-1)|b|+|b|}. \end{aligned}$$

By the recursion equation for $((D_{i,j}), (P_{i,j}), (Q_{i,j}))$ and for $\left((D^k_{i,j}), (P^k_{i,j}), (Q^k_{i,j})\right)$, we immediately get

$$M_{i^\bullet,j(r)} \geq M^k_{i^\bullet,l|b|+j(r)}.$$

Proceeding this way, one gets that there are k, l such that $M_{i^\bullet,j^\bullet} \geq M^k_{i^\bullet,l|b|+j^\bullet}$, which is a contradiction.

2. The traceback is infinite. Then the traceback must be cyclic, i.e., there are $r < r'$ such that

$$M(r) = M(r') \qquad \text{and} \qquad j(r) = j(r').$$

By definition of the traceback and the conditions $|b| > 1$, $\alpha \geq 0$ and $\beta > 0$, there must be a positive value $v \in \mathbb{R}$ with $v > 0$ such that

$$M(r)_{i^\bullet,j(r)} = M(r')_{i^\bullet,j(r')} + v,$$

which is a contradiction to $(M(r), j(r)) = (M(r'), j(r'))$.

■

The previous does not directly lead to a dynamic programming (DP) algorithm. The problem is with the $(Q_{i,j})$ matrix, where we need to know $Q_{i,|b|}$ before we can define $Q_{i,1}$. Hence, we have to modify the DP algorithm to work into two passes (which is shown in Algorithm 3.1). An application example, comparing the matrix yielded by the standard Needleman–Wunsch algorithm [NW70] applied to two copies of b, and the matrix as generated by the Wraparound Dynamic Programming, is shown in Figure 3.6.

In the following, we want to use $D^{(m)}_{i,j}$ to indicate the values of $D_{i,j}$ after the pass m, where $m = 1$ or $m = 2$. We can show that if the last row of the matrices in the first and second passes agree, then we have found the final matrices. If this is not true, then we need additional passes.

LEMMA 3.19
If for all $1 \leq i \leq |a|$

$$\begin{aligned} & \exists 0 < j < |b| \; D^{(1)}_{i,j} = D^{(2)}_{i,j}, \\ & \exists 0 < j < |b| \; P^{(1)}_{i,j} = P^{(2)}_{i,j}, \\ & \exists 0 < j < |b| \; Q^{(1)}_{i,j} = Q^{(2)}_{i,j}, \end{aligned} \tag{3.19}$$

then

$$D^{(m)}_{i,j} = D_{i,j}.$$

Algorithm 3.1 Wraparound Dynamic Programming

Initialization:
{
  for $i = 0$ to $|a|$ {
    $D_{i,0} = g(i)$
    $Q_{i,0} = \infty$
  }
  for $j = 0$ to $|b|$ {
    $D_{0,j} = g(j)$
    $P_{0,j} = \infty$
  }
}

Pass 1: {
  for $i = 1$ to $|a|$
    for $j = 1$ to $|b|$ {

$$P_{i,j} = \min \left\{ \begin{array}{l} D_{i-1,j} + g(1) \\ P_{i-1,j} + \beta \end{array} \right\}$$

$$Q_{i,j} = \min \left\{ \begin{array}{l} D_{i,j-1} + g(1) \\ P_{i,j-1} + \beta \end{array} \right\}$$

$$D_{i,j} = \min \left\{ \begin{array}{l} D_{i-1,j-1} + w(a_i, b_j) \\ \texttt{if } j > 1 \texttt{ then } \infty \texttt{ else } D_{i-1,|b|} + w(a_i, b_j) \texttt{ endif} \\ P_{i,j} \\ Q_{i,j} \end{array} \right\}$$

    }
}

Pass 2: {
  for $i = 1$ to $|a|$
    for $j = 1$ to $|b|$ {

$$P_{i,j} = \min \left\{ \begin{array}{l} D_{i-1,j} + g(1) \\ P_{i-1,j} + \beta \end{array} \right\}$$

$$Q_{i,j} = \min \left\{ \begin{array}{l} D_{i,j-1} + g(1) \\ \texttt{if } j > 1 \texttt{ then } \infty \texttt{ else } D_{i,|b|} + g(1) \texttt{ endif} \\ P_{i,j-1} + \beta \\ \texttt{if } j > 1 \texttt{ then } \infty \texttt{ else } P_{i,|b|} + \beta \texttt{ endif} \end{array} \right\}$$

$$D_{i,j} = \min \left\{ \begin{array}{l} D_{i-1,j-1} + w(a_i, b_j) \\ \texttt{if } j > 1 \texttt{ then } \infty \texttt{ else } D_{i-1,|b|} + w(a_i, b_j) \texttt{ endif} \\ P_{i,j} \\ Q_{i,j} \end{array} \right\}$$

    }
}

		A	C	C	G	A	C	C	G
	0.00	1.00	2.00	3.00	4.00	5.00	6.00	7.00	8.00
A	1.00	0.00	1.00	2.00	3.00	4.00	5.00	6.00	7.00
T	2.00	1.00	1.00	2.00	3.00	4.00	5.00	6.00	7.00
C	3.00	2.00	1.00	1.00	2.00	3.00	4.00	5.00	6.00
G	4.00	3.00	2.00	2.00	1.00	2.00	3.00	4.00	5.00
G	5.00	4.00	3.00	3.00	2.00	2.00	3.00	4.00	4.00
G	6.00	5.00	4.00	4.00	3.00	3.00	3.00	4.00	4.00
A	7.00	6.00	5.00	5.00	4.00	3.00	4.00	4.00	5.00
C	8.00	7.00	6.00	5.00	5.00	4.00	3.00	4.00	5.00
T	9.00	8.00	7.00	6.00	6.00	5.00	4.00	4.00	5.00
G	10.00	9.00	8.00	7.00	6.00	6.00	5.00	5.00	4.00
T	11.00	10.00	9.00	8.00	7.00	7.00	6.00	6.00	5.00

		A	C	C	G
	0.00	1.00	2.00	3.00	4.00
A	1.00	0.00	1.00	2.00	3.00
T	2.00	1.00	1.00	2.00	3.00
C	3.00	2.00	1.00	1.00	2.00
G	4.00	2.00	2.00	2.00	1.00
G	5.00	2.00	3.00	3.00	2.00
G	6.00	3.00	3.00	4.00	3.00
A	7.00	3.00	4.00	4.00	4.00
C	8.00	4.00	3.00	4.00	5.00
T	9.00	5.00	4.00	4.00	5.00
G	10.00	5.00	5.00	5.00	4.00
T	11.00	5.00	6.00	6.00	5.00

$$\text{Alignment:} \quad \begin{array}{lcl} a & = & \text{ATCGGGACTGT} \\ bb & = & \text{A-C-CGACCG-} \end{array}$$

Figure 3.6 Comparison of Needleman–Wunsch and Wraparound Needleman–Wunsch. The cost is 1 for substitutions, and 1 for indels. The task is to find the minimal distance between a = ATCGGGACTGT and an unknown number of copies of b = ACCG. The optimum is found by two copies. The figure shows the distance matrix for Needleman–Wunsch applied to a and bb, and the distance matrix for Wraparound Needleman–Wunsch applied to a and b. One can see that the traceback of the first matrix is just folded into the second matrix. Note that the number of wraparound steps in the traceback determines the number of copies of b used.

PROOF After pass 2, we know that $(D^{(2)}_{i,j})$ satisfies (3.11)–(3.13). If (3.19) holds, then especially

$$\begin{aligned} D^{(1)}_{i,|b|} &= D^{(2)}_{i,|b|}, \\ P^{(1)}_{i,|b|} &= P^{(2)}_{i,|b|}, \\ Q^{(1)}_{i,|b|} &= Q^{(2)}_{i,|b|}. \end{aligned}$$

By the definition of the algorithm, this implies

$$\begin{aligned} Q^{(2)}_{i,1} &= \min \left\{ \begin{array}{ll} D^{(1)}_{i,0} + g(1) & \\ D^{(1)}_{i,|b|} + g(1) & \text{(wraparound)} \\ Q^{(1)}_{i,0} + \beta & \\ Q^{(1)}_{i,|b|} + \beta & \text{(wraparound)} \end{array} \right\} \\ &= \min \left\{ \begin{array}{ll} D^{(2)}_{i,0} + g(1) & \\ D^{(2)}_{i,|b|} + g(1) & \text{(wraparound)} \\ Q^{(2)}_{i,0} + \beta & \\ Q^{(2)}_{i,|b|} + \beta & \text{(wraparound)} \end{array} \right\}. \end{aligned}$$

This gives (3.15), and, with a similar argument, we also get (3.14). The same can be done for the $(D^{(2)}_{i,j})$ matrix. By Lemma 3.18, this proves the claim. ■

The cost for duplications can easily be incorporated by adding the duplication extension cost every time a wraparound step is performed.

3.3.3 Similarity Methods

Let $a = a_1 \dots a_n$ and $b = b_1 \dots b_m$ be words over a finite alphabet Σ. Rather than minimizing the *distance* between a, b, we *maximize* the similarity between a, b. Instead of a cost function $w(x, y)$, we consider a similarity function $s(x, y)$, where generally $s(x, y)$ is positive if $x = y$, and negative if $x \neq y$ (the score for gap $s(x, -)$ and $s(-, y)$ may be different than the score $s(x, y)$ of a mismatch). *Transition* is the substitution of a purine (adenine or guanine) for a purine, or a pyrimidine (cytosine or thymine) for a pyrimidine. *Transversion* is the substitution of a purine for a pyrimidine or vice versa. Such information can be incorporated into the similarity function s. The best scoring global alignment between a, b (without special treatment of gap, i.e. without context sensitivity) must have score

$$S(a, b) = \max \left\{ \sum_{i=1}^{L} s(a_i^*, b_i^*) \;\middle|\; \text{over all } a^*, b^*, \text{ provided that for no } i \text{ are both } a_i^* \ b_i^* \text{ equal to '}-\text{'} \right\}.$$

Here, $L \leq n + m$, and a^*, b^* are sequences in $\Sigma \cup \{-\}$, where $-$ represents a gap ($-$ in the a sequence represents insertion into the b sequence, while $-$ in the b sequence represents deletion from the a sequence; this is clear if one imagines a sequence of edit operations which produce b from a).

A dynamic algorithm to compute $S(a, b)$ goes as follows. Let

$$S_{i,j} = S(a_1, \ldots, a_i, b_1, \ldots, b_j).$$

Letting δ be the gap penalty, we consider the linear gap function $g(n) = \delta(n)$ Then

$$\begin{aligned} S_{0,0} &= 0, \\ S_{0,j} &= j\delta, \\ S_{i,0} &= i\delta, \\ S_{i,j} &= \max\{S_{i-1,j-1} + s(a_i, b_j), S_{i-1,j} + \delta, S_{i,j-1} + \delta\}. \end{aligned}$$

The similarity score computation takes time $O(nm)$, and either defining *pointers* or using *tracebacks* produces the sequence alignment with this score.

T. Smith and M. Waterman [SW81] modified the previous algorithm to produce *local alignments*. Define the local similarity measure

$$H(a, b) = \max\{S(a_k \ldots a_i, b_\ell \ldots b_j) \mid 1 \leq k \leq i \leq n, 1 \leq \ell \leq m\}.$$

In computing the similarity score for local alignments, it is necessary that the expected local similarity score between random sequences is negative (this condition is guaranteed for instance by the PAM matrices). Otherwise, a local alignment could always be extended to a longer local alignment and one could expect a higher score. This condition is considered in detail Durbin *et al.* [DEKM98]. To determine $H(a, b)$, compute

$$H_{i,j} = \max\{0, S(a_k \ldots a_i, b_\ell \ldots b_j) \mid 1 \leq k \leq i, 1 \leq \ell \leq j\}$$

by dynamic programming, where

$$\begin{aligned} H_{0,0} &= 0, \\ H_{0,j} &= 0, \\ H_{i,0} &= 0, \\ H_{i,j} &= \max\{0, H_{i-1,j-1} + s(a_i, b_j), H_{i-1,j} + \delta, H_{i,j-1} + \delta\}. \end{aligned}$$

Then the score of the best scoring subsequence of a aligned with subsequence of b is

$$H(a, b) = \max\{H_{i,j} \mid 1 \leq i \leq n, 1 \leq j \leq m\}.$$

3.4 Multiple Sequence Alignment

The problem of multiple sequence alignment is, given k sequences of length at most n, to determine an optimal alignment of all sequences. Suitably formulated and with k, n arbitrary, this problem is known to be NP-complete. There are many approaches towards multiple sequence alignment, of which we mention only a few.

For the formal description of multiple sequence alignment, let Σ be the alphabet, and $\Sigma' = \Sigma \cup \{-\}$ be Σ extended by the gap character $-$. An alignment for k sequences $S_1 \ldots S_k$ is given by a character matrix

$$A = (A_{ij})_{1 \leq i \leq k, 1 \leq j \leq K}$$

over the alphabet Σ' with the property that S_i can be obtained from $A_{i1} \ldots A_{iK}$ by removing the gaps. Thus, the matrix A represents the alignment of the k sequences $S_1, \ldots, S_k$ in K columns.

The cost function for multiple sequence alignment is a straightforward extension of the cost function for pairwise sequence alignment. In both cases, the costs are given to the individual columns of the alignment. Thus, a cost function for aligning sequences $S_1 \ldots S_k$ is a function $w : \Sigma^k \to \mathbb{R}$. The *cost* $D(A)$ of an alignment A with K columns is given by

$$D(A) = \sum_{1 \leq j \leq K} w(A_{1j}, \ldots, A_{kj}).$$

Different multiple sequence alignment problems can be formulated by specifying how the function $w(c_1, \ldots, c_k)$ is composed.

3.4.1 *Dynamic Programming*

In a straightforward manner, it is possible to extend the dynamic programming techniques from pairwise alignment to the alignment of k sequences. For instance, when trying to align three sequences, $a_1 \ldots a_n$, $b_1 \ldots b_m$ $c_1 \ldots c_\ell$, drawn from the finite alphabet Σ, define

$$D_{i,j,k} = \min \left\{ \begin{array}{l} D_{i-1,j-1,k-1} + w(a_i, b_j, c_k) \\ D_{i,j-1,k-1} + w(-, b_j, c_k) \\ D_{i-1,j,k-1} + w(a_i, -, c_k) \\ D_{i-1,j-1,k} + w(a_i, b_j, -) \\ D_{i-1,j,k} + w(a_i, -, -) \\ D_{i,j-1,k} + w(-, b_j, -) \\ D_{i,j,k-1} + w(-, -, c_k) \end{array} \right\},$$

where $w(x, y, z)$ is the cost of comparing $x, y, z \in (\Sigma \cup \{-\})$. This situation corresponds to a linear gap function, and one could define $w(x, y, z)$ to be the sum of pairwise costs $w(x, y) + w(x, z) + w(y, z)$ or alternatively define a new measure, such as

$$w(x, y, z) = \begin{cases} 0 & \text{if } x = y = z, \\ 1 & \text{if 2 of the 3 symbols are the same,} \\ 2 & \text{if all 3 symbols are distinct.} \end{cases} \tag{3.20}$$

The determination of $D_{n,m,\ell}$ then requires $O(nm\ell) = O(n^3)$ time, if $n \geq m, \ell$. Similarly, multiple alignment for k sequences by dynamic programming would take n^k time, where n is the maximum sequence length, when using a linear gap function.

3.4.2 *Gibbs Sampler*

In [LAB+93], an interesting application of the method of Gibbs sampler was developed for local multiple sequence alignment. The generic idea concerns alignment without gaps, although a modification of the method allows for gaps.

Suppose that we have m sequences, of possibly differing lengths. Let w be a fixed window size. After convergence, Algorithm 3.2 determines a local sequence alignment of all windows. In Algorithm 3.2, various window sizes are tried, and the optimal scoring local alignment is output.

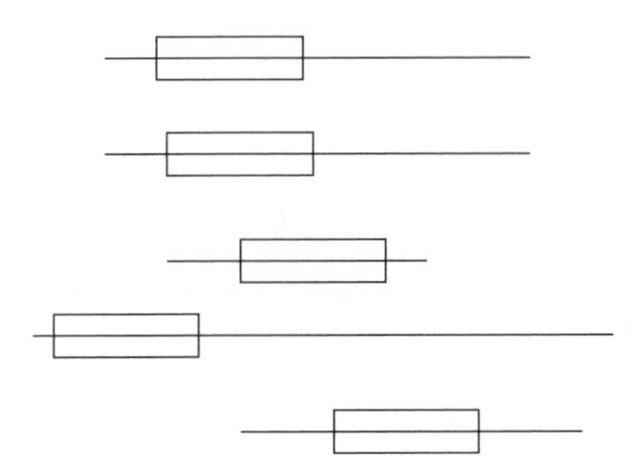

Figure 3.7 Multiple sequence alignment using Gibbs sampler.

Algorithm 3.2 Lawrence, Altschul, Boguski,Liu, Neuwald, Wootton [LAB+93]

---

1. Identify random size w segments, or windows, from m amino acid sequences. (See Figure 3.7 for an illustration.)
2. Choose one of the m proteins at random and temporarily call it P, where $r_1, \ldots, r_n$ is its amino acid sequence.
3. Define a $20 \times w$ frequency matrix Q where

$$Q_{r,j} = \frac{N_{r,j}}{m-1}$$

and $N_{r,j}$ is the number of occurrences of residue r at position j among the $m-1$ remaining proteins. (To avoid a probability of 0 in the case that $N_{r,j} = 0$, *pseudocounts* are added – see [LAB+93] for details.)
4. For $i = 1$ to $|P| - w + 1$ compute

$$p_i = \prod_{j=1}^{w} Q_{r_{i+j},j},$$

which is the probability that the subsequence of P of size w beginning at position i is generated by Q.
5. Choose the starting position of the window in P randomly according to probability p_i. In other words, define cumulative probabilities q_i by

$$\begin{aligned} q_0 &= 0, \\ q_1 &= p_1, \\ q_i &= q_{i-1} + p_i = \sum_{j \leq i} p_j \end{aligned}$$

for $1 \leq i \leq n - w + 1$. Generate random $z \in (0, 1)$, determine i_0, the smallest $1 \leq i \leq n - w$ for which $z \leq q_i$, and set the new starting position for the window in P at position i_0.
6. If convergence has occurred, then stop, else return to Step 2.

3.4.3 Maximum-Weight Trace

Kececioglu [Kec91, Kec93] introduced a special formalization of multiple sequence alignment, the *complete maximum-weight trace (CMWT)* formalization. The CMWT formalization handles a subclass of multiple sequence alignment problems, which can roughly be described as multiple sequence alignment problems that merges pairwise sequence alignments. The most prominent instance of this subclass is the sum-of-pairs multiple sequence alignment problem, which was introduced by Carrillo and Lipman [CL88]. We will first define the sum-of-pairs multiple sequence alignment problem. Later, we define the CMWT.

In the *sum-of-pairs* multiple sequence alignment problem with linear gap penalties, an alignment for n sequences $S_1, \ldots, S_n$ is given as usual by a character matrix

$$A = (A_{ij})_{1 \leq i \leq n, 1 \leq j \leq K}$$

over the alphabet Σ' with the property that S_i can be obtained from $A_{i1} \ldots A_{iK}$ by removing the gaps. The cost function is composed of pairwise cost functions $w_p : \Sigma' \times \Sigma' \to \mathbb{R}$. The *cost* $D(A)$ of an alignment A is given by

$$D(A) = \sum_{i<i'} \sum_{1 \leq j \leq K} w_p(A_{ij}, A_{i'j}).$$

Of course, this is equivalent to $D(A) = \sum_{1 \leq j \leq K} \sum_{i<i'} w_p(A_{ij}, A_{i'j})$, and is hence a special case of the general multiple sequence alignment problem, where the cost for a column is given by $w(a_1, \ldots, a_n) = \sum_{i<i'} w_p(A_{ij}, A_{i'j})$.

The sum-of-pairs problem (with linear gap penalties) is one of the motivations for the CMWT problem. In this problem, the letters of the strings $S_i = s_{i1} \ldots s_{in_i}$ are considered to be the set of vertices $V = V_1 \uplus \ldots \uplus V_n$[5] of a complete n-partite graph $G = (V, E)$ (i.e., G satisfies that for every $v \in V_i$ and $v' \in V_j$, we have $e = (v, v') \in E$ if and only if $i \neq j$). G is called the *complete alignment graph* for the sequences $S_1, \ldots, S_n$. It represents all possible multiple sequence alignments for the sequences $S_1, \ldots, S_n$. An *alignment graph* G' is a subgraph of the complete alignment graph. Alignment graphs can be used to restrict the search for a multiple sequence alignment to a subset of all possible alignments. For example, let S_1 be AACG and S_2 be AGG. Then the complete alignment graph for AACG and AGG is the 2-partite graph

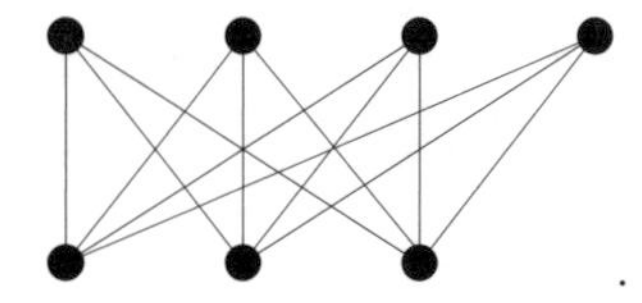

With every edge $e \in E$, there is a positive weight $w(e)$ associated. An alignment A for the sequences $S_1, \ldots, S_n$ *realizes* an edge $e = (s_{ij}, s_{i'j'}) \in E$ of an alignment graph $G = (V, E)$ for the sequences $S_1, \ldots, S_n$ if the jth character of S_i and the j'th character of $S_{i'}$ are aligned in A. For example, consider the alignment

```
A A C G
A - G G
```

[5] Where $\uplus$ is the disjoint union

Then this alignment realizes three edges, indicated by straight lines:

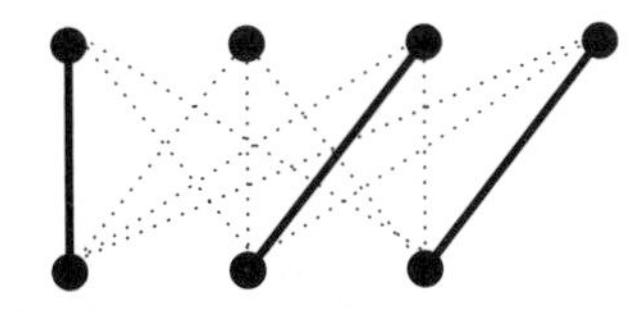

Given an alignment A, the set of all edges realized by A is called the *trace* of A. A set $T \subset E$ of edges is called a *trace* if it is the trace of some alignment A. Given the weight function w, the *weight* of a trace T is $\sum_{e \in T} w(e)$.

PROBLEM 3.20 ((COMPLETE) MAXIMUM-WEIGHT TRACE) Let $S_1, \ldots, S_n$ be sequences, let $G = (V, E)$ be the complete alignment graph for $S_1, \ldots, S_n$, and let w be a weight function. The *complete maximum-weight trace problem* is to find a trace $T \subset E$ that has maximal weight (under w). The *maximum-weight trace problem* is defined analogously for an alignment graph $G = (V, E)$ for $S_1, \ldots, S_n$.

A remaining problem is that not any subset of edges is a trace (i.e., not every subset of E corresponds to a real alignment). Consider again the two sequences AACG and AGG, and consider the following subset of edges indicated by straight lines:

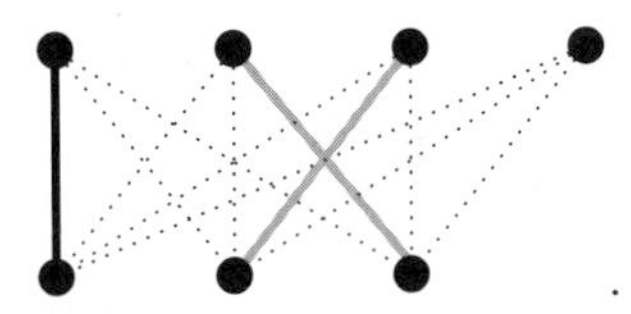

The problem are the two crossing edges indicated in gray above.

In order to characterize set of edges realized by alignments, we must define extended alignment graphs. We need some definitions. A binary relation $\leq$ is a *partial order* if it is (1) *reflexive* (i.e., $a \leq a$), (2) *antisymmetric* (i.e., $a \leq b$ and $b \leq a$ implies $a = b$), and (3) *transitive* (i.e., $a \leq b$ and $b \leq c$ implies $a \leq c$). A binary relation $<$ is a *strict partial order* if it is (1) *irreflexive* (i.e., $a \not< a$), and (2) *transitive* (i.e., $a < b$ and $b < c$ implies $a < c$).

Given sequences $S_1, \ldots, S_n$ with $S_i = s_{i1} \ldots s_{in_i}$, one defines the *extended alignment graph* $G = (V, E, \prec)$ for $S_1, \ldots, S_n$ to be a triple such that (V, E) is an alignment graph for $S_1, \ldots, S_n$, and $\prec$ is defined by

$$\prec = \{(s_{ij}, s_{ij+1}) \mid 1 \leq i \leq n \wedge 1 \leq j < n_i\}.$$

With $\prec^*$, we denote the transitive closure[6] of $\prec$, i.e.,

$$\prec^* = \{(s_{ij}, s_{ij'}) \mid 1 \leq i \leq n \wedge 1 \leq j < j' \leq n_i\}.$$

Note that $\prec^*$ is a strict partial order of V.

Using the extended alignment graph, one can characterize traces. A *connected component* of a graph $G = (V, E)$ is a $\subseteq$-maximal set $V' \subseteq V$ such that for all vertices $v, v' \in V'$ there is a path of edges in E connecting v and v'. For any two subsets $X, Y \subseteq V$, we define

$$X \lhd Y \text{ if and only if } \exists v \in X\, \exists v' \in Y : v \prec v'.$$

---

[6] Given a binary relation $\prec$, the transitive closure of $\prec$ is a binary relation $\prec^*$ such that $x \prec^* x'$ if there is a sequence of elements $x = x_1 \ldots x_k = x'$ with $x_1 \prec x_2 \prec \ldots \prec x_{k-1} \prec x_k$

We define $\lhd^*$ to be the transitive closure of $\lhd$. Hence, $X \lhd^* Y$ if and only if $\exists v \in X \exists v' \in Y : v \prec^* v'$. For the sequences AACG and AGG the extended complete alignment graph is

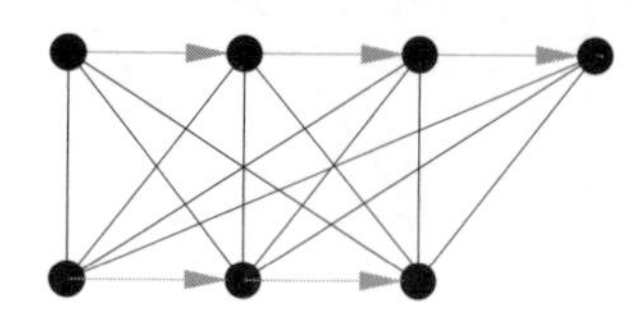

where we have indicated the edges for $\prec$ by arrows.

THEOREM 3.21 (KECECIOGLU)
Let $S_1, \dots, S_n$ be sequences, and let $G = (V, E, \prec)$ be the extended alignment graph for $S_1, \dots, S_n$. Then a subset $T \subseteq E$ is a trace if and only if $\lhd^*$ is a strict partial order on the connected components of $G' = (V, T)$.

PROOF For the first direction, let T be the trace of an alignment A with K columns. By the definition of an alignment, we know that $(s_{ij}, s_{kl}) \in T$ if and only the jth position of S_i and the lth position of S_k are aligned in one column. Hence, any connected component of $G' = (V, T)$ consists of the vertices that are aligned in one column, i.e., $V = X_1 \uplus \dots \uplus X_K$, where X_r is the connected component of $G' = (V, T)$ corresponding to the rth column of the alignment A. We have to show that $X_r \lhd^* X_s$ implies $r < s$, since then $\lhd^*$ must be a strict partial order on $X_1, \dots, X_K$. If $X_r \lhd^* X_s$, then there is a s_{ij} in X_r such that there is a $j' > j$ with $s_{ij'}$ in X_s, which implies by the definition of an alignment that column s must appear after column r.

For the second direction, let T be a subset of E such that $\lhd^*$ is a strict partial order on the connected components of $G' = (V, T)$. Let $\lhd^*_{\text{tot}}$ be any strict total order of the connected components of $G' = (V, T)$ extending $\lhd^*$ (which must exist since $\lhd^*$ is a strict partial order). Let $X_1, \dots, X_K$ be an enumeration of the connected components of $G' = (V, T)$ such that $r < s$ implies $X_r \lhd^*_{\text{tot}} X_s$. Then we define the alignment A of $S_1, \dots, S_n$ in K rows by

$$A_{ij} = \begin{cases} s_{il} & \text{if } s_{il} \in X_j, \\ - & \text{otherwise.} \end{cases}$$

We have to show that A is an alignment. First, we have to show that A is properly defined; i.e., we have to show that $s_{il} \in X_j$ implies that there is no other $s_{il'}$ such that $s_{il'} \in X_j$. Suppose that there were $s_{il} \in X_j$ and $s_{il'} \in X_j$ with $l \neq l'$. Without loss of generality, we can assume that $l < l'$, which implies that $s_{il} \prec^* s_{il'}$ and therefore $X_j \lhd^* X_j$, which is a contradiction to the assumption that $\lhd^*$ is a strict partial order. Second, we have to show that we achieve the string S_i if we ignore the gap symbol $-$ in the string $A_{i1} \dots A_{iK}$. Let w be the string that results from ignoring the gap symbol $-$ in $A_{i1} \dots A_{iK}$. Since every s_{ij} must occur in one of the X_r, we know the word w is composed of some permutation of the symbols in S_i. Hence, we have to show that the permutation is the identity, i.e., we have to show that the symbols occur in correct order.

So assume that there is a s_{il} such that there is some s_{ik} with $l < k$ that appears earlier in the string $A_{i1} \dots A_{iK}$ than s_{il}. By the definition of A, we know that there are X_r and X_s such that $s_{ik} \in X_r$, $s_{il} \in X_s$, and $r < s$. This implies $X_r \lhd^* X_s$. On the other hand, $l < k$ implies $s_{il} \prec^* s_{ik}$ and therefore $X_s \lhd^* X_r$. But $X_s \lhd^* X_r$ and $X_r \lhd^* X_s$ is contradictory to the assumption that $\lhd^*$ is a strict partial order. ■

In [RLM+97, KLM+99], the complete maximum weight trace problem was generalized and coded in the framework of integer linear programming.

3.4.4 Hidden Markov Models

In a later chapter, we will describe hidden Markov models (HMM), a widely used machine learning technique, based on solid mathematical foundations of maximum likelihood and expectation maximization. In [BCHM94, KBM+94, EMD95, Edd95] HMMs have been described for multiple sequence alignment; moreover HMM multiple sequence alignment software has been independently developed by various groups, including P. Baldi, Y. Chauvin and V. Mittal-Henkle (HMMpro at `www.netid.com`), S. Eddy (HMMER `www.genetics.wustl.edu/ eddy`), and R. Hughey, K. Karplus et al. (SAM `www.cse.ucsc.edu/research/compbio/sam.html`).

The underlying idea is that by viewing the sequences to be aligned as a *training set* of observations, which deviate from an ancestor sequence by a stochastic process, a stochastic model is computed, which is most likely to have generated the sequences in the training set. This approach leads to an algorithm for multiple sequence alignment that scales linearly in the number of sequences, rather than exponentially; however, determining the alignment with *global maximum* likelihood, rather than *local maximum*, can require exponential time.

3.4.5 Steiner Sequences

In [KLT97], a problem somewhat related to multiple sequence alignment is considered. In the laboratory technique of *single-molecule DNA sequencing*, single-stranded DNA is cut, a single base at a time. This base then flows down a microscopic tube at high speed past an optical sensor, which detects the type of base. This technique is error-prone, due to sputtering, especially at the beginning and end of the DNA molecule being cut. By repeating the process many times, we produce a collection of N erroneous copies of an original DNA sequence $s_1, \ldots, s_n$, which is to be determined. A heuristic to determine $s_1, \ldots, s_n$ is given in Algorithm 3.3.

The algorithm's runtime is of order

$$\frac{N}{3}n^3 + \frac{N}{3^2}n^3 + \frac{N}{3^3}n^3 + \cdots + \frac{N}{3^{\log_3 N}}n^3 \leq \frac{3}{2}Nn^3 = \Omega(Nn^3).$$

Algorithm 3.3 can be appropriately modified to produce a general multiple sequence alignment, by constructing a tree with branching factor 3 and depth $\log_3 N$, where the N sequences to be aligned are at the leaves, and intermediate nodes are Steiner sequences of the children. From the tree, it is clear how to produce the multiple sequence alignment. The resulting multiple sequence alignment could, however, have the following undesired property. Suppose that $s_1, \ldots, s_N$ are the N sequences to be aligned, and that s_i and s_j are identical (or almost identical), where $1 \leq i \ll j \leq N$. Then it could happen, because of grouping together of every 3 adjacent sequences in producing the intermediate Steiner sequences, that s_i and s_j are misaligned in the resulting multiple sequence alignment. A heuristic remedy might be to first apply pairwise sequence alignment in order to sort the sequences, thus producing $s_{\sigma(1)}, \ldots, s_{\sigma(N)}$ for some permutation $\sigma \in S_N$, with the property that adjacent sequences in the sorted order are similar to each other.

How well does the heuristic perform in determining the original sequence $s_1, \ldots, s_n$ from

Algorithm 3.3 Kececioglu, Li, Tromp [KLT97]

1. Given N erroneous copies of an original DNA sequence to be determined, let $M = N$.
2. Partition the M sequences into $M/3$ groups of three sequences.
3. Apply dynamic programming with cost function $w(x, y, z)$ given in (3.20) to determine an optimal alignment of the three sequences U, V, W in each group. From this alignment, define the *consensus sequence* S obtained by taking the majority symbol in each column, and then removing all occurrences of '$-$' (i.e., it can happen that '$-$' is the majority symbol of a column, in which case this is removed from S). This consensus sequence S is a *Steiner sequence* for U, V, W, meaning that $S \in \{A, C, G, T\}^*$ satisfies

$$D(S, U) + D(S, V) + D(S, W) = \min \{D(T, U) + D(T, V) + D(T, W) : T \in \{A, C, G, T\}^*\}.$$

4. The previous step yields $M/3$ Steiner sequences. If $M = 3$, then stop and output the resulting sequence. Else let $M = M/3$, and return to Step 2.

erroneous copies produced by single-molecule DNA sequencing, and how many erroneous copies are required to determine $s_1, \ldots, s_n$ with high probability? Kececioglu, Li and Tromp answer these questions, subject to the hypotheses that

1. the original DNA sequence $s_1, \ldots, s_n$ is a random sequence (technically *Kolmogorov random*, or *incompressible*, essentially related to maximal entropy), and
2. insertion, deletion and substitution are equally likely.

The former assumption is biologically unrealistic, especially for DNA sequences of interest (genes, promoter sequences, etc.), while the latter assumption seems reasonable, given the physical device used for single-molecule DNA sequencing. Under these assumptions, it is shown in [KLT97] that Step 3 of Algorithm 3.3 reduces an original error rate of ϵ to almost ϵ^2.

It suffices to reduce the error rate to less than $1/n$, since the expected number of errors in a sequence of length n will be less than $n \cdot 1/n = 1$, and the original sequence will have been determined. To this end, let c be such that $\epsilon = 1/c$, and note that the least k such that $\epsilon^{2^k} = \frac{1}{c^{2^k}} < 1/n$ is $\lceil \log_2 \log_2 n - \log_2 \log_2 c \rceil = O(\log_2 \log_2 n)$. Assuming that the resulting $M/3$ Steiner sequences in each pass of the algorithm are random (this has no analytic justification), Kececioglu, Li, and Tromp then argue that it suffices to take $N = 3^k = 3^{\log_2 \log_2 n} = (\log_2 n)^{\log_2 3} \approx \log_2^{1.5849} n$, which is a small number of erroneous copies.

3.5 Genomic Rearrangements

Using dynamic programming methods for sequence alignment, we have shown how genetic distance between two species can be measured using sequence alignment (global or local, using distance or similarity methods) of two homologous DNA or amino acid sequences. *Chromosomal rearrangement* events are much rarer than pointwise mutations, and hence can possibly be used to determine similarity between vastly different organisms.

1. *Intra*-chromosomal events: *inversion* of a contiguous segment of genes (from $5' \to 3'$ to $3' \to 5'$, or vice versa); *duplication*, possibly caused by transposons, leading to pseudogene families such as ALU; *transposition*, where a portion of the chromosome is broken and placed elsewhere in the same chromosome.
2. *Inter*-chromosomal events: *reciprocal translocation*, where the end segments (not including the centromere) of two chromosomes break off and exchange positions (this can happen during crossover); *chromosomal duplication*, where the number of chromosomes in the genome is doubled (this happened in the evolution of a wild grass into wheat); *fission*, where one chromosome is broken into two; *fusion*, where two chromosomes fuse into one.

An even higher level of abstraction was considered by Ferretti, Nadeau, and Sankoff [FNS96], in picturing a chromosome as an *unordered set* of genes. Define a *synteny* as a partition of the genome into distinct sets of genes, or chromosomes. Two genes are said to be *syntenic* if they both lie on the same chromosome. From physical genome maps, current knowledge about the placement of genes on chromosomes is far greater than more detailed knowledge about the order of genes, their orientation, etc.; hence the algorithmic study of synteny seems particularly appropriate.

If we consider that synteny sets (chromosomes, as unordered sets of genes) can be transformed only by the operations of fission, fusion, and reciprocal translocation, then the following questions immediately come to mind.

- Given synteny sets of current organisms, can one construct likely synteny sets of ancestral organisms?
- How many chromosomes did ancestral species have?
- Which genes were on which chromosomes of ancestral species?
- Do phylogenetic trees constructed from comparing synteny sets of different species resemble those constructed from sequence alignment data?

In order to answer such questions, we must compute the *syntenic distance* between two species, i.e., the minimal number of fission, fusion and reciprocal translocation events necessary to transform one synteny set into another.

For example, suppose the genome $\mathcal{G}$ consists of genes a, b, c, d, e, f, g, h and is partitioned into two chromosomes A, B, where $A = \{a, b, c, d\}$ and $B = \{e, f, g, h\}$. Suppose the genome $\mathcal{H}$ consists of the same genes a, b, c, d, e, f, g, h and is partitioned into three chromosomes C, D, E, where $C = \{a, b\}$, $D = \{c, e, f\}$, and $E = \{d, g, h\}$. Then a fission of A into $\{a, b\}$ and $\{c, d\}$, followed by a translocation of $\{c, d\}$ and $\{e, f, g, h\}$ into $\{c, e, f\}$, and $\{d, g, h\}$ transforms genome $\mathcal{G}$ into genome $\mathcal{H}$. See Figure 3.8 for an illustration.

As another example, suppose the genome $\mathcal{G}$ consists of genes a, b, c, p, q, r, x, y and is partitioned into three chromosomes C_1, C_2, C_3, where $C_1 = \{x, y\}$, $C_2 = \{p, q, r\}$, and $C_3 = \{a, b, c\}$. Suppose the genome $\mathcal{H}$ consists of (different) genes a, b, p, q, r, x, y, z and is partitioned into two chromosomes D_1, D_2, where $D_1 = \{p, q, x\}$ and $D_2 = \{a, b, r, y, z\}$. The common genes between genomes $\mathcal{G}$ and $\mathcal{H}$ are a, b, p, q, r, x, y, and the syntenic distance between $\mathcal{G}$ and $\mathcal{H}$ is 2, since a translocation of $\{p, q, x\}, \{a, b, r, y\}$ produces $\{x, y\}, \{a, b, p, q, r\}$, and a fission of $\{a, b, p, q, r\}$ produces $\{p, q, r\}, \{a, b\}$. thus

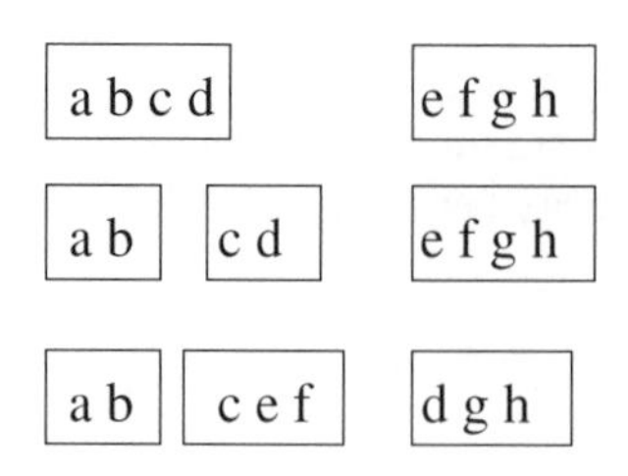

Figure 3.8 Fission and translocation of synteny sets.

transforming the genome $\mathcal{H}$ into $\mathcal{G}$. Note that we only consider those genes common to both genomes.

It is known that the problem of determining syntenic distance is NP-complete. Certain other measures of chromosomal rearrangement distance have been shown to be NP-complete, where, however, a variant of the problem can sometimes be shown to be polynomial-time-computable. In some cases, using heuristics, branch-and-bound, or other means, efficient algorithms for approximate distance have been developed. The interested reader should consult the references [KS95, HP95, FNS96, SM97] for more information on this subject.

3.6 Locating Cryptogenes and Guide RNA

A recent application of sequence alignment methods was given by A. von Haeseler *et al.* [vHBS$^+$92]. DNA is transcribed into pre-edited mRNA, whereupon the intron transcriptions are removed. In some instances, before producing the final mRNA, the pre-edited mRNA is *edited* – pointwise, certain nucleotides are added or removed. The original genes, called *cryptogenes*, may thus be significantly different than the cDNA obtained by retranscribing the edited mRNA back into DNA. One editing operation can shift the entire reading frame, so it is truly amazing that nature has evolved such an intricate mechanism to produce proteins. There are many documented cases of editing found in plants, fungi, and animals, but for mitochondrial mRNA in the case of kinetoplastid protozoa (such as trypanosomes), *guide RNA* (gRNA) has been found that forms, over a small region, a (reverse complementary) template for the desired nucleotide sequence. The gRNA forms a short *anchor region* with the pre-edited mRNA downstream of the edited region (i.e., on the $3'$ end of the pre-edited mRNA) and forms a perfect hybrid (allowing G, U base pairs) with the mRNA. The anchor region is terminated by a mismatch, which starts the *guiding region* of the guide RNA. This region guides the editing of the cryptogene. There is low sequence similarity of the guiding region with the corresponding region of the pre-edited mRNA of the cryptogene, but a perfect match (allowing G, U base pairs) with the edited mRNA. See Figure 3.9 for details.

More complicated still, there are cases of *chaining* of guide RNAs in the sense that a first gRNA anchors itself on the pre-edited RNA, permits the editing over a small region, and the newly edited portion is reverse complementary to a second gRNA responsible for further changes. This permits a linear application of desired editing. See [AH94, ES99] for a description of known mechanisms for RNA editing.

In [vHBS$^+$92], algorithms derived from Smith–Waterman local sequence alignment are given for finding potential gRNA genes from a given cryptogene, and for finding potential cryptogenes from similar proteins. Since Smith–Waterman alone does not place real gRNA genes among the highest scoring sequences found, certain rules concerning the formation

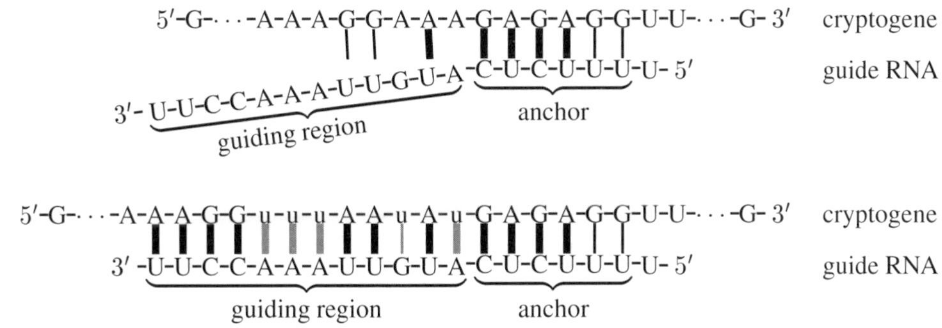

5′-G-···-A-A-A-G-G-A-A-A-G-A-G-A-G-G-U-U-···-G- 3′ cryptogene
3′- U-U-C-C-A-A-A-U-U-G-U-A-C-U-C-U-U-U-U- 5′ guide RNA
guiding region anchor

5′-G-···-A-A-A-G-G-u-u-u-A-A-u-A-u-G-A-G-A-G-G-U-U-···-G- 3′ cryptogene
3′ -U-U-C-C-A-A-A-U-U-G-U-A-C-U-C-U-U-U-U- 5′ guide RNA
guiding region anchor

Figure 3.9 Cryptogene and guide RNA before and after editing. The inserted uracils are indicated by lower-case us. Usual Watson–Crick base pairs are indicated by thick lines, G–U bonds by thin lines.

of an *anchor* region between gRNA and pre-edited mRNA are given. Later a Markov chain model is developed, for which the expected length of gRNA is computed. This computation agrees fairly well with the data.

The following algorithm accounts for pointwise uracil additions, as documented in the trypanosome data. For the application to gRNA, we match the cryptogene with the gRNA gene. For simplicity, we consider the matching of the RNA transcripts of both the gRNA gene and the cryptogene (instead of the DNA genes themselves; of course, one has to use the DNA version for a database search). Recall that in RNA, there are no steric interferences between guanine and uracil, and so we consider the weaker G–U bond in addition to the Watson–Crick pairs A–U with 2 hydrogen bonds and C–G with 3 hydrogen bonds. Consider the pre-edited cryptogene transcript as the a sequence, and the gRNA as the b sequence. Thus, the pairs $(-, A)$ and $(-, G)$ corresponds to insertions in the cryptogene a. Since U base-pairs with both A and G, this correspond to the insertion of U as described above. This suggests the similarity function

$$\sigma(x, y) = \begin{cases} c_1 & \text{if } (x, y) \in \{(A, U), (U, A)\}, \\ c_2 & \text{if } (x, y) \in \{(C, G), (G, C)\}, \\ c_3 & \text{if } (x, y) \in \{(G, U), (U, G)\}, \\ c_4 & \text{if } (x, y) \in \{(-, A), (-, G)\}, \\ -\infty & \text{if } (x, y) \in \{(-, U), (-, C), (A, -), (C, -), (G, -)\}, \\ c_5 & \text{if } (x, y) \in \{(U, -)\}, \\ -\infty & \text{otherwise.} \end{cases}$$

Note that c_4 corresponds to the editing instance of adding uracil to the pre-edited mRNA, while c_5 corresponds to the editing instance of deleting uracil from the pre-edited mRNA. If one models no uracils being deleted, then one sets $c_5 = -\infty$. This model *disallows* bulges and interior loops, since we do not account for any pairing of bases that do not form a Watson–Crick or G–U bond. Thus, the last line in definition of σ is $-\infty$.

For gRNA search in the organism *L. tarentolae* (4 known cryptogenes at the time of [vHBS+92]), the similarity function was defined by

$$\sigma(x,y) = \begin{cases} 500 & \text{if } (x,y) \in \{(A,U),(U,A)\}, \\ 1000 & \text{if } (x,y) \in \{(C,G),(G,C)\}, \\ 1 & \text{if } (x,y) \in \{(G,U),(U,G)\}, \\ 0 & \text{if } (x,y) \in \{(-,A),(-,G)\}, \\ -\infty & \text{if } (x,y) \in \{(-,U),(-,C),(A,-),(C,-),(G,-)\}, \\ -\infty & \text{if } (x,y) \in \{(U,-)\}, \\ -\infty & \text{otherwise.} \end{cases}$$

The algorithm of [vHBS+92] is reported to be robust, in that setting $c_1 = 1, c_2 = 3, c_3 = 1$ does not change the results of search.

3.6.1 *Anchor and Periodicity Rules*

Since known gRNAs did not have the highest scores for the algorithm, the following rules were introduced in [vHBS+92]. These rules concern formation of an *anchor region* (where gRNA binds to pre-edited mRNA) and the periodicity of uracil inserts. Only insertion of uracil was considered.

Rules

1. There are at least 5 base pairs at the $3'$ end of cryptogene region, including G, U pairs. This is the *anchor* region.
2. There can be no more than 3 G, U base pairs in the anchor.
3. There must be at least 4 contiguous Watson–Crick base pairs in the anchor.
4. No more than 8 contiguous Us may be inserted.
5. There are no more than 3 base pairs between successive U inserts.
6. There must be at least 1 base pair at the $5'$ end of the cryptogene (end of editing).

The rule applications were incorporated into the dynamic programming algorithm, providing better (but still not optimal) results.

3.6.2 *Search for Cryptogenes*

Until now, knowledge of the cryptogene has been assumed, and we have searched for gRNA genes. One can search for cryptogenes by finding similarity between unknown cryptogenes and homologous proteins $y_1 \ldots y_m$ (y_is are amino acids) of closely related organisms. Here, similarity between two amino acids is given by the Dayhoff PAM 250 matrix, δ_N is the cost for a nucleotide gap, while δ_A is the cost for an amino acid gap. Again, we use the RNA transcript of the cryptogene for simplicity.

Since we compare a nucleotide sequence $a_1 \ldots a_n$ with an amino acid sequence $y_1 \ldots y_m$, we have always to compare a nucleotide codon with an amino acid in the matching case. Note that this implies that we explicitly model the reading frame (and the shifting of the reading frame caused by insertion of Us). Hence, define

$$H_{i,j} = \max \left\{ \begin{array}{lll} 0 & & \\ H_{i-1,j} & - & \delta_N \\ H_{i,j-1} & - & \delta_A \\ H_{i-3,j-1} & - & s(a_{i-2}a_{i-1}a_i, y_j) \\ H_{i-2,j-1} & - & s(ua_{i-1}a_i, y_j) \\ H_{i-2,j-1} & - & s(a_{i-1}ua_i, y_j) \\ H_{i-2,j-1} & - & s(a_{i-1}a_iu, y_j) \\ H_{i-1,j-1} & - & s(uua_i, y_j) \\ H_{i-1,j-1} & - & s(ua_iu, y_j) \\ H_{i-1,j-1} & - & s(a_iuu, y_j) \\ H_{i,j-1} & - & s(uuu, y_j) \end{array} \right\}.$$

Again, we use a lower-case u to indicated insertion of Us in the transcript of the cryptogene.

3.7 Expected Length of gRNA in Trypanosomes

The problem with the previous approaches is that they do not rank known guide RNAs among the highest scoring ones. For this reason, [vHBS+92] developed a Markov chain to model guide RNAs that respect the rules given in Section 3.6.1 (with the exception of the rule concerning 8 contiguous uracils) in order to estimate the expected maximum length of gRNA from given parameters. It would be interesting to write a program to output the expected length without going through the following statistical analysis.

The general concept is as followed. Since a sequence is a guide RNA only if there is a corresponding cryptogene, one generates the guide RNA together with the cryptogene. The simplest way to describe the construction of guide RNAs is to start with the anchor and proceed into the guiding region. Since we want to use the standard left-to-right direction for strings, this implies that we write down alignments of cryptogene and guide RNAs in the reverse direction compared to Figure 3.9 (see Figure 3.10, where we have also specified the editing events).

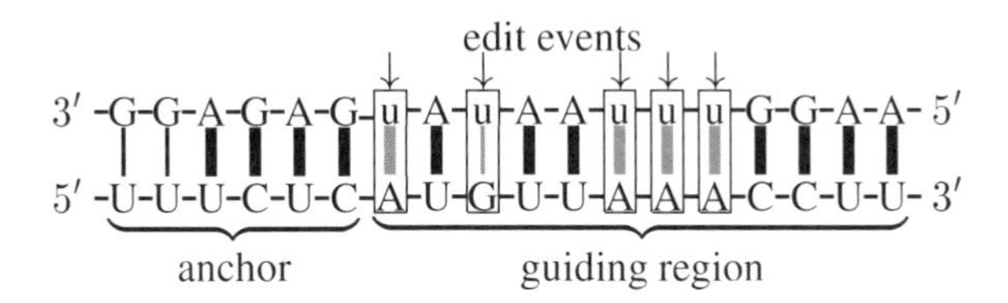

Figure 3.10 Reverse alignment (together with edit events).

Recall that $a = a_1 \dots a_n$ is the transcript of the cryptogene, and $b = b_1 \dots b_m$ the guide RNA. Sequences are modeled as being generated from independent, identically distributed nucleotides. Set

$$p_\alpha = Pr[a_i = \alpha]$$

and

$$q_\alpha = Pr[b_i = \alpha]$$

for $\alpha \in \{A, C, G, U\}$ (since we use the RNA transcripts again). Let

$$p_1 = p_A q_U + p_U q_A + p_C q_G + p_G q_C$$

be the probability of a Watson–Crick base pair, and

$$p_2 = p_G q_U + p_U q_G$$

be the probability yielding a G, U base pair. Then the probability of a non-pairing is

$$Pr[\text{mispair}] = 1 - p_1 - p_2.$$

Let L_a be the length of the anchor region, and L_e the length of the edited region. The length of the guide RNA is clearly $L_a + L_e$. Elementary probability yields an estimate for maximum length of L_a, while the method of Perron–Frobenius yields an approximation for the expected maximum length of L_e.

The anchor must begin after a mismatch, then consist of a region with base pairs, allowing at most three G, U base pairs, and requiring at least 4 contiguous Watson–Crick base pairs. Here is an example of a mismatch followed by an anchor:

$$XWWWNWWWNWWWNWWWW,$$

where X denotes mismatch (before start of anchor), W denotes Watson–Crick base pair, N denotes G, U base pair. Note there are 3 G, U base pairs in the anchor (the maximum allowed), and that the anchor length is 16. If the length of the guide RNA is shorter than 16, then not all possible strings containing less than 3 Ns have at least 4 contiguous Ws (Watson–Crick base pairs). For example, the string of $XWWNWWNWWNWW$ of length 11 does not satisfy this condition, whereas the string $XWWNNNWWWWNWW$ of the same length has 4 contiguous Ws. If, on the other hand, the anchor has length at least 16, then we know that the condition 'less than 3 Ns' implies that there are at 4 contiguous Ws. This implies that for anchors of length at least 16,

$$Pr[L_a \geq \ell] \geq (1 - p_1 - p_2) \sum_{k=0}^{3} \binom{\ell}{k} p_1^{\ell-k} p_2^k.$$

From the binomial theorem, we have the approximation

$$(p_1 + p_2)^\ell = \sum_{k=0}^{\ell} \binom{\ell}{k} p_1^{\ell-k} p_2^k \geq \sum_{k=0}^{3} \binom{\ell}{k} p_1^{\ell-k} p_2^k,$$

so that

$$Pr[L_a \geq \ell] \approx (1 - p_1 - p_2)(p_1 + p_2)^\ell;$$

hence

$$Pr[L_a = \ell] \approx (1 - p_1 - p_2)^2 (p_1 + p_2)^\ell$$

where the extra $(1 - p_1 - p_2)$ term comes from having a mismatch at the end. In this case, the mismatch signals the start of editing.

We continue with the description of the guiding region of the gRNA, which starts with the mismatch that ends the anchor region. Note that, by definition, a RNA sequence is a gRNA if there is a corresponding cryptogene whose edited mRNA aligns to the gRNA. Hence, we have to generate (at random) alignments of gRNA to cryptogenes.

We have to consider the edit event in more detail first. An edit event has two basic properties: (1) editing can take place only if the gRNA has G or A (since these can base-pair with U), and (2) there is a mismatch if editing is not performed. Hence, the possible edit events are

?-u-A-... ?-u-A-... ?-u-G-... ?-u-G-... ?-u-C-...
?-A-?-... ?-G-?-... ?-A-?-... ?-G-?-... ?-A-?-...

These events correspond to the following mismatches with the pre-edited mRNA of the cryptogene:

?-A-... ?-A-... ?-G-... ?-G-... ?-C-...
?-A-... ?-G-... ?-A-... ?-G-... ?-A-...

respectively. Thus, when generating at random an edit event, we have to take into account that there is no base-pairing with the next nucleotide in the cryptogene. For this reason, we add the next nucleotide that has to appear in the (pre-edited) cryptogene to the edit event. The states

A→ A→ G→ G→ C→
u u u u u
A, G, A, G, A

characterize edit events, where the following nucleotides in the cryptogene are A, A, G, G, C, respectively. The states

$*$ ×1, $*$ ×2, $*$ ×3
$*$, $*$, $*$

represent base-pairing in the alignment. They are used to count the number of matches after an edit event. The Moore automaton for generating alignments of gRNA and cryptogene is given in Figure 3.11. Recall that a Markov model is a stochastic Moore automaton. Note that the automaton restricts the number of successive matches to 3, which accounts for Rule 5.

For example, one possible sequence of states generated by the automaton is

A→ A→ G→G→G→
u * u * * u u u *
A * G * * G A A *

This sequence of states represents a set of possible guiding region alignments. For example, this set includes the alignment

u A u A G u u u G
A U G U C G A A C

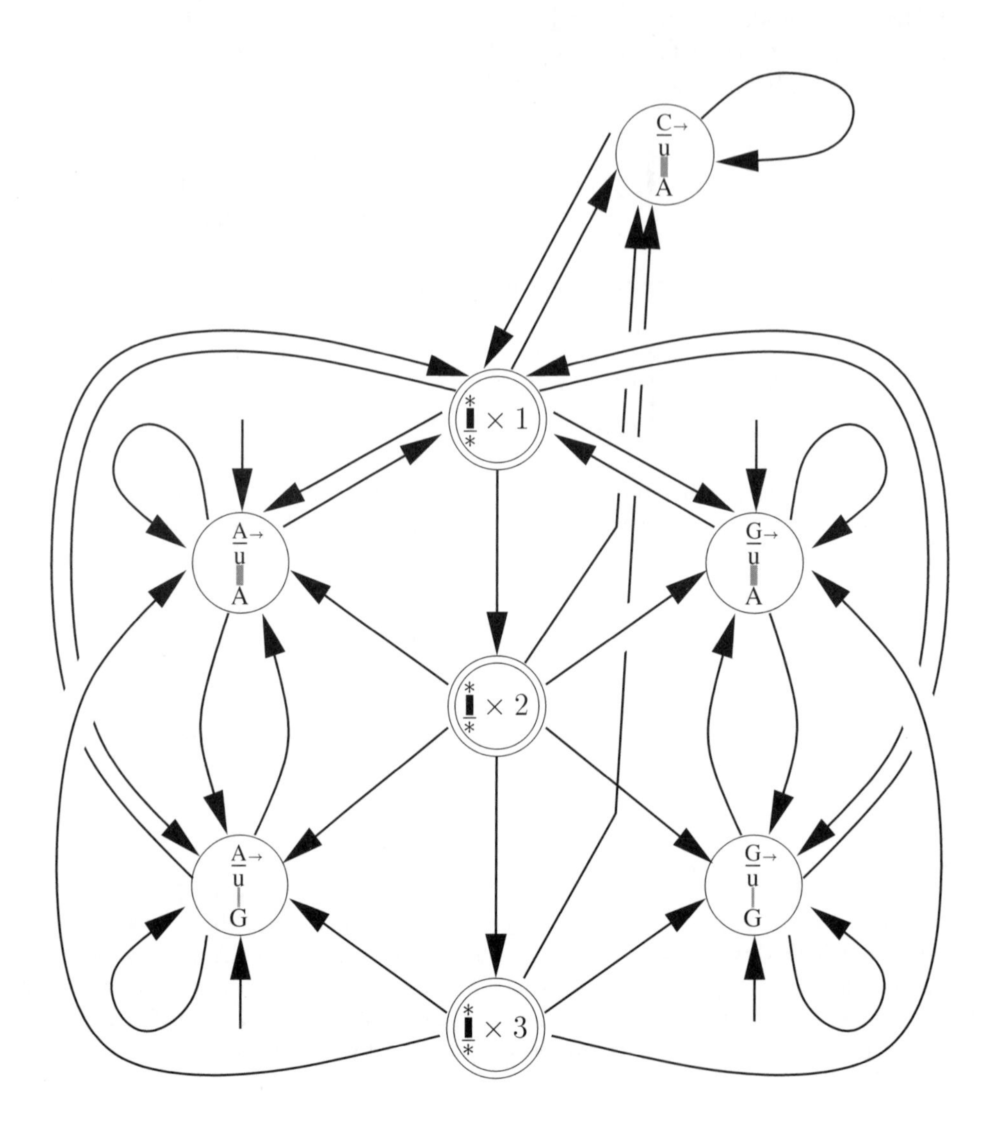

Figure 3.11 Moore automaton for generating alignment of guide RNA *and* edited cryptogene for the guiding region.

but excludes the alignment

$$\begin{array}{ccccccccc} & & & \downarrow & & & & & \\ u & A & u & C & G & u & u & u & G \\ | & | & | & | & | & | & | & | & | \\ A & U & G & G & C & G & A & A & C \end{array}$$

The reason is that the first C in the above sequence (indicated by an arrow) does not satisfy the conditions we have stated above for edit events.

We continue by adding the probabilities to the Moore automaton given in Figure 3.11, thus generating the Markov chain. Recall that a Markov chain is a stochastic process producing a sequence of states $q_0 q_1 \ldots q_t q_{t+1} \ldots$, where q_t is the state at time t. For the initial state q_0, we know that it must be an edit state (since the guiding region starts with an edit state). Consider for example the probability

$$Pr\left[q_0 = \begin{smallmatrix} \underline{A}\to \\ u \\ | \\ A \end{smallmatrix}\right].$$

This is the probability that we see a symbol A in the cryptogene (albeit it appears only later in the alignment), and a symbol A in the guide RNA, i.e., $p_A q_A$. Similarly, we get

$$Pr\left[q_0 = \begin{smallmatrix} \underline{A}\to \\ u \\ | \\ G \end{smallmatrix}\right] = p_A q_G, \quad Pr\left[q_0 = \begin{smallmatrix} \underline{G}\to \\ u \\ | \\ A \end{smallmatrix}\right] = p_G q_A, \quad Pr\left[q_0 = \begin{smallmatrix} \underline{G}\to \\ u \\ | \\ G \end{smallmatrix}\right] = p_G q_G,$$

and

$$Pr\left[q_0 = \begin{smallmatrix} \underline{C}\to \\ u \\ | \\ A \end{smallmatrix}\right] = p_C q_A.$$

Since a guiding region does not start with a match, we get

$$Pr\left[q_0 = \begin{smallmatrix} * \\ \blacksquare \\ * \end{smallmatrix} \times 1\right] = 0, \quad Pr\left[q_0 = \begin{smallmatrix} * \\ \blacksquare \\ * \end{smallmatrix} \times 2\right] = 0, \quad Pr\left[q_0 = \begin{smallmatrix} * \\ \blacksquare \\ * \end{smallmatrix} \times 3\right] = 0.$$

By π, we denote the distribution for the initial state, i.e.,

$$\pi = (p_A q_A, p_A q_G, p_G q_A, p_G q_G, p_C q_A, 0, 0, 0).$$

Next, we have to calculate the transition probabilities. We want to describe two kinds of transitions in greater detail. The first are transitions from a match state into an edit state. For instance, consider

$$Pr\left[q_{t+1} = \begin{smallmatrix} \underline{A}\to \\ u \\ | \\ G \end{smallmatrix} \middle| q_t = \begin{smallmatrix} * \\ \blacksquare \\ * \end{smallmatrix} \times 1\right].$$

Then this is again the probability that we see an A in the cryptogene, and an G in the guide RNA, i.e., $p_A q_G$.

The second kind of transition is a transition from an edit state to a match state, e.g.,

$$Pr\left[q_{t+1} = \overset{*}{\underset{*}{\blacksquare}} \times 1 \,\middle|\, q_t = \overset{\underline{\text{A}}\rightarrow}{\overset{\text{u}}{\underset{\text{G}}{|}}}\right]. \tag{3.21}$$

In order to investigate this probability, we have to consider all alignments that are allowed for any state sequence of the form

$$\cdots \overset{\underline{\text{A}}\rightarrow}{\overset{\text{u}}{\underset{\text{G}}{|}}} \overset{*}{\underset{*}{\blacksquare}} \times 1 \cdots$$

Since we know already that the next letter in the cryptogene must be A (by the definition of the edit state), and since we know that a match must follow, we know that the allowed alignments are of the form

$$\cdots \overset{\text{u}}{\underset{\text{G}}{|}} \overset{\text{A}}{\underset{\text{U}}{\blacksquare}} \cdots$$

Since the only new character we determine is the U in the guiding region, we get that the transition probability given in (3.21) is q_U.

The transition probabilities are summarized in matrix M given in Table 3.1. Note that the matrix M is *substochastic*. Given this matrix M, we know that the probability for a guiding region of length k is given by the probability for a sequence of states of length k generated by M that begins with an edit state and ends with a match state (according to Rule 6 in Section 3.6.1). Edit states correspond to rows $1 \ldots 5$, and match states to rows $6, 7$ and 8. Thus, we get

$$Pr[L_e = k] = \sum_{i=1}^{5} \sum_{j=6}^{8} \pi_i M_{i,j}^k \left(1 - \sum_{s=1}^{8} M_{j,s}\right),$$

where the term $\sum_{s=1}^{8}(1 - M_{j,s})$ accounts for the probability for terminating the Markov chain with the jth state. Since M^k is positive for $k \geq 4$ (i.e. all entries are positive), the Perron–Frobenius theory can be applied to approximate M^k and therefore $Pr[L_e = k]$ (see [vHBS+92] for details).

3.8 Exercises

1. Write a program that, given a nucleotide sequence s_0 along with probabilities p_i, p_d, p_s respectively for the insertion, deletion and substitution of nucleotides, generates new sequences $s_1, s_2, \ldots$, where s_{i+1} is obtained from s_i by means of one insertion, deletion or substitution. Keep track of the position where the pointwise mutation took place, so as to produce the historically correct alignment between s_0 and any s_n. Experiment with different probabilities p_i, p_d, p_s.

   (a) In the first step, develop a program that reads a DNA sequence using a command-line argument and then repeatedly allows the user to determine the type of mutation and the mutation site. For example, `i 3 A` could mean to insert the nucleotide `A` in the third position, while `d 2` could mean to delete the second

Table 3.1 Transition matrix M of the Markov chain (whose graph is given in Figure 3.11).

	A→ u ▮ A	A→ u | G	G→ u ▮ A	G→ u | G	C→ u ▮ A	* ▮×1 *	* ▮×2 *	* ▮×3 *
A→ u ▮ A	q_A	q_G	0	0	0	q_U	0	0
A→ u | G	q_A	q_G	0	0	0	q_U	0	0
G→ u ▮ A	0	0	q_A	q_G	0	$q_U + q_C$	0	0
G→ u | G	0	0	q_A	q_G	0	$q_U + q_C$	0	0
C→ u ▮ A	0	0	0	0	q_A	q_G	0	0
* ▮×1 *	$p_A q_A$	$p_A q_G$	$p_G q_A$	$p_G q_G$	$p_C q_A$	0	$p_1 + p_2$	0
* ▮×2 *	$p_A q_A$	$p_A q_G$	$p_G q_A$	$p_G q_G$	$p_C q_A$	0	0	$p_1 + p_2$
* ▮×3 *	$p_A q_A$	$p_A q_G$	$p_G q_A$	$p_G q_G$	$p_C q_A$	0	0	0

nucleotide, and `s 4 C` could mean to substitute a `C` in the fourth position. After each step, the program should output the total number of mutations effected and produce an alignment between the original input sequence and the current mutated sequence.

HINT An approach along the lines of *bucket sort* can be made, where if the original input sequence has length n, one keeps track of an array `a[ ]` of $n+1$ arrays of character. Thus `a[0]` might be the array of characters inserted *before* the first nucleotide of the original sequence, while `a[1]` consists of that character in the first position of the original sequence, together with other characters inserted between the first and second character of the original sequence, etc.

(b) Extend the program so that random mutations will be carried out in random positions of the sequence, where the user enters as command-line arguments the DNA sequence and the total number of mutations to be carried out (i.e. total number of generations). Your program should be so written that if no number of mutations is entered, then the program goes into the interactive phase as in part (1a) Experiment with different values for the probabilities of insertion, deletion and substitution.

2. Implement Gotoh's algorithm to determine the sequence alignment distance between pairs of nucleotide sequences (you need not implement the backtracking algorithm to determine the alignment, but only determine the distances). Using command line arguments, your program takes a given input file consisting of nucleotide sequences, and outputs a distance matrix $D = (d_{i,j})$, where $d_{i,j}$ is the sequence alignment distance between the ith and jth sequences. This program can be used in combination with phylogeny algorithms discussed in Chapter 4.
3. Implement Gotoh's algorithm for sequence alignment, with backtracking to construct the alignment between two input sequences, given in command line arguments.
4. Write a program to create a dot plot when comparing two amino acid or nucleotide sequences. Given two input sequences $a_1, \ldots, a_n, b_1, \ldots, b_m$, your program should output all pairs (i, j) for which $1 \leq i \leq n, 1 \leq j \leq m$ and $a_i = b_j$. Using a plotting program such as `gnuplot`, you should then produce a 2-dimensional plot with a dot in the position $(n + 1 - i, j)$ for each output pair (i, j). Before the advent of dynamic programming sequence alignment methods, such dot plots were used by biologists, who detected possible local alignments by the presence of a diagonal line of dots from the top left towards the bottom right in the dot plot. Test your program on the proteins Ste6 and Cdc25 from yeast, where you can find the yeast genome at `www.mips.biochem.mpg.de`.
5. Align the Epstein–Barr and HIV genomes, using BLAST and your algorithm.
6. Prove that the Smith–Waterman local alignment algorithm is correct. You need not take apply a special gap penalty, so that

$$H_{i,j} = \max\{0, S(a_k, \ldots, a_i, b_\ell, \ldots, b_j) \mid 1 \leq k \leq i, 1 \leq \ell\}$$

where

$$\begin{aligned} H_{0,0} &= 0, \\ H_{0,j} &= 0, \\ H_{i,0} &= 0, \\ H_{i,j} &= \max\{0, H_{i-1,j-1} + s(a_i, b_j), H_{i-1,j} + \delta, H_{i,j-1} + \delta\}. \end{aligned}$$

Define

$$H(a, b) = \max\{S(a_k, \ldots, a_i, b_\ell, \ldots, b_j) \mid 1 \leq k \leq i \leq n, 1 \leq \ell \leq m\},$$

and show that the optimal alignment of a subsequence of a with a subsequence of b has score

$$H(a, b) = \max\{H_{i,j} \mid 1 \leq i \leq n, 1 \leq j \leq m\}.$$

7. Using the software SAM of the University of California at Santa Cruz for building hidden Markov models, construct a multiple alignment of 10 tRNAs from *M. jannaschii*. You should create a file called `trna10.seq` consisting of the 10 tRNAs, as follows:

```
; This is a comment followed by ID and sequence!!
TRNA1
GGGGAUGUAGCUCAGAGGCCCCGGGUUCGAUCCCCGGCA
;
TRNA2
GGCCAUGUAGCUCAGCCCCGGGUUCGAUCCCCGGCA
;
```

Type the following commands:

(a) buildmodel test -train trna10.seq
(b) buildmodel test -alphabet RNA -train trna10.seq -seed 0
(c) align2model trna10 -i test.mod -db trna10.seq
(d) prettyalign trna10.a2m -l90 > trna10.pretty
(e) buildmodel test -train trna10.seq -seed 0 -printFrequencies 1
(f) drawmodel test.mod test.ps
(g) hmmscore test -i test.mod -db trna10.seq

You can then view the postscript file using *ghostview*. Now, use SAM to perform sequence alignment for 2 sequences, and compare the result with your implementation of Gotoh's algorithm. Note that SAM does *not* necessarily yield the same alignment obtained by dynamic programming.

8. Implement a version of sequence alignment which allows the user to highlight and color (hence weight) certain portions of a nucleotide sequence (or amino acid sequence). This project could use Motif with X-Windows, or Java.

   A biologist who enters a nucleotide (or amino acid) sequence might be particularly interested in finding sequences with BLAST that come very close to matching in a particular region, because of knowledge of biological function of that region. Ideally, the biologist user should be able to select a region with the mouse, and give a certain color to that region, where different colors represent different strengths of desired match.

   Using the usual dynamic programming algorithm for sequence alignment, this is a simple task. Namely, letting $d_{i,j}$ represent the distance between prefix $x_1, \ldots, x_i$ of word x and prefix $y_1, \ldots, y_j$ of word y, write the recurrence relation

$$d_{i+1,j+1} = \min\{d_{i,j} + m_{i+1}(x_{i+1}, y_{j+1}), d_{i+1,j} + g_{j+1}, d_{i,j+1} + g_{i+1}\}.$$

   The usual recurrence relation has a function $m(x_{i+1}, y_{j+1})$ for the cost of a match between x_{i+1} and y_{j+1}, where 0 is for a match, and some penalty for a mismatch. Here the penalty for a mismatch depends on the position $i+1$ in the first sequence x (this sequence is the sequence the biologist is trying to apply BLAST to find homologs).

9. Use Monte Carlo with simulated annealing to perform local multiple sequence alignment; i.e. choose at random one of the sequences to be aligned, shift by a random amount, and accept or reject according to whether it is energetically favorable or unfavorable according to the Metropolis criterion.

10. Write a program that carries out a sequence of pointwise mutations in an input nucleotide sequence, and then generates a 'historically correct' alignment between the original input sequence and the output mutated sequence. (Since the program

keeps track of where the pointwise mutations were made, such a historically correct alignment is possible.)

3.9 Appendix: Maximum-Likelihood Estimation for Pair Probabilities

Let Σ_{amino} be the one-letter alphabet of amino acids, and let $p_1, \dots, p_m$ be an arbitrary enumeration of the pair probabilities $\{p_{AB} \mid A, B \in \Sigma_{\text{amino}}\}$. Let s and s' be two sequences of length n aligned without gaps, and let n_i be the number of times the ith amino acid pair occurs in the alignment of s and s'. Then $\sum_{i=1}^{m} n_i = n$. We want to estimate the parameters p_i from the alignment of s and s'. We will handle the simple case where for all i with $1 \leq i \leq m$ we have $n_i > 0$, i.e., all possible pairings occur at least once in the alignment of s and s'. The description of the case that there some pairings are missing would be too long.

Applying the maximum-likelihood approach as defined from Chapter 2, we want to maximize the likelihood $L_D(M) = Pr[D|M]$ of the model M, given the data D. In our case, the data D is given by the sequence alignment, and the model M by the pair probabilities $p_1, \dots, p_m$, where $m = 20 \cdot 20 = 400$. Thus, we want to maximize

$$f(p_1, \dots, p_m) = \prod_{i=1}^{m} p_i^{n_i}$$

subject to the restrictions

$$\sum_{i=1}^{m} p_i = 1 \tag{3.22}$$

and

$$\forall 1 \leq i \leq m: \;\; p_i \geq 0. \tag{3.23}$$

The restriction (3.22) can directly be handled using a Lagrange multiplier (in contrast to the restrictions in (3.23)). We will calculate the extremum of f under the restriction given in (3.22). As we will see, the solution to this problem also satisfies the other restrictions in (3.23) and is hence a solution of the complete problem.

For this purpose, let

$$g(p_1, \dots, p_m) = \sum_{i=1}^{m} p_i - 1.$$

We have to find the maximum of

$$\begin{aligned} F(p_1, \dots, p_n, \lambda) &= f(p_1, \dots, p_n) + \lambda g(p_1, \dots, p_n) \\ &= \prod_{i=1}^{m} p_i^{n_i} + \lambda \left(\sum_{i=1}^{m} p_i - 1 \right), \end{aligned}$$

where λ is the Lagrange multiplier. Under our assumption that n_i is greater 0 for all

$1 \leq i \leq m$, the partial derivations of F are

$$\frac{\partial F}{\partial p_i} = n_i p_i^{n_i - 1} \prod_{j \neq i} p_j^{n_j} + \lambda$$

$$= n_i \frac{\prod_{j=1}^{m} p_j^{n_j}}{p_i} + \lambda \qquad (3.24)$$

and

$$\frac{\partial F}{\partial \lambda} = \sum_{i=1}^{m} p_i - 1. \qquad (3.25)$$

Setting (3.24) equal to 0, we get

$$p_i = -n_i \frac{\prod_{j=1}^{m} p_j^{n_j}}{\lambda}. \qquad (3.26)$$

Setting (3.25) equal to 0 gives us $1 = \sum_{i=1}^{m} p_i$, which yields, with (3.26),

$$1 = \sum_{i=1}^{m} -n_i \frac{\prod_{j=1}^{m} p_j^{n_j}}{\lambda}, \qquad (3.27)$$

and therefore

$$\lambda = \sum_{i=1}^{m} -n_i \prod_{j=1}^{m} p_j^{n_j} = \prod_{j=1}^{m} p_j^{n_j} \sum_{i=1}^{m} -n_i = -n \prod_{j=1}^{m} p_j^{n_j}.$$

With (3.26), this yields

$$p_i = -n_i \frac{\prod_{j=1}^{m} p_j^{n_j}}{-n \prod_{j=1}^{m} p_j^{n_j}} = \frac{n_i}{n}.$$

The solution $p_i = \frac{n_i}{n}$ satisfies the additional restriction $p_i \geq 0$, and it is easy to check that this is really a maximum of f (under the restriction $g(p_1, \ldots, p_n) = 0$).

Acknowledgments and References

Pairwise sequence alignment by the method of dynamic programming is an old topic, investigated in computer science under the name of *edit distance* (Wagner–Fischer, Masek–Paterson [MP80]) and for applications in biology (Needleman–Wunsch, Sellers, etc.). Chapter 9 of Waterman [Wat95] covers pairwise sequence alignment, as do most other texts on computational biology. Various algorithmic approaches for multiple sequence alignment have been developed, including the application of Gibbs samplers [LAB+93, RT98], hidden Markov models [BCHM94, KBM+94, EMD95], and a method due to [LAK89] not given in the text. The recurrence relation for the tandem duplication was drawn from [Ben97]. Wraparound dynamic programming was introduced in [MM89] (and independently in [FLSS92]). The determination of a DNA nucleotide sequence from polylogarithmically many erroneous copies obtained by single-molecule DNA sequencing was drawn from [KLT97]. The reader should consult the references [FNS96, KS95, HP95, SM97] for more information on chromosal rearrangement distance, and [vHBS$^+$92] concerning guide RNA. Durbin *et al.* [DEKM98] is a good reference for the description of scoring matrices, and some of the description on scoring matrices was drawn from this book. For more details on the maximum-weight trace problem, consider [Kec91, Kec93]. Integer linear programming approaches to solve this problem (and extensions) are given in [RLM$^+$97, KLM$^+$99, LMR99, Rei99].

4

All About EVE

> So the Lord God caused a deep sleep to fall upon the man, and while he slept took one of his ribs and closed up its place with flesh: and the rib which the Lord God had taken from the man he made into a woman and brought her to the man. . . . The man called his wife's name Eve, because she was the mother of all living. (Genesis 2,3, *The Holy Bible*, Meridian Books, 1962).

4.1 Introduction

From the remarkable paleoanthropological work of Louis and Mary Leakey, Donald Johanson, and others, there emerges a generally accepted picture of the evolution of modern man, whose hominid ancestors diverged from ape lines around 5–6 million years ago. As contrasted with the *multiregional model*,[1] the *Out of Africa Model* posits the evolution of a line (or tree) of hominids, who adapted to the savannah life in Africa under the then prevailing dryer climatic conditions, and who left Africa not more than 1 million years ago. D. Johanson discovered the fossil remains of a female *Australopithecus* dated at 3.5 million years, whom he named *Lucy*.[2] The Leakeys' fossil remains of *Homo habilis* from the Great Rift Valley in Ethiopia were dated at around 2 million years.[3] *Homo erectus* emerged in Africa around 1 million years ago, was unusually tall (about 6 feet), had a smaller birth canal than modern man, but because of this and the smaller pelvic rotation, had a better structure for bipedalism and could run faster than modern man [JJE94]. Modern *Homo sapiens* is thought to have emerged between 100 000 and 60 000 years ago. By a new technique of *thermal luminescence*, cave art (Lascaux and other caves) have been accurately dated, some only at 12 000 years ago. Though the Lascaux cave art, being European, is better known to the public at large, certain caves in Australia have been found with similar cave art and dated at around 30 000 years. It is generally accepted that *Cro Magnon* man, responsible for European cave art, is our immediate ancestor *H. sapiens*, and was distinct from *Neanderthal* man, the latter having become extinct around 30 000 years ago. Indeed, DNA amplification and sequence alignment by S. Pääbo an

---

[1] The multiregional model posits the evolution of *Homo sapiens* from a convergence of various distinct hominid lines in different geographic regions.

[2] The Beatles' song 'Lucy in the sky with diamonds' was playing in Johanson's camp when he returned after his discovery, hence the name *Lucy*.

[3] Certain paleoanthropologists now tend to believe that *Homo habilis*, rather than being a distinct species of hominid, is rather a 'catch-all' for a number of hominid fossil fragments.

co-workers supports this view. However, it is believed that *Neanderthal* lived in some areas at the same time as *Cro Magnon*, since nearby cave sites in Israel indicate the presence of both species 35 000 years ago.

Against this background of painstaking paleoanthropological work, R.M. Cann, M. Stoneking, and A. Wilson [CSW87] carried out a statistical analysis of the mitochondrial DNA extracted from the placental tissue of 147 women of different races and from different countries. By constructing a phylogenetic tree under the assumption of a constant molecular clock, Alan Wilson's group (University of California at Berkeley) concluded that modern man emerged from Africa roughly a mere 200 000 years ago, and that race differences arose some 50 000 years ago! This sensational discovery caught the public attention, when on 11 January 1988, *Newsweek* featured a cover article on the *Mitochondrial Eve hypothesis*. As reported in [JJE94], A. Templeton (Washington University in St Louis), later obtained 100 distinct trees, all at most 2 steps away from Wilson's tree, and *all* of which support a non-African hypothesis. Moreover, the molecular clock has been claimed by some specialists not to be constant, and Wilson's clock not to be accurate, leading to an analysis that supports a common ancestor of modern man from 100 000 to 1 million years ago.

Human mitochondrial DNA has a circular double-stranded form consisting of about 16 500 base pairs, and is known to contain genes for coding 13 proteins, 22 tRNA genes, and 2 rRNA genes [GT94]. Mitochondria are inherited in a growing mammalian zygote only from the egg, hence an individual's mitochondria comes from his/her mother, the mother's mitochondria come from her mother, etc.[4] Mitochondrial DNA, mtDNA, is known to have pointwise mutation substitution rates roughly 10 times faster than nuclear DNA. To simulate and test the conclusions of Wilson's group and those of other biologists, one needs knowledge (or mathematical models) of molecular evolution rates as well as computer algorithms for generation of phylogenetic trees. That is the topic of this chapter.

We begin with a discussion of mathematical models for pointwise substitution in nucleotide sequences, and then go on to develop three types of algorithm for the construction of phylogenetic trees (clustering methods, maximum likelihood, and quartet puzzling). The phylogenetic tree construction methods introduced in this chapter are only an introduction to the field. In particular, there are numerous other approaches that are not dealt with in this text (such as neighbor joining and parsimony), but are adequately covered in the literature. Moreover, to the best of our knowledge, the central question of describing mathematical conditions satisfied by those trees, which are historically correct phylogenetic trees, seems yet unanswered. Of course, this is not a question for computational biology, which is concerned with algorithm development, but for molecular biology and population genetics. The situation as of a few years ago was summarized by an expert of population genetics in the following statement:

> At the present time, however, the theoretical foundations of many tree-making methods are not well established from the point of view of population genetics. (M. Nei, *Molecular Evolutionary Genetics*, 1987 [Nei87])

---

[4] The ovum is substantially more complex than the spermatozoon, and requires a correspondingly higher investment of an organism's resources. For instance, human females produce on the order of 400 eggs throughout a lifetime, while human males produce on the order of 7 million sperm per hour.

4.2 Rate of Evolutionary Change

Given n distinct species, all of which evolved from a common ancestor, there is historically a unique phylogenetic tree. This is perhaps not quite accurate, since a species, rather than having a clear-cut, unequivocal taxonomic identity, is really a probability distribution on a gene pool.

A *taxon*, sometimes called *operational taxonomic unit* or *OTU*, is an entity (such as species, amino acid sequence, nucleotide sequence, language, etc.) whose distance or similarity from other entities can be measured. In computational biology, we are mostly concerned with taxa of nucleotide and amino acid sequences, but one could attempt to construct a phylogenetic tree for North American Indian languages, for instance. When constructing a phylogenetic tree based on data from protein or nucleotide sequence comparisons, one can apply sequence alignment algorithms to obtain a distance measure $d_{i,j}$ between two taxa. We are interested principally in the topology of the phylogenetic tree, and to a lesser extent in the branch lengths. In assigning branch lengths to phylogenetic trees, one must consider as well whether evolutionary rate is constant. It is to these considerations that we turn first.

The determination of a phylogenetic tree for the evolution of species from amino acid or nucleotide sequence comparisons depends on measurements and assumptions concerning *rate of evolutionary change*, i.e. a *molecular clock*. This is not a simple task when comparing homologous sites in two sequences, since over long periods of time, there can be *back mutations* (the site in one sequence may originally contain A, later be mutated to G, then again be mutated back to A) as well as *parallel mutations* (homologous sites in the two sequences undergo the same mutation). Following [Nei87], in Sections 4.2.1 and 4.2.2, we look at well-known approaches for mathematically modeling the evolution of amino acid and nucleotide sequences, subject to the restriction that only *substitutions* will be considered.

4.2.1 Amino Acid Sequences

From a multitude of studies, it has become clear that the rate of amino acid substitution varies between organisms, and between protein classes, presumably because of the *selection pressure* of protein functionality, though the overall rate is roughly constant (the number of substitutions is roughly linear in time). Letting λ denote the amino acid substitution rate per site per year, it has been shown that λ for guinea pig insulin is roughly 5.3×10^{-9}, compared with a rate of 0.33×10^{-9} for other organisms,[5] a variation by a factor of more than 10. Also, fibrinopeptides[6] have a substitution rate of 9×10^{-9} per site per year, as compared with a rate of 1×10^{-11} for histone H4,[7] while the rates for cytochrome c and for hemoglobin lie between these two extremes. Changes in the hydrophobic core of a protein are often not tolerated, since the tertiary structure usually changes radically upon substitution of a hydrophobic by a hydrophilic amino acid.[8]

---

[5] p. 54 of [Nei87]

[6] Fibrinopeptides are involved in producing *fibrogen*, which is converted to fibrin for blood clotting.

[7] Histones are proteins around which DNA is wrapped in the nucleosomes. The chromatids are dark banded regions when seen under a microscope.

[8] A good example is the class HS70 of heat shock chaperones of 70 kilodaltons, whose hollow, cylindrically shaped core consists of hydrophobic amino acids. When an incorrectly folded protein (often signaled by hydrophobic residues on the outside) lies within the cylindrical core of HS70, a conformational change occurs to produce an environment allowing the protein to refold correctly.

How are amino acid substitution rates computed? One technique proceeds by determining the amino acid sequence for the same protein of two distantly related species, for which geological data suggests a time of divergence. Define *unit evolutionary time* T_u, the average time to produce one substitution per 100 amino acids, by

$$T_u = \frac{1}{100\lambda}.$$

For histone H4, it has been determined that there are only 2 differences in a sequence of 105 amino acids, when comparing calf and pea. Since it is thought that plants and animals diverged roughly 1 billion years ago, T_u is thus estimated at between 0.5 to 1×10^9, yielding

$$\lambda \approx \frac{1}{100T_u} \approx 10^{-11}.$$

Let X and Y be homologous proteins of the same length n. Let n be the length of the same protein X, Y isolated from two distantly related species. Letting n_d be the number of differences between homologous amino acid sites, the probability p of an amino acid substitution occurring at a given site of either X or Y can be estimated by

$$p \approx \frac{n_d}{n}.$$

Note that $1 - p$ is the probability that no substitution has occurred at a given site of either X or Y. Over a long period of time, because of backward and parallel mutations, this may be a serious underestimate of p. A second approximation of p can be derived by assuming that amino acid substitution at a given site in a protein is a Poisson process. Let Z be a random variable counting the number of mutations over time t at a fixed site for an amino acid sequence having substitution rate λ per site per year. Then the probability $Pr[Z = k]$ that k substitutions have occurred over time t at a given site satisfies

$$Pr[Z = k] = P_{\lambda t}(k) = e^{-\lambda t}\frac{(\lambda t)^k}{k!}.$$

The probability $P_{\lambda t}(0)$ that no substitution occurs at a given site of X is then $e^{-\lambda t}$; hence the probability that no substitution occurs at the same homologous site in X and Y is

$$q = 1 - p = e^{-2\lambda t} \approx 1 - \frac{n_d}{n}.$$

From this, the total number of substitutions $d = \lambda t$ occurring at a fixed site satisfies

$$d = \frac{-\ln q}{2},$$

and a first approximation of d can be obtained by

$$d \approx -\frac{\ln\left(1 - \frac{n_d}{n}\right)}{2}.$$

A more serious criticism of this approach altogether is that the Poisson process model assumes that λ does not depend on the residue site in a protein. As mentioned above, however, hydrophilic substitutions in a hydrophobic core usually lead to a dysfunctional protein, so it is clear that substitution rate *is* site-dependent. Another example is that substitution of glycine, the smallest amino acid, by another amino acid often disrupts the function of the protein.

For these reasons, the amino acid substitution matrix methods of Dayhoff [DSO78] (PAM matrices), Henikoff [HH92, HH93] (BLOSUM matrices), and more recently of Wei, Altman and Chang [WAC97] have been proposed. See Sections 3.2 and 3.9 on scoring matrices for a more detailed treatment of the PAM matrices. By comparing sequences of various classes of proteins (hemoglobins, cytochrome c, fibrinopeptides, etc.) Dayhoff devised an empirical estimation of the probability $p_{i,j}$ that during one evolutionary time unit T_u,[9] amino acid residue i will be substituted by residue j. The substitution matrix obtained, $M = (p_{i,j})$, is called the *PAM1 matrix*, since there is one *point accepted mutation* per 100 residues. Let $\pi = (\pi_1, \ldots, \pi_{20})^{\mathrm{T}}$ be the amino acid composition frequency of a given polypeptide, expressed as a column vector.[10] Denoting M^t as the tth matrix power of M, it follows that the amino acid frequencies t time units later (recall that the time unit is T_u) are given by the matrix product

$$\pi M^t$$

of the row vector π with the matrix M^t. Often one uses the PAM250 (i.e. M^{250}) matrix when comparing two proteins using sequence alignment algorithms. Note that under this model, the probability p that a substitution at a given site of X or Y has occurred during t time units is given by

$$p = 1 - \sum_{i=1}^{20} p_{i,i}^{(2t)} \pi_i$$

where $p_{i,i}^{(2t)}$ is the ith diagonal entry of M^{2t}.[11]

A recent technique has been proposed by Wei, Altman, and Chang [WAC97], who introduced the WAC matrix, created by measuring the physico-chemical properties in radial shells up to 10 Å centered around a given amino acid, thus constituting a description of the 'micro-environment' of the amino acid. The WAC similarity matrix has integer entries ranging between -5 and 4, where $WAC(a, b) = 4$ when amino acids a, b are identical, and -5 for very dissimilar amino acids.

4.2.2 *Nucleotide Sequences*

Nucleotide substitution must be treated somewhat differently than amino acid substitutions, because of redundancy in the genetic code. *Synonymous* or *neutral substitutions* are nucleotide substitutions that leave the expressed amino acid unchanged. Glancing at the genetic code, it is clear that most substitutions to the third position are synonymous. Moreover, from computer simulations, it appears that the genetic code is optimized so that a single non-synonymous nucleotide substitution is likely to change an amino acid into a related amino acid (both hydrophobic, for instance). Since natural selection acts on protein functionality, rather than occurrence of particular nucleotides, it has been argued that a molecular clock should be

---

[9] Recall that T_u is the average time required for one substitution to occur in 100 residues of a protein.

[10] $(\pi_1, \ldots, \pi_{20})$ is a row vector and $(\pi_1, \ldots, \pi_{20})^{\mathrm{T}}$ is its transpose, a column vector.

[11] The factor $2t$ appears rather than t, since one must consider both X and Y. Dayhoff's PAM matrices are symmetric; going from X to Y is modeled by going from X at time t backwards to time 0 and then forwards to Y at time t.

based on the substitution rate in the third codon position, which, according to [Nei87] has been shown by Kimura to be as high as the amino acid substitution rate for fibrinopeptides.

As is the case for amino acids, it appears that nucleotide substitution rate is species dependent, perhaps because of better DNA repair mechanisms in certain species. For instance, it has been estimated that the nucleotide substitution rate λ per site per year for nuclear DNA of higher primates is 1.3×10^{-9}, while rodents and sea urchins have the higher rate of 6.6×10^{-9}. Mammalian mitochondrial DNA, mtDNA, consists of about 16 500 base pairs, and is known to evolve about 10 times as fast as mammalian nuclear DNA, with rates of 10^{-8} per site per year. *Transitional mutations* (purine–purine i.e. A $\leftrightarrow$ G, and pyrimidine–pyrimidine, i.e. C $\leftrightarrow$ T) are known to account for 92% of the total substitutions in mammalian mtDNA, with only 8% for *transversional mutations* (purine–pyrimidine).[12] Chloroplast DNA, cpDNA, consists of about 150 000 base pairs, and has a much slower mutation rate than mtDNA, where one study estimated 1.1×10^{-9} per site per year.

We want to give a more appropriate modeling of the evolutionary process for nucleotides. In the definition of the PAM matrices, one assumes a discrete Markov chain, where the 1 PAM matrix is the transition matrix of the Markov chain. The parameters are estimated from close homologs using local sequence alignment. In this estimation, it is assumed that the two sequences are generated using one application of the transition matrix. This implies the assumption that for the aligned sequences, no multiple substitutions had occurred. Furthermore, it assumes that the evolutionary distance of more distantly related sequences can be modeled by n-times iteration of the Markov chain, which would only allow evolutionary distances that are multiples of the evolutionary distance used for setting up the 1 PAM matrix. Of course, both assumption are simplifications.

Thus, we have to find a better evolutionary model. Both assumptions (no multiple substitutions, discrete evolutionary time) can be avoided if one models the evolutionary process as a continuous Markov process instead of a discrete Markov chain. The definition of a Markov process is completely analogous to the definition of a Markov chain, except that the transition matrix is substituted by a matrix of transition probability functions, which depend on the time parameter t. For the formal definition, we need a 'stochastic function' $X(t)$ with a real-valued time parameter $t \geq 0$, and which can take values from a state space $\{1, \ldots, n\}$. A (time-homogeneous) *Markov process* for this stochastic function $X(t)$ is a triple $(Q, \pi, \mathbf{P}(t))$, where $Q = \{1, \ldots, n\}$ is a set of states, $\pi = (\pi_1, \ldots, \pi_n)$ is the initial distribution (i.e., $\pi_i = Pr[X(0) = i]$), and $\mathbf{P}(t)$ is an $n \times n$ matrix

$$\mathbf{P}(t) = \begin{pmatrix} p_{1,1}(t) & \ldots & p_{1,n}(t) \\ \vdots & \ddots & \vdots \\ p_{n,1}(t) & \ldots & p_{n,n}(t) \end{pmatrix}$$

of transition probability functions. These transition probabilities have to satisfy the Markov property:

$$\begin{aligned} Pr[X(t+s) = j \mid X(s) = i] &= Pr[X(t) = j \mid X(0) = i] \qquad \text{(MP)} \\ &= p_{i,j}(t). \end{aligned}$$

[12] Both the higher mutation rate for mtDNA and the much higher rate of transitional versus transversional mutations seems specific to mammals, and is not shared by *Drosophila* or by plant mtDNA. Note that plant mtDNA is much more complex than animal mtDNA.

The Markov property (MP) directly gives us $p_{i,j}(t) \geq 0$ and $\sum_{1\leq j\leq n} p_{i,j}(t) = 1$. Using an argument analagous to the case of (discrete) Markov chains, one gets

$$\mathbf{P}(t+s) = \mathbf{P}(t)\mathbf{P}(s). \tag{MP$'$}$$

Analogous to Markov chains, we define $p_i(t) = Pr[X(t) = i]$, which is given by $p_i(t) = \sum_{k=1}^{n} \pi_k p_{k,i}(t)$. A distribution $\pi^* = (\pi_1^*, \ldots, \pi_n^*)$ is called *stationary* for $\mathbf{P}(t)$ if

$$\pi^* \mathbf{P}(t) = \pi^*$$

for all t.

A common assumption (which we assume also for the case of nucleotide sequences) is that the functions $p_{i,j}(t)$ are right-continuous at $t = 0$:

$$\lim_{t\to 0+} p_{i,j}(t) = \begin{cases} 1 & \text{if } i = j \\ 0 & \text{otherwise.} \end{cases} \tag{C}$$

With this assumption, one can show that the derivative at 0 exists. This assumption allows one to define $\mathbf{P}(0)$ as the identity matrix:

$$\mathbf{P}(0) = \mathbf{I}. \tag{C'}$$

In the case of nucleotide sequences, the set of all states is $Q = \{1, 2, 3, 4\}$, which corresponds to the set of nucleotides $\{A, C, G, T\}$. The stochastic variable $X(t)$ is the nucleotide at a specific site, and the corresponding Markov matrix is

$$\begin{aligned}
\mathbf{P}(t) &= \begin{pmatrix} p_{1,1}(t) & \ldots & p_{1,4}(t) \\ \vdots & \ddots & \vdots \\ p_{4,1}(t) & \ldots & p_{4,4}(t) \end{pmatrix} \\
&= \begin{pmatrix} Pr[A \mid A, t] & Pr[C \mid A, t] & Pr[G \mid A, t] & Pr[T \mid A, t] \\ Pr[A \mid C, t] & Pr[C \mid C, t] & Pr[G \mid C, t] & Pr[T \mid C, t] \\ Pr[A \mid G, t] & Pr[C \mid G, t] & Pr[G \mid G, t] & Pr[T \mid G, t] \\ Pr[A \mid T, t] & Pr[C \mid T, t] & Pr[G \mid T, t] & Pr[T \mid T, t] \end{pmatrix},
\end{aligned}$$

where $Pr[\sigma \mid \tau, t]$ is short for $Pr[X(t) = \sigma \mid X(0) = \tau]$.

Next, we need a simple way to generate the transition probability functions $p_{i,j}(t)$. We will show, that these functions can be defined using a 4×4 matrix of substitution rates. For this purpose, consider the matrix of derivations $\mathbf{P}'(t)$, which we define as follows:

$$\mathbf{P}'(t) = \begin{pmatrix} \frac{\partial p_{1,1}}{\partial t}(t) & \ldots & \frac{\partial p_{1,4}}{\partial t}(t) \\ \vdots & \ddots & \vdots \\ \frac{\partial p_{4,1}}{\partial t}(t) & \ldots & \frac{\partial p_{4,4}}{\partial t}(t) \end{pmatrix}.$$

Then we have

$$\begin{aligned}
\mathbf{P}'(t) &= \lim_{\Delta t\to 0} \frac{\mathbf{P}(t+\Delta t) - \mathbf{P}(t)}{\Delta t} && (4.1) \\
&= \lim_{\Delta t\to 0} \frac{\mathbf{P}(t)\mathbf{P}(\Delta t) - \mathbf{P}(t)\mathbf{I}}{\Delta t} && \text{(by (MP}'\text{))} \\
&= \mathbf{P}(t) \lim_{\Delta t\to 0} \frac{\mathbf{P}(\Delta t) - \mathbf{P}(0)}{\Delta t} \\
&= \mathbf{P}(t)\, \mathbf{\Lambda}, && (4.2)
\end{aligned}$$

where $\mathbf{\Lambda}$ is the matrix $\lim_{\Delta t \to 0} \frac{\mathbf{P}(\Delta t) - \mathbf{I}}{\Delta t}$. That is, $\mathbf{\Lambda}$ is the matrix

$$\mathbf{\Lambda} = \begin{pmatrix} -\lambda_1 & \lambda_{1,2} & \lambda_{1,3} & \lambda_{1,4} \\ \lambda_{2,1} & -\lambda_2 & \lambda_{2,3} & \lambda_{3,4} \\ \lambda_{3,1} & \lambda_{3,2} & -\lambda_3 & \lambda_{3,4} \\ \lambda_{4,1} & \lambda_{4,2} & \lambda_{4,3} & -\lambda_4 \end{pmatrix},$$

where

$$-\lambda_i = \frac{\partial p_{i,i}}{\partial t}(0) = \lim_{\Delta t \to 0} \frac{p_{i,i}(\Delta t) - 1}{\Delta t}$$

and

$$\lambda_{i,j} = \frac{\partial p_{i,j}}{\partial t}(0) = \lim_{\Delta t \to 0} \frac{p_{i,j}(\Delta t) - 0}{\Delta t} \qquad \text{for } i \neq j.$$

Now $\lambda_{i,j}$ can easily be interpreted as the rate per site and year for substituting nucleotide i into nucleotide j as follows. For small Δt, we have

$$p_{i,j}(\Delta t) = \lambda_{i,j}\Delta t + o(\Delta t) \approx \lambda_{i,j}\Delta t,$$

where the term $o(\Delta t)$ accounts for multiple substitutions (with the property that $o(\Delta t)$ converges faster to 0 than Δt).[13] Analogously, we can conclude that λ_i is also a substitution rate, but this time the rate with which we substitute *out* of nucleotide i. Since any substitution out of i must go into some $j \neq i$, it is natural to conjecture that

$$\lambda_i = \sum_{j \neq i} \lambda_{i,j}. \tag{4.3}$$

This can be shown as follows:

$$\begin{aligned} -\lambda_i + \sum_{j \neq i} \lambda_{i,j} &= \lim_{\Delta t \to 0} \frac{-1 + p_{i,i}(\Delta t)}{\Delta t} + \sum_{j \neq i} \lim_{\Delta t \to 0} \frac{p_{i,j}(\Delta t)}{\Delta t} \\ &= \lim_{\Delta t \to 0} \frac{-1 + p_{i,i}(\Delta t) + \sum_{j \neq i} p_{i,j}(\Delta t)}{\Delta t} = \lim_{\Delta t \to 0} \frac{-1+1}{\Delta t} = 0. \end{aligned}$$

The remaining question is why the matrix of substitution rates $\mathbf{\Lambda}$ determines the matrix $\mathbf{P}(t)$, as we have claimed above. The reason is that equation (4.2) is nothing other than a linear differential equation:

$$\mathbf{P}'(t) = \mathbf{P}(t)\,\mathbf{\Lambda}.$$

---

[13] More precisely, this follows from the Taylor expansion at 0, which is

$$\begin{aligned} p_{i,j}(\Delta t) &= p_{i,j}(0) + \frac{\partial p_{i,j}}{\partial t}(0)\Delta t + \frac{1}{2!}\frac{\partial^2 p_{i,j}}{\partial t}(0)(\Delta t)^2 + \ldots \\ &= 0 \qquad + \lambda_{i,j}\Delta t \qquad + \frac{1}{2!}\frac{\partial^2 p_{i,j}}{\partial t}(0)(\Delta t)^2 + \ldots \quad . \end{aligned}$$

Analogously to the one-dimensional case, this is solved by $e^{\mathbf{\Lambda} t}$, where $e^{\mathbf{\Lambda} t}$ is defined by the Taylor expansion

$$e^{\mathbf{\Lambda} t} = \sum_{n=0}^{\infty} \frac{\mathbf{\Lambda}^n t^n}{n!}, \tag{4.4}$$

where we define $\mathbf{\Lambda}^0$ as $\mathbf{I}$. It is easy to check that this is a solution of the differential equation.

To find a closed form for $e^{\mathbf{\Lambda} t}$, we turn to explicit forms of rate matrices. A more general one, which will be used in Section 4.4, considers a nucleotide-specific rate u_j for substituting to nucleotide j. Under this assumption, we get for all i, j that $u_j = \lambda_{i,j}$. By (4.3), this gives us the following rate matrix:

$$\mathbf{\Lambda} = \begin{pmatrix} -(u_2+u_3+u_4) & u_2 & u_3 & u_4 \\ u_1 & -(u_1+u_3+u_4) & u_3 & u_4 \\ u_1 & u_2 & -(u_1+u_2+u_4) & u_4 \\ u_1 & u_2 & u_3 & -(u_1+u_2+u_3) \end{pmatrix}. \tag{4.5}$$

Now consider the matrix $P = \mathbf{I} + \mathbf{\Lambda}/u$, where $u = u_1 + u_2 + u_3 + u_4$. Simple calculation shows that P is of the form

$$P = \begin{pmatrix} \pi_1 & \pi_2 & \pi_3 & \pi_4 \\ \pi_1 & \pi_2 & \pi_3 & \pi_4 \\ \pi_1 & \pi_2 & \pi_3 & \pi_4 \\ \pi_1 & \pi_2 & \pi_3 & \pi_4 \end{pmatrix},$$

where $\pi_i = \frac{u_i}{u}$. Note that $\pi_1 + \pi_2 + \pi_3 + \pi_4 = 1$, which implies that P is stochastic. Furthermore, it is already stationary, since

$$\begin{aligned} P^2 &= \begin{pmatrix} \pi_1\pi_1 + \pi_1\pi_2 + \pi_1\pi_3 + \pi_1\pi_4 & \cdots & \pi_4\pi_1 + \pi_4\pi_2 + \pi_4\pi_3 + \pi_4\pi_4 \\ \vdots & \ddots & \vdots \\ \pi_1\pi_1 + \pi_1\pi_2 + \pi_1\pi_3 + \pi_1\pi_4 & \cdots & \pi_4\pi_1 + \pi_4\pi_2 + \pi_4\pi_3 + \pi_4\pi_4 \end{pmatrix} \\ &= \begin{pmatrix} \pi_1(\pi_1+\pi_2+\pi_3+\pi_4) & \cdots & \pi_4(\pi_1+\pi_2+\pi_3+\pi_4) \\ \vdots & \ddots & \vdots \\ \pi_1(\pi_1+\pi_2+\pi_3+\pi_4) & \cdots & \pi_4(\pi_1+\pi_2+\pi_3+\pi_4) \end{pmatrix} \\ &= P. \end{aligned}$$

Hence $P^n = P$ for $n \geq 1$. This gives us the solved form for $e^{\mathbf{\Lambda} t}$ as follows:

$$\begin{aligned} e^{\mathbf{\Lambda} t} &= e^{-u(\mathbf{I}-P)t} = e^{-ut\mathbf{I}} e^{utP} \\ &= \left[\sum_{n=0}^{\infty} \frac{\mathbf{I}^n(-ut)^n}{n!}\right] \left[\sum_{n=0}^{\infty} \frac{P^n(ut)^n}{n!}\right] \\ &= \mathbf{I}\left[\sum_{n=0}^{\infty} \frac{(-ut)^n}{n!}\right] \left\{\mathbf{I} + P\left[\sum_{n=1}^{\infty} \frac{(ut)^n}{n!}\right]\right\} \\ &= e^{-ut}[\mathbf{I} + P(e^{ut} - 1)] \\ &= e^{-ut}\mathbf{I} + P(e^{-ut}e^{ut} - e^{-ut}) \\ &= e^{-ut}\mathbf{I} + P(1 - e^{-ut}) \end{aligned}$$

Writing down $\mathbf{P}(t) = e^{-ut}\mathbf{I} + P(1 - e^{-ut})$ componentwise, we get

$$p_{i,j}(t) = e^{-ut}\delta_{i,j} + (1 - e^{-ut})\pi_j, \tag{4.6}$$

where the Kronecker δ-function is defined by

$$\delta_{i,j} = \begin{cases} 1 & \text{if } i = j \\ 0 & \text{otherwise.} \end{cases}$$

The model depicted in (4.5) subsumes the model by Jukes and Cantor [JC69], who introduced one of the first statistical models for nucleotide substitution. The Jukes–Cantor model is generated by setting $u_1 = u_2 = u_3 = u_4 = \alpha$, which gives us the rate matrix

$$\mathbf{\Lambda} = \begin{pmatrix} -3\alpha & \alpha & \alpha & \alpha \\ \alpha & -3\alpha & \alpha & \alpha \\ \alpha & \alpha & -3\alpha & \alpha \\ \alpha & \alpha & \alpha & -3\alpha \end{pmatrix}.$$

Then $u = 4\alpha$ and $\pi_1 = \pi_2 = \pi_3 = \pi_4 = \frac{1}{4}$. Hence, by (4.6), we have

$$\begin{aligned} p_{i,i}(t) &= e^{-4\alpha t} + \frac{1}{4}(1 - e^{-4\alpha t}) \\ &= \frac{1}{4}(1 + 3e^{-4\alpha t}) \end{aligned}$$

and

$$p_{i,j}(t) = \frac{1}{4}(1 - e^{-4\alpha t}) \qquad \text{for } i \neq j.$$

This gives us the possibility to determine the evolutionary process by a small number of parameters as needed for the rate matrix $\mathbf{\Lambda}$. Of course, the remaining problem is to determine these parameters, which have to be estimated from data.

4.3 Clustering Methods

Given n taxa, where the distance $d_{i,j}$ between taxa i, j is available, how can one determine the best fitting phylogenetic tree for the data? In principle, one might attempt to determine the minimum number of mutations yielding intermediate ancestors for all possible tree topologies having n leaves. Theorem 4.1 indicates that this method is infeasible, since the number of possible *binary* trees is exponential. First, we introduce some definitions to clarify concepts.

A *directed graph* $G = (V, E)$ consists of a set V of *nodes* or *vertices* and a set $E \subseteq V \times V$ of directed edges. Here $(i, j) \in E$ means that there is a directed edge from vertex i to j; i.e. $i \rightarrow j$. A graph is *undirected* if the edge relation is symmetric; i.e. for all vertices i, j we have $(i, j) \in E$ if and only if $(j, i) \in E$. In this case, we may think of E as consisting of unordered pairs $\{i, j\}$ corresponding to the undirected edges between i and j. A directed graph is *connected* if between any two distinct nodes there is a directed path. A directed graph is *acyclic* if it does not contain a cycle, i.e. a path $(v_0, \ldots, v_m)$ where $v_0 = v_m$, $m \geq 1$, and for all $i < m$ it is the case that $(v_i, v_{i+1}) \in E$.

A *tree* is an undirected, connected, acyclic graph. Nodes x, y are *immediate neighbors* if there is an undirected edge between x and y. From this definition, it easily follows that there is a unique path between any two distinct *nodes* or *vertices* in a tree. A *rooted tree* has a distinguished node, called the *root*. The *parent* of node y in a tree T is that unique node x which lies immediately before y on the path from root r to y. Node y is the *child* of x exactly if x is the parent of y. An *ancestor* of y is any node on the path from root r to y. A *leaf* or *external node* of a rooted tree T has no children. Non-leaf nodes are called *internal* nodes. The *depth* of a tree is one less than the maximal number of nodes on a path from the root to a leaf. An *ordered* tree is a rooted tree such that the children of internal nodes are ordered – there is a difference whether a child is the leftmost child, or second child, etc.

A *phylogenetic tree* on n taxa is a tree with leaves labeled by $1, \ldots, n$. Let T_1, T_2 be phylogenetic trees on n taxa. Then $T_1 = (V_1, E_1)$ and $T_2 = (V_2, E_2)$, $E_1 \subseteq V_1 \times V_1$ and $E_2 \subseteq V_2 \times V_2$, and the leaves of both T_1 and T_2 are labeled by $1, \ldots, n$. A function $f : V_1 \rightarrow V_2$ is an *isomorphism* between T_1 and T_2 if the following conditions are satisfied:

1. f is one-to-one and onto.
2. $x \in V_1$ is a leaf of T_1 labeled by i if and only if $f(x) \in V_2$ is a leaf of T_2 labeled by i (i.e. f respects leaf labels).
3. $(x, y) \in E_1$ if and only $(f(x), f(y)) \in E_2$ (i.e. f respects the edge relation).

Phylogenetic trees are isomorphic if there exists an isomorphism between them. When phylogenetic trees are counted in Theorem 4.1 and Corollary 4.2, the number of phylogenetic trees is counted up to isomorphism.

A tree is *binary* if every node has at most two children, otherwise it is *multifurcating*. For phylogeny, one usually considers unordered binary trees with the property that every internal node has exactly two children. Though multifurcating trees perhaps model biological reality better, constructing binary phylogenetic trees meeting some sort of optimality criterion is sufficiently complicated, so that only binary trees are considered.

It is well known that $b(n)$, the number of rooted, ordered binary trees having n nodes, satisfies the recurrence relation

$$b(n) = \begin{cases} 1 & \text{if } n = 0, \\ \sum_{k=0}^{n-1} b(k)b(n-k-1) & \text{if } n \geq 1, \end{cases}$$

since in the inductive case a tree of n nodes can be constructed from a left subtree of k nodes, a root, and a right subtree of $n - k - 1$ nodes. The solution of $b(n)$ is the Catalan number $\frac{1}{n+1}\binom{2n}{n}$ (see [Knu73] for details). Similarly, one can count the number of rooted and unrooted binary phylogenetic trees.

THEOREM 4.1 (L.L. CAVALLI-SFORZA AND A.W.F. EDWARDS [CSE67])
There are

$$1 \cdot 3 \cdot 5 \cdots (2n-3) = \prod_{i=2}^{n} (2i-3) = \frac{(2n-3)!}{2^{n-2}(n-2)!} = \Omega\left(\left(\frac{2n}{3}\right)^{n-1}\right)$$

many rooted, binary phylogenetic trees on n taxa.

PROOF This is by induction on the number of leaves (or taxa) $n \geq 2$. When $n = 2$, there is a unique rooted, binary phylogenetic tree having 2 leaves, depicted below.

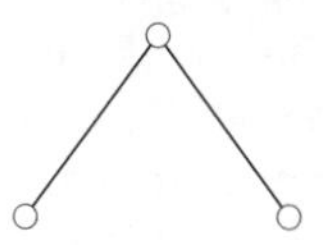

Let $\mathcal{T}_n$ denote the collection of rooted, binary phylogenetic trees having n leaves, and suppose that $t(n) = |\mathcal{T}_n|$. By Exercise 1, any rooted, binary tree having n leaves must have $n-1$ internal nodes. By the same exercise, any binary tree has one fewer edges than nodes. Thus any rooted, binary tree having n leaves has $n + (n-1) - 1 = 2n - 2$ many edges. A tree in $\mathcal{T}_{n+1}$ can be constructed from a tree in $\mathcal{T}_n$ by attaching the new $(n+1)$th leaf to a new interior node created in the middle of an edge of a tree from $\mathcal{T}_n$ as follows:

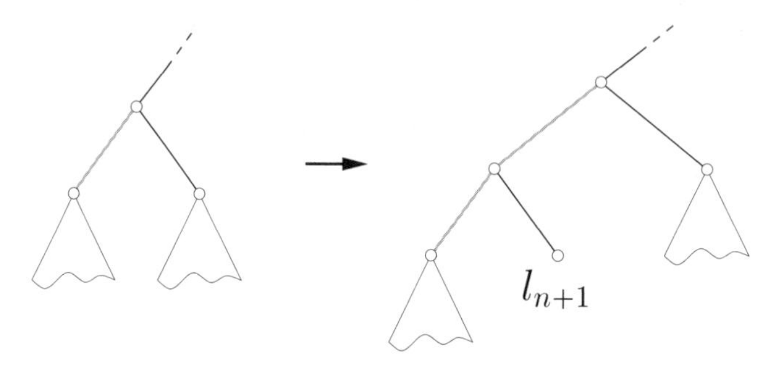

This yields $t(n)(2n-2)$ possible choices. Also, one can attach the new $(n+1)$th leaf to a new root, where the other child of the new root is the old root of a tree in $\mathcal{T}_n$. This yields $t(n)$ additional trees, hence altogether $t(n)(2n-1)$ trees, so establishing the recurrence relation

$$t(n+1) = \begin{cases} 1 & \text{if } n = 1, \\ t(n)(2n-1) & \text{if } n > 1, \end{cases}$$

with solution

$$t(n) = 1 \cdot 3 \cdot 5 \cdots (2n-3).$$

Finally, note that

$$2 \cdot 4 \cdot 6 \cdots (2n-4) = 2^{n-2}[1 \cdot 2 \cdot 3 \cdots (n-2)],$$

and so

$$t(n) = \frac{(2n-3)!}{2^{n-2}(n-2)!}.$$

Stirling's approximation $n! \sim \left(\frac{n}{e}\right)^n \cdot \sqrt{2\pi n}$ yields the lower bound $\Omega((\frac{2n}{3})^{n-1})$. ■

The 3 rooted, binary phylogenetic trees having 3 leaves are depicted in Figure 4.1. An unrooted binary phylogenetic tree is assumed to satisfy the condition that every node has degree either 1 or 3, where leaves have degree 1, and internal nodes have degree 3. See Figure 4.2 for the unique unrooted, binary phylogenetic tree having 3 leaves.

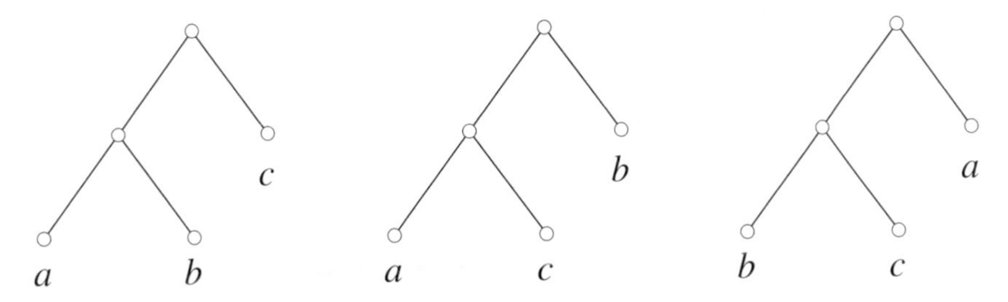

Figure 4.1 The three rooted trees having three leaves.

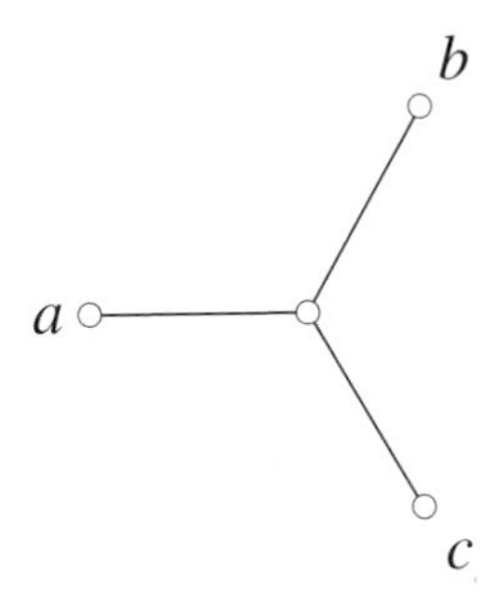

Figure 4.2 The unique unrooted tree having three leaves.

COROLLARY 4.2 (L.L. CAVALLI-SFORZA AND A.W.F. EDWARDS [CSE67])
There are

$$1 \cdot 3 \cdot 5 \cdots (2n-5) = \prod_{i=3}^{n} (2i-5) = \frac{(2n-5)!}{2^{n-3}(n-3)!} = \Omega\left(\left(\frac{2n}{3}\right)^{n-2}\right)$$

many unrooted, binary phylogenetic trees having n leaves.

PROOF The number $t(n)$ of rooted, binary trees having n leaves equals the number of unrooted, binary trees having $n+1$ leaves. Specifically, given the rooted, binary tree T with n leaves and root r, attach a new leaf node x to r (above r). This transforms T into an unrooted binary phylogenetic tree (r now has degree 3) with $n+1$ leaves. This transformation defines a $1-1$ correspondence between rooted trees with n leaves and unrooted trees with $n+1$ leaves. The corollary now follows by replacing n by $n-1$ in Theorem 4.1. ■

4.3.1 Ultrametric Trees

Recall that a distance or *metric* ρ satisfies the axioms $\rho(i,j) \geq 0$ with equality if and only if $i = j$; $\rho(i,j) = \rho(j,i)$ (symmetry); $\rho(i,k) \leq \rho(i,j) + \rho(j,k)$ (triangle inequality). Assume that positive edge weights are assigned to a tree T. If the value $d_{i,j}$ of the distance function between all leaves i, j of T is simply the sum of edge weights along the path connecting i and j (i.e. the *path length* between i, j), then d is called an *additive tree metric*. If, moreover, the path length from the root r of tree T to every leaf of T is identical, then d is called an *ultrametric*. Assume for an instant that only nucleotide substitutions are allowed (no insertions or deletions) and that the evolution rate is constant across all taxa for which we wish to construct a phylogenetic tree (e.g. the Jukes–Cantor or Kimura model). Then the expected number of substitutions having occurred from the root to any leaf is the same; i.e. a constant evolution rate for nucleotide substitution gives rise to an ultrametric tree.

In [Wat95], it is shown that for every ultrametric d, and for all taxa (i.e. leaves) i, j, k, of the three distances $d(i,j), d(i,k), d(j,k)$, two are equal and not less than the third (the *ultrametric* or 3-point condition). For instance, $d(i,j) \leq d(i,k) = d(j,k)$. In [Wat95], it is also shown that every additive metric satisfies the *4-point condition*, where the latter states that for all taxa (i.e. leaves) i, j, k, ℓ, of the three sums $S_1 = d(i,j) + d(k,\ell)$, $S_2 = d(i,k) + d(j,\ell)$, $S_3 = d(i,\ell) + d(j,k)$, two are equal and not less than the third. For instance, $S_1 \leq S_2 = S_3$.

Clustering algorithms attempt to repeatedly cluster the data by grouping the closest elements. Apart from phylogeny construction, clustering algorithms are currently used to group similar results from gene expression microarray experiments. When pairs are repeatedly amalgamated, this clustering technique is called the *pair group method* PGM. A simple algorithm for molecular data where sequence alignment distance between sequences has been determined in a distance matrix D is known as UPGMA (unweighted pair group method with arithmetic mean). Under the hypothesis of an ultrametric, one can show that UPGMA always correctly constructs the original topology.

The clustering methods presented in this section do not require that distance defined from the distance matrix D be a metric. However, one can show that the following UPGMA method produces the correct tree topology for an ultrametric, and that the Farris transformed UPGMA method transforms an additive metric d into an ultrametric e, and hence produces the correct tree topology. While UPGMA methods do not yield realistic branch lengths, a later method of Fitch–Margoliash does.

Algorithm 4.1 can be implemented using linked lists with pointers, and produces a rooted tree. To prove that the distances associated with the internal nodes are well-defined, we define the height $height(e)$ of a node e to be the distance from e to its leaves as defined in step 3c of the UPGMA algorithm. We have to show that for any node e that is generated by joining the nodes c, d, we have $height(e) \geq height(c)$ and $height(e) \geq height(d)$. This is a consequence of the following proposition.

PROPOSITION 4.3
Let e be a node generated by joining clusters c and d in step 3b of Algorithm 4.1. Let C be the set of all clusters immediately before joining c and d to e. Then

$$\forall t \in C : height(e) \geq height(t).$$

PROOF This is by induction. The base case that e was the first cluster generated by joining is trivial since the height of leaves is 0. For the induction step, let e be the node

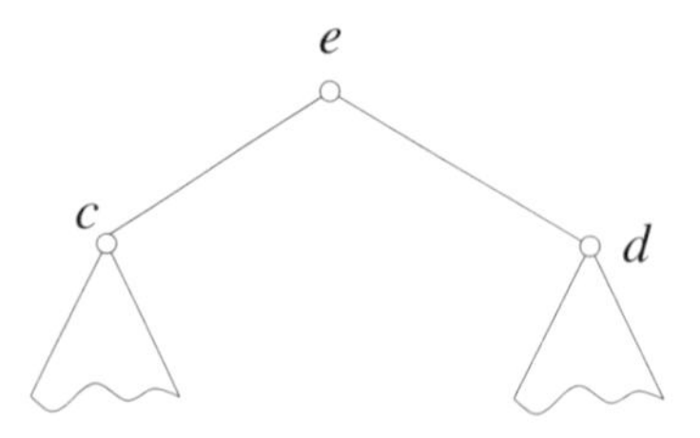

generated by joining c, d, and there is at least one additional cluster that is not a singleton set. For the leaves, the claim holds trivially. Let t be the node that was joined in the step directly

Algorithm 4.1 UPGMA

INPUT: $n \times n$ distance matrix D

1. Initialize set $\mathcal{C}$ to consist of n initial singleton clusters $\{1\}, \ldots, \{n\}$
2. Initialize function $dist(c, d)$ on $\mathcal{C}$ by defining for all $\{i\}$ and $\{j\}$ in $\mathcal{C}$

$$dist(\{i\}, \{j\}) = D(i, j)$$

3. Repeat $n - 1$ times
   (a) determine pair c, d of clusters in $\mathcal{C}$ such that $dist(c, d)$ is minimal; define

$$d_{\text{min}} = dist(c, d)$$

   (b) define new cluster $e = c \cup d$; define

$$\mathcal{C} = \mathcal{C} - \{c, d\} \cup \{e\}$$

   (c) define a node with label e and daughters c, d, where the e has distance $\frac{d_{\text{min}}}{2}$ to its leaves (i.e. the leaves of the subtree rooted at e).
   (d) define for all $f \in \mathcal{C}$ with $f \neq e$

$$dist(e, f) = dist(f, e) = \frac{dist(c, f) + dist(d, f)}{2}$$

---

before joining c and d. If we can show that

$$height(c) \geq height(t),$$

then the claim follows from the induction hypothesis. Since t was generated by joining, we know that t is of the form

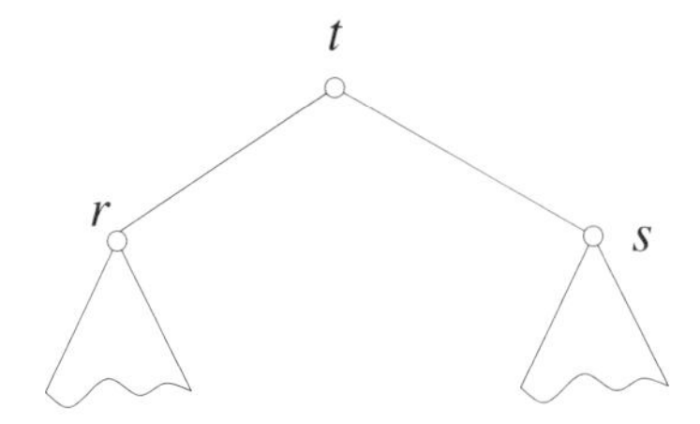

where r, s are the clusters joined in the creation of t. Let $\mathcal{C}^{\text{before } t}$ be the set of clusters immediately before joining r and s to t. Since $height(e) = \frac{dist(c,d)}{2}$ and $height(t) = \frac{dist(r,s)}{2}$, we have to show that

$$dist(c, d) \geq dist(r, s). \tag{4.7}$$

We have two cases:

1. t is different from c and d. Then

$$\mathcal{C} = \{c, d, t\} \cup \mathcal{C}^{\text{rest}}$$

and

$$\mathcal{C}^{\text{before } t} = \{c, d, r, s\} \cup \mathcal{C}^{\text{rest}}.$$

By the minimality of $dist(r, s)$ when joining r, s to t, we get (4.7) directly.

2. t is one of c or d. Without loss of generality, we can assume that t equals c, i.e., e is of the form

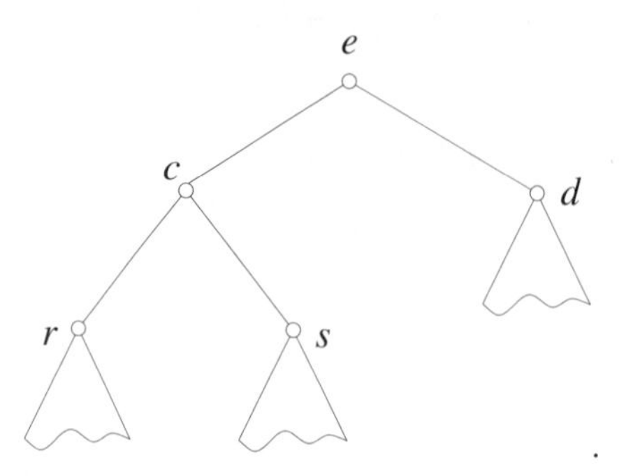

Note that our assumption (that $c = t$ was generated immediately before generating e) implies that c was joined later than d. Hence,

$$\mathcal{C} = \{c, d\} \cup \mathcal{C}^{\text{rest}}$$

and

$$\mathcal{C}^{\text{before } t} = \{r, s, d\} \cup \mathcal{C}^{\text{rest}}.$$

By the minimality of $dist(r, s)$ when joining r, s to c, we know that

$$dist(r, d) \geq dist(r, s) \quad \text{and} \quad dist(s, d) \geq dist(r, s).$$

Hence,

$$\begin{aligned} dist(c, d) &= \frac{dist(r, d) + dist(s, d)}{2} \\ &\geq \frac{dist(r, s) + dist(r, s)}{2} = dist(r, s), \end{aligned}$$

which proves (4.7).

■

One can always choose as label of cluster c the smallest integer $1 \leq i \leq n$ belonging to c. This allows one to update the distance matrix D directly, as follows. Suppose that the labels of the two closest clusters are i_0, j_0 and that $i_0 < j_0$. Then updating the distance matrix D can be done as follows:

```
for (k = 0; k < n; k++)
     if ( (k ≠ j0) && (k ≠ i0) && (D[i0][k] ≠ 0) )
          E[i0][k] = (D[i0][k] + D[j0][k]) / 2;
     else
```

```
        E[i0][k]  =  0;

    for (k = 0; k < n; k++)
        D[i0][k]  =  D[k][i0]  =  E[i0][k];
        // copy E into D and symmetrify

    for (k = 0; k < n; k++)
        D[j0][k]  =  D[k][j0]  =  0;
        // erase j0-th row and column of D
```

A modification of this algorithm is to weight the clusters by their size, yielding Algorithm 4.2. Note that if the cluster size of pairs amalgamated is roughly the same, then WPGMA is essentially UPGMA.

Algorithm 4.2 WPGMA

INPUT: $n \times n$ distance matrix D

1. Initialize set $\mathcal{C}$ to consist of n initial singleton clusters $\{1\}, \ldots, \{n\}$
2. Initialize function $dist(c, d)$ on $\mathcal{C}$ by defining for all $\{i\}$ and $\{j\}$ in $\mathcal{C}$

$$dist(\{i\}, \{j\}) = D(i, j)$$

3. Repeat $n - 1$ times
   (a) determine pair c, d of clusters in $\mathcal{C}$ such that $dist(c, d)$ is minimal; define

$$d_{\min} = dist(c, d)$$

   (b) define new cluster $e = c \cup d$; define

$$\mathcal{C} = \mathcal{C} - \{c, d\} \cup \{e\}$$

   (c) define a node with label e and daughters c, d, where the e has distance $\frac{d_{\min}}{2}$ to its leaves.
   (d) define for all $f \in \mathcal{C}$ with $f \neq e$

$$dist(e, f) = dist(f, e) = \frac{|c| dist(c, f) + |d| dist(d, f)}{|c| + |d|}$$

In the case of WPGMA, we can define the distance function on cluster without using a recursive definition.

PROPOSITION 4.4
Let $dist(e, f)$ be the distance defined by WPGMA for clusters e and f. Then

$$dist(e, f) = \frac{1}{|e||f|} \sum_{k \in e, l \in f} D(k, l).$$

PROOF This is by induction. The base case holds by definition. For the induction step, let e be defined by $e = c \cup d$. Then we have

$$
\begin{aligned}
dist(e,f) &= \frac{|c|dist(c,f) + |d|dist(d,f)}{|c|+|d|} \\
&\overset{\text{Ind.Hyp.}}{=} \frac{|c|\frac{1}{|c||f|}\sum_{i\in c, l\in f} D(i,l) + |d|\frac{1}{|d||f|}\sum_{j\in d, l\in f} D(j,l)}{|c|+|d|} \\
&= \frac{\frac{1}{|f|}\left[\sum_{i\in c, l\in f} D(i,l) + \sum_{j\in d, l\in f} D(j,l)\right]}{|c|+|d|} \\
&= \frac{\frac{1}{|f|}\sum_{k\in c\cup d, l\in f} D(k,l)}{|c|+|d|} = \frac{1}{|e||f|}\sum_{k\in e, l\in f} D(k,l)
\end{aligned}
$$

■

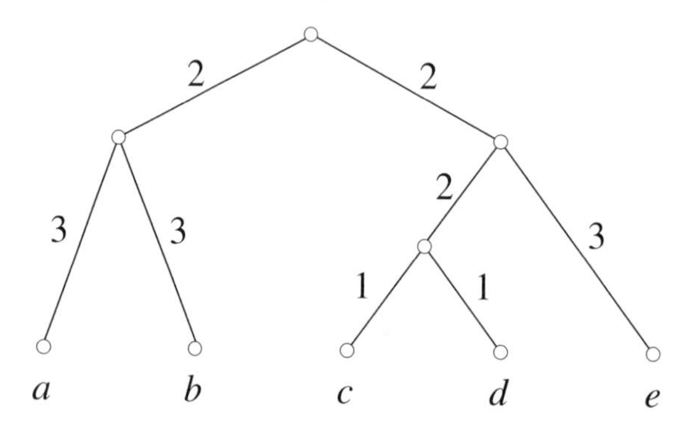

Figure 4.3 Ultrametric topology.

In Figure 4.3, we give an example of an ultrametric topology whose corresponding distance table and matrix is shown in Table 4.1. In Figure 4.4, we show the correctly reconstructed topology and path lengths using the UPGMA algorithm applied to the distance matrix given in Table 4.1. We use a presentation of trees that is commonly used by phylogenetic programs. The internal nodes are labeled with the distance of the node to the leaves (the height of the node). Note that this kind of representation implies an ultrametric topology.

4.3.2 Additive Metric

Consider the additive metric with distance matrix in Figure 4.6 and topology given in Figure 4.5. Note that the ultrametric or 3-point condition is violated for this example. Namely, $d_{a,b} = 9$, $d_{a,e} = 11$, $d_{b,e} = 18$, and it is not the case that two of these distances are equal and *not* less than the third. However, the 4-point condition is satisfied. For instance, letting $i = a$, $j = c$, $k = d$ and $\ell = e$, we have that $S_1 = d(i,j) + d(k,\ell) = 6 + 19 = 25$, $S_2 = d(i,k) + d(j,\ell) = 14 + 11 = 25$, $S_3 = d(i,\ell) + d(j,k) = 11 + 12 = 23$, and it is the case that two are equal and not less than the third. One readily verifies that the 4-point condition is valid for all $\binom{5}{4} = 5$ possible choices of 4 points among a, b, c, d, e.

Table 4.1 Distance table for the ultrametric topology given in Figure 4.3. To the right we have shown the matrix notation of the distance table, which will be used in the following.

	a	b	c	d	e
a	0	6	10	10	10
b	6	0	10	10	10
c	10	10	0	2	6
d	10	10	2	0	6
e	10	10	6	6	0

$$\begin{pmatrix} 0 & 6 & 10 & 10 & 10 \\ 6 & 0 & 10 & 10 & 10 \\ 10 & 10 & 0 & 2 & 6 \\ 10 & 10 & 2 & 0 & 6 \\ 10 & 10 & 6 & 6 & 0 \end{pmatrix}$$

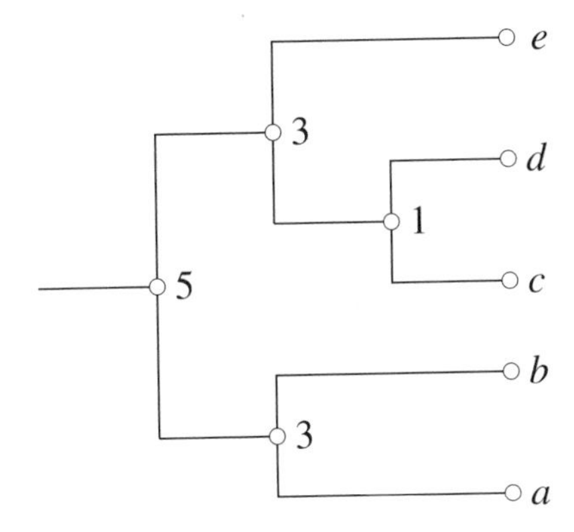

Figure 4.4 Output of UPGMA. The labels of the internal nodes are the distances to the leaves.

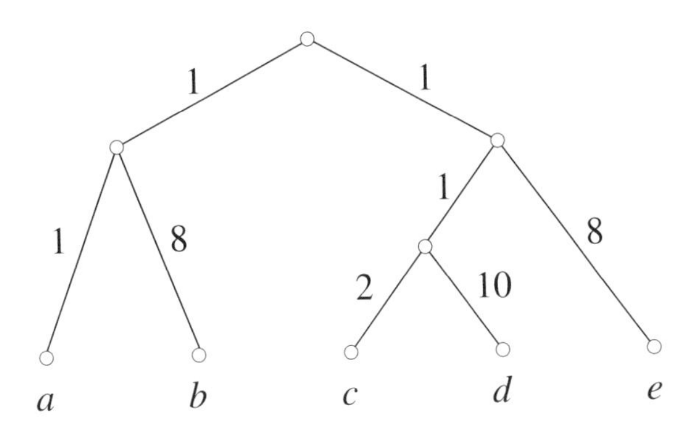

Figure 4.5 Additive metric topology.

$$\begin{pmatrix} 0 & 9 & 6 & 14 & 11 \\ 9 & 0 & 13 & 21 & 18 \\ 6 & 13 & 0 & 12 & 11 \\ 14 & 21 & 12 & 0 & 19 \\ 11 & 18 & 11 & 19 & 0 \end{pmatrix}$$

Figure 4.6 Distance matrix for additive, non-ultrametric metric.

Applying UPGMA to this example yields a tree with the incorrect topology and incorrect branch lengths, as shown in Figure 4.7. However, since the metric is additive (i.e. the distance between taxa i, j is the sum of edge distances along a path from i to j), the modification of UPGMA given in Theorem 4.5, due to Farris [Far77], and later independently discovered by [KKBM79], will compute the correct topology.

THEOREM 4.5 (FARRIS TRANSFORMED DISTANCE METHOD)
Let T be a phylogenetic tree with ancestor r and leaves (current taxa) $1, \ldots, n$, and suppose that the distance $d_{i,j}$ between nodes i, j of T is the path length of the unique path connecting i, j. Define the transformed distance

$$e_{i,j} = \frac{d_{i,j} - d_{i,r} - d_{j,r}}{2} + \overline{d}_r$$

where $\overline{d}_r = (\sum_{i=1}^n d_{i,r})/n$ is the average distance between r and the leaves. Then UPGMA applied to the transformed distance matrix generates the correct topology of T.

PROOF Let $lca(i, j)$ denote the *least common ancestor* of leaves i, j. We claim that

$$e_{i,j} = \overline{d}_r - d_{r,lca(i,j)}.$$

Indeed, from the property that distance is path length between nodes in T, it follows that

$$d_{i,j} = d_{i,r} + d_{j,r} - 2d_{r,lca(i,j)},$$

so

$$\frac{d_{i,j} - d_{i,r} - d_{j,r}}{2} = -d_{r,lca(i,j)},$$

and hence $e_{i,j} = \overline{d}_r - d_{r,lca(i,j)}$. From this, it is immediate that the matrix $(e_{i,j})$ is an ultrametric: the distance from the root r to any leaf i is $e_{r,i} = \overline{d}_r - d_{r,lca(r,i)}$, which is

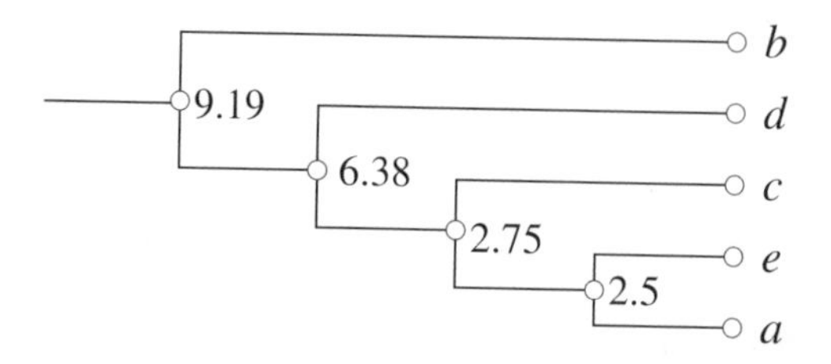

Figure 4.7 UPGMA incorrectly reconstructed topology.

$$\begin{pmatrix} 0 & 6.2 & 7.2 & 7.2 & 7.2 \\ 6.2 & 0 & 7.2 & 7.2 & 7.2 \\ 7.2 & 7.2 & 0 & 5.2 & 6.2 \\ 7.2 & 7.2 & 5.2 & 0 & 6.2 \\ 7.2 & 7.2 & 6.2 & 6.2 & 0 \end{pmatrix}$$

Figure 4.8 Transformed distance matrix $E = (e_{i,j})$.

$\overline{d}_r$, since the least common ancestor of r and i is r, so $d_{r,lca(r,i)} = 0$. Thus we have the same distance from the root to any leaf, and applying UPGMA to the transformed distance matrix yields the correct topology. ■

Continuing with the example of Figure 4.6, we compute from Figure 4.5 that $\overline{d}_r = \frac{2+9+4+12+9}{5} = 7.2$ and so the transformed distance matrix $E = (e_{i,j})$, where

$$e_{i,j} = \frac{d_{i,j} - d_{i,r} - d_{j,r}}{2} + \overline{d}_r = \overline{d}_r - d_{r,lca(i,j)}$$

has the values appearing in Figure 4.8, yielding the correct topology in Figure 4.9.

What does this method yield if the ancestor r is not known, nor are distances from ancestor to leaves? In [Nei87], Nei argues for the use of the Farris transformed distance matrix method by taking r to be a known outgroup. When comparing n taxa, one could determine that taxon whose average distance from all others is a maximum, and define this taxon to be the *external reference* or outgroup. This is the approach in our implementation of the Farris transformed method (either UPGMA or WPGMA can be then applied to the transformed distance matrix). Note that without being able to stipulate the root r, the program sometimes correctly captures the topology and sometimes not.

If we do not know the correct tree topology, then how can we tell the most likely topology? There have been different approaches: parsimony (minimize the total number of substitutions or distance for intermediate ancestors), tree distance (a modification of sequence alignment distance, where trees in place of sequences are aligned), neighbor joining, and maximum-likelihood methods. In the next section, we pursue Felsenstein's method of maximum-likelihood tree construction, while leaving the reader to other texts or the original literature

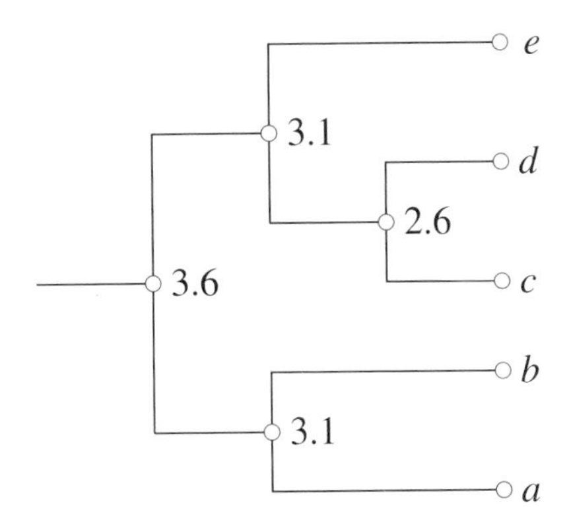

Figure 4.9 Farris transform method.

for parsimony, tree distance, and neighbor joining methods.

Suppose that we know the true phylogenetic tree T for n taxa. How can we measure the deviation of a tree T', output from a phylogenetic tree algorithm, from T? The tree T' is constructed from a given input distance matrix, and gives rise to an extended distance matrix $(d_{i,j})$ for $1 \leq i, j \leq N$, where $N = 2n - 1$ is the number of nodes (leaves and interior nodes) for both T and T' (see Exercise 1). Thus it is natural to compare the *root mean square deviation* (RMSD)

$$\sqrt{\frac{2\sum_{1 \leq i < j \leq n}(d_{i,j} - e_{i,j})^2}{n(n-1)}}$$

between the extended distance matrices for T and T'.

According to [Nei87], Totero first investigated RMSD as above defined, while Fitch, Margoliash, and Farris suggested alternate, easily computable statistical measures for the deviation of tree topology and branch length error between a candidate phylogenetic tree and the correct phylogenetic tree. All of these measures are computationally simpler than a tree alignment algorithm. In a later chapter, we shall see a different application of RMSD for quantifying the effect of the hydrophobic force in protein folding.

4.3.3 Estimating Branch Lengths

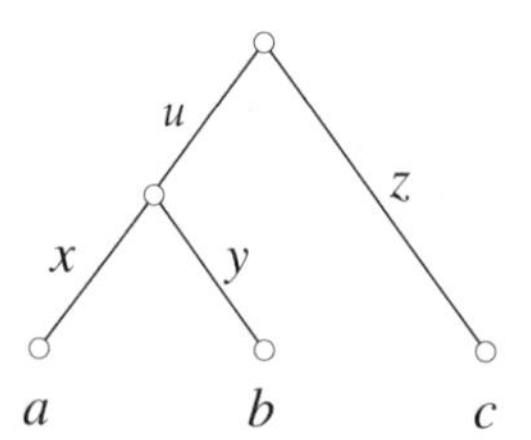

Figure 4.10 Observation for branch lengths.

The previous distance methods assume a constant evolutionary rate of change, and according to [Nei87], under this assumption, when there is a large stochastic error in distance measurements, the UPGMA method performs better than other techniques in yielding the best approximation to tree topology. In [FM67], Fitch and Margoliash developed a method, applicable under non-constant rate of evolution, for estimating branch lengths. This method is based on a simple observation. For Figure 4.10, $d_{a,b} = x + y$, $d_{b,c} = u + y + z$ and $d_{a,c} = u + x + z$; hence

$$\begin{aligned} x &= \frac{d_{a,b} + d_{a,c} - d_{b,c}}{2}, \\ y &= \frac{d_{a,b} + d_{b,c} - d_{a,c}}{2}, \\ u + z &= \frac{d_{a,c} + d_{b,c} - d_{a,b}}{2}. \end{aligned}$$

The Fitch–Margoliash method (Algorithm 4.3) modifies WPGMA only in branch length determination, and hence produces the same topology as WPGMA. This method repeatedly

Algorithm 4.3 Fitch–Margoliash

INPUT: $n \times n$ distance matrix D

1. Initialize the set $\mathcal{C}$ to consist of n singleton clusters $\{1\}, \ldots \{n\}$.
2. While number of clusters > 2 do
   (a) determine closest pair a, b of clusters
   (b) let C consist of all points in all other clusters
   (c) determine average distances

$$d_{a,C} = \frac{\sum_{i \in a} \sum_{j \in C} d_{i,j}}{|a||C|}$$

   and

$$d_{b,C} = \frac{\sum_{i \in b} \sum_{j \in C} d_{i,j}}{|b||C|}$$

   (d) create parent node $P(a, b)$ of a, b obtained by amalgamating clusters a, b and define branch lengths x from $P(a, b)$ to a and y from $P(a, b)$ to b by

$$x = \frac{d_{a,b} + d_{a,C} - d_{b,C}}{2},$$
$$y = \frac{d_{a,b} + d_{b,C} - d_{a,C}}{2}.$$

determines the closest two clusters a, b, temporarily grouping all other clusters into c, and uses the above observation to determine the branch lengths x, y.

This method produces an unrooted tree, whose topology is identical to that obtained by UPGMA. To produce a rooted tree, Nei [Nei87] suggests when there are 3 remaining clusters a, b, c in the above algorithm, where c is the outlying group, that one define the root to be at distance

$$\frac{x + y + 2(u + z)}{4}$$

from $P(a, b)$ and at distance

$$u + z - \frac{x + y + 2(u + z)}{4}$$

from c.

4.4 Maximum Likelihood

In 1981, J. Felsenstein [Fel81] applied the method of *maximum likelihood* to construct phylogenetic trees from DNA sequence data whose likelihood is a (local) maximum. Starting from a given initial topology, the idea is to use the maximum-likelihood method to determine optimal branch lengths for the topology, then to make local modifications to the topology and again optimize new branch lengths, etc. New species or taxa are added one, by one, each time optimizing branch lengths and topologies obtained by local modifications. For an unrooted tree T having m edges, as in the construction in the proof of Theorem 4.1, there are m possibilities of adjoining a new leaf taxon.

As Felsenstein observes, parsimony methods (not treated in our discussion) yield incorrect tree topologies when the amount of evolutionary change is sufficiently divergent in different branches. Felsenstein's model assumes the following:

- n nucleotide sequences are given, each of the same length m.
- In constructing a phylogenetic tree, *no* insertions or deletions have occurred.
- The evolutionary process is a reversible Markov process $\mathbf{P}(t)$, whose substitution rate matrix $\mathbf{\Lambda} = \mathbf{P}'(0)$ is given by nucleotide-specific substitution rates u_1, u_2, u_3 and u_4, i.e., the substitution rate matrix $\mathbf{\Lambda}$ has the form as given in (4.5).

Analogously to the case of Markov chains, a Markov process $\mathbf{P}(t)$ with stationary distribution π is called *reversible* if

$$\pi_i p_{i,j}(t) = \pi_j p_{j,i}(t) \tag{4.8}$$

for all states i, j and all times t. The assumption that the evolutionary process is a reversible Markov process is crucial to the pulley principle, given below, necessary for efficient computation.

Let $u = u_1 + u_2 + u_3 + u_4$ and $\pi_i = \frac{u_i}{u}$. Note that π_i can be interpreted as nucleotide frequencies. As we have shown in Section 4.2, the assumption of a Markov process with nucleotide specific substitution rates implies a transition probability function $p_{i,j}(t)$ of the form

$$p_{i,j}(t) = e^{-ut}\delta_{i,j} + (1 - e^{-ut})\pi_j,$$

where the Kronecker δ-function is defined by

$$\delta_{i,j} = \begin{cases} 1 & \text{if } i = j \\ 0 & \text{otherwise.} \end{cases}$$

While it may be intuitively clear that frequencies in π are stationary, because they have evolved and appear to be a limiting phenomenon in homeostasis, what must be shown is that for all t,

$$(\pi_1, \ldots, \pi_n)\mathbf{P}(t) = (\pi_1, \ldots, \pi_n).$$

In other words, we must show that for each j, $\sum_i \pi_i p_{i,j}(t) = \pi_j$. Fixing j, we have that

$$\begin{aligned} \sum_i \pi_i p_{i,j}(t) &= \left[\sum_i \pi_i (1 - e^{-ut})\pi_j\right] + \pi_j e^{-ut} \\ &= \left[\pi_j (1 - e^{-ut}) \sum_i \pi_i\right] + \pi_j e^{-ut} \\ &= \pi_j. \end{aligned}$$

So for the transition matrix $\mathbf{P}(t)$, nucleotide frequencies $\pi_1, \pi_2, \pi_3, \pi_4$ are then indeed stationary.

We prove that one can place the root of tree $t^l(d_1, \ldots, d_k)$ at an arbitrary distance between the nodes 1, 2 without affecting the likelihood. That is, letting a_i^l be the lth site of a_i, then

$$L = Pr[a_1^l, \ldots, a_n^l \mid t^l(d_1, d_2, \ldots, d_k)] = Pr[a_1^l, \ldots, a_n^l \mid t^l(d_1', d_2', \ldots, d_k)],$$

where

$$d_1' = d_1 - x, \qquad d_2' = d_2 + x$$

holds for arbitrary $0 \le x \le d_1$. Note first that $p_{i,k}(s+t) = \sum_j p_{i,j}(s)p_{j,k}(t)$. From equations (4.9) and (4.10), we have the following. Let s_0, s_1, and s_2 denote the states for the nodes 0, 1, and 2, respectively. Then

$$\begin{aligned}
L &= \sum_{s_0} \pi_{s_0} \left[\sum_{s_1} p_{s_0,s_1}(d_1)L_{1,s_1}\right]\left[\sum_{s_2} p_{s_0,s_2}(d_2)L_{2,s_2}\right] \\
&= \sum_{s_1}\sum_{s_2} L_{1,s_1}L_{2,s_2} \sum_{s_0} \pi_{s_0} p_{s_0,s_1}(d_1) p_{s_0,s_2}(d_2) \\
&= \sum_{s_1}\sum_{s_2} L_{1,s_1}L_{2,s_2} \sum_{s_0} \underbrace{\pi_{s_1} p_{s_1,s_0}(d_1)}_{\text{reversible Markov}} \; p_{s_0,s_2}(d_2) \\
&= \sum_{s_1}\sum_{s_2} L_{1,s_1}L_{2,s_2}\pi_{s_1} \sum_{s_0} p_{s_1,s_0}(d_1) p_{s_0,s_2}(d_2) \\
&= \sum_{s_1}\sum_{s_2} \pi_{s_1} L_{1,s_1}L_{2,s_2} p_{s_1,s_2}(d_1 + d_2).
\end{aligned}$$

Thus the likelihood depends only on the sum $d_1 + d_2$, and so the root may be placed anywhere between nodes 1 and 2 without affecting the likelihood. Generalizing this observation, one can prove that the root can be placed anywhere in the tree without affecting the likelihood. ■

Thus, by setting x to d_1 in the last lemma, we can simply omit the root. The tree in Figure 4.11 is then equivalently represented by the unrooted tree in Figure 4.12, which is called a *4-taxon tree*.

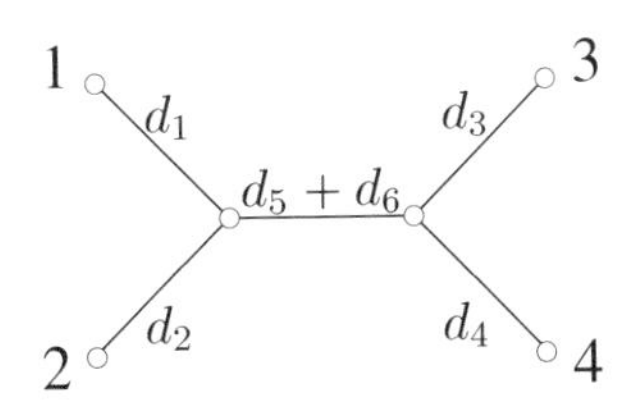

Figure 4.12 Equivalent representation of the tree in Figure 4.11.

COROLLARY 4.7
Let t be the tree for one site of given taxa, and let

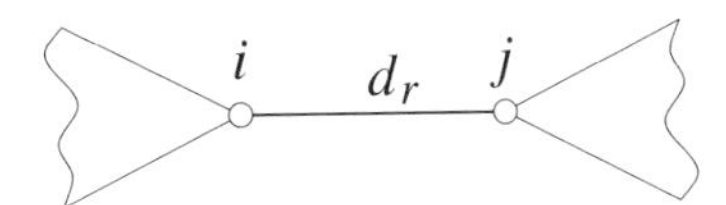

be an arbitrary edge of t. *Then*

$$L(t) = \sum_{s_i}\sum_{s_j} \pi_{s_i} L_{i,s_i} L_{j,s_j} p_{s_i,s_j}(d_r).$$

PROOF This follows directly from the proof of the Pulley Principle by assuming the root to be between i and j. ■

Now we can maximize one single edge

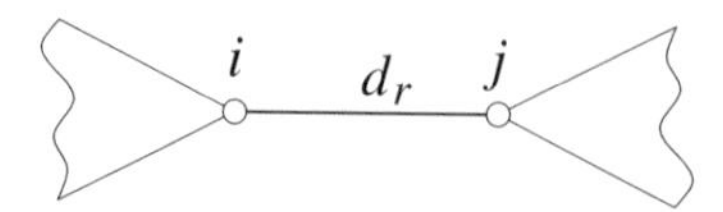

in a given topology t, keeping the other distances in t fixed. Recall that, letting $u = 1$,

$$p_{i,j}(t) = e^{-t}\delta_{i,j} + (1 - e^{-t})\pi_j.$$

By the last corollary, we get the following for the likelihood L for the tree at a single site:

$$\begin{aligned}
L &= \sum_{s_i}\sum_{s_j} \pi_{s_i} L_{i,s_i} L_{j,s_j} p_{s_i,s_j}(d_r) \\
&= \sum_{s_i}\sum_{s_j} \pi_{s_i} L_{i,s_i} L_{j,s_j} \left[e^{-d_r}\delta_{s_i,s_j} + (1 - e^{-d_r})\pi_{s_j}\right] \\
&= \sum_{s_i}\sum_{s_j} \pi_{s_i} L_{i,s_i} L_{j,s_j} e^{-d_r}\delta_{s_i,s_j} + \sum_{s_i}\sum_{s_j} \pi_{s_i} L_{i,s_i} L_{j,s_j}(1 - e^{-d_r})\pi_{s_j} \\
&= e^{-d_r}\sum_{s} \pi_s L_{i,s} L_{j,s} + (1 - e^{-d_r})\sum_{s_i}\sum_{s_j} \pi_{s_i} L_{i,s_i} L_{j,s_j}\pi_{s_j} \\
&= e^{-d_r}A + (1 - e^{-d_r})B,
\end{aligned}$$

where A and B abbreviate the appropriate expressions. This holds independently for each site $l = 1, \ldots, m$ of the n nucleotide sequences being compared. The likelihood for the lth site is written

$$A_l q + B_l p,$$

where

$$q = e^{-d_r}, \qquad p = 1 - q,$$

and

$$\begin{aligned}
A_l &= \sum_{s} \pi_s L_{i,s} L_{j,s}, \\
B_l &= \sum_{s_i}\sum_{s_j} \pi_{s_i} L_{i,s_i} L_{j,s_j}\pi_{s_j}.
\end{aligned}$$

The total likelihood over all m independent sites is thus

$$L = \prod_{l=1}^{m} A_l q + B_l p.$$

We need to solve for d_r in order to maximize the likelihood L, or, equivalently, to maximize the log likelihood

$$\ln\ L = \sum_{l=1}^{m} \ln(A_l q + B_l p).$$

To find the maximal value for p, we solve for p the equation

$$\frac{\partial \ln L}{\partial p}(p) = \sum_{l=1}^{m} \frac{B_l - A_l}{A_l q + B_l p} = 0. \tag{4.11}$$

Now

$$m = \sum_{l=1}^{m} \frac{A_l q + B_l p}{A_l q + B_l p} = \sum_{l=1}^{m} \frac{B_l - (B_l - A_l)q}{A_l q + B_l p}.$$

From (4.11), we obtain

$$m = \sum_{l=1}^{m} \frac{B_l}{A_l q + B_l p}.$$

Multiplying both sides by $\frac{p}{m}$, we get

$$p = \frac{1}{m} \sum_{l=1}^{m} \frac{B_l p}{A_l q + B_l p}. \tag{4.12}$$

This yields an iteration formula

$$p^{(k+1)} = \frac{1}{m} \sum_{l=1}^{m} \frac{B_l p^{(k)}}{A_l q^{(k)} + B_l p^{(k)}}$$

where $q^{(k)} = 1 - p^{(k)}$. A calculation shows that

$$\sum_{l=1}^{m} A_l q^{(k+1)} + B_l p^{(k+1)} \geq \sum_{l=1}^{m} A_l q^{(k)} + B_l p^{(k)},$$

so the likelihood increases to a maximum likelihood where $p = \lim_k p^{(k)}$ satisfies (4.12). Felsenstein's algorithm is in fact a special case of the EM algorithm (expectation maximization) of Dempster *et al.* [DLR77, Wu83] and so is guaranteed to converge under these conditions. If $\lim_k p^{(k)} = 0$ then there is no stationary point for likelihood for given topology, so consider a rearrangement.

Now, in turn, apply this technique to estimate each of the distances in t. Keep on cycling through until values no longer change. Since the likelihood always increases, we cannot enter an endless cycle. The final stationary value satisfies (4.12), but is not guaranteed to be a global maximum (it could be a *local* maximum). Now the likelihood for given topology is computed.

4.4.4 *Determining the Topology*

The intent is then to choose that tree (model) with maximum likelihood. Letting $\mathcal{T}_n$ be the set of all binary trees having n leaves, note that

$$\sum_{t \in \mathcal{T}_n} L(t)$$

is not necessarily 1.

In order to construct an unrooted phylogenetic tree for n given taxa, start with the first two taxa, then repeatedly add the next taxon in the following manner. After k species have been added, the current tree has k external nodes, and hence $k-2$ internal nodes,[14] so as the number of edges is one less than the total number of nodes, there are

$$k + (k-2) - 1 = 2k - 3$$

many segments in an unrooted tree in which one could add the next node. Apply the previously described maximum-likelihood method to determine the optimal placement of the next node, and continue. Felsenstein states that this method examines

$$2n^2 - 9n + 8 = O(n^2)$$

different topologies for the n species. Note that there is no guarantee that this is the best topology among the exponentially many possible trees, and that the algorithm is dependent on the *order* in which successive taxa are added to the tree. Nevertheless, it is moderately comforting that there is a local application of the maximum-likelihood principle.

4.5 Quartet Puzzling

Felsenstein's method of using expectation maximization and maximum likelihood to construct a phylogenetic tree T has the advantage of being based on the appealing and mathematically well-founded notion of maximum likelihood; i.e., the conditional probability of the species data given T is a maximum over all trees. When there are a large number of species, Felsenstein's method takes too much computation time. In [SvH96], K. Strimmer and A. von Haeseler proposed a new phylogenetic algorithm, based on an initial application of Felsenstein's method for 4-taxon trees or *quartet trees*, followed by a clever yet simple method of combining the quartet trees to produce an overall tree. Given n taxa along with a distance matrix, the overall structure of the algorithm consists of the following three steps:

1. Compute $\binom{n}{4}$ maximum-likelihood trees for all possible quartets.
2. Combine the quartet trees into an n-taxon tree, which tries to respect the neighbor relation of all quartet trees. This is known as the quartet *puzzling* step.
3. Repeat steps (1) and (2) many times, and output the majority consensus tree, defined later. The consensus tree is multifurcating (not simply bifurcating).

The first two steps of the preceding are given in pseudocode in Algorithm 4.4. For a quartet tree, such as that illustrated in Figure 4.13, where a, b have a common parent, as do c, d, we write the *neighbor relation* $N(a, b; c, d)$ to indicate the fact that a and b are nearest neighbors, as are c and d. Note that a quartet topology uniquely determines one neighbor relation, and a neighbor relation identifies uniquely a quartet topology.

---

[14] Recall that the tree is unrooted here, and that by *internal node* we mean a node whose degree is exactly 3.

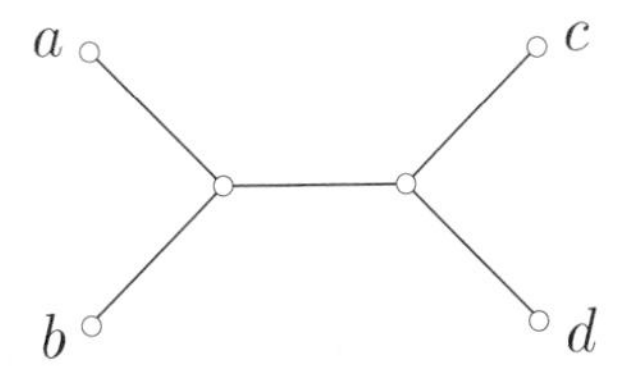

Figure 4.13 4-taxon or quartet tree.

Before we present the quartet puzzling algorithm, we want to motivate the use of quartet trees and quartet recombination. We use the terminology as used for example in [KLJ98, BJK+99]. Let t be any unrooted, binary phylogenetic tree, and let $\{a, b, c, d\}$ be any quartet of leaf labels of t. Then t *induces* the neighbor relation $N(a, b; c, d)$ if a, b and c, d reside in disjoint subtrees, i.e., if the paths between a, b and c,d are disjoint. Note that there is exactly one neighbor relation induced for every quartet. Note, furthermore, that with the neighbor relation, we induce also the corresponding quartet topology.

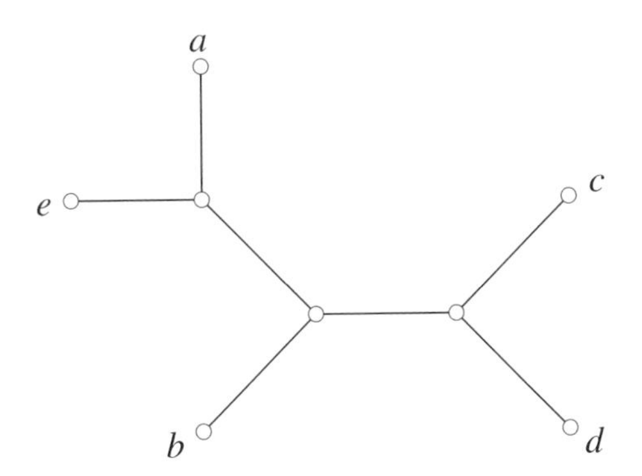

Figure 4.14 Original topology for 5 taxa.

Consider the topology in Figure 4.14, and the quartet $\{a, b, c, d\}$. Then the induced neighbor relation is $N(a, b; c, d)$, which corresponds to the quartet tree as shown in Figure 4.13. The quartet trees induced by t for all $\binom{5}{4}$ possible quartets of the label set $\{a, b, c, d, e\}$ are shown in Figure 4.15.

Now it is well known that, given the set of all induced quartet trees, one can reconstruct the original topology [Bun71]. Many phylogenetic algorithms are known in the literature that use the recombination of quartet trees to reconstruct a phylogeny – the method of quartet puzzling, due to Strimmer and von Haeseler, is one of many such reconstruction algorithms. See [KLJ98, BJK+99] for a detailed discussion. Note that we can reconstruct the phylogeny completely only if the set of induced quartet trees is error-free. Since this is not the case in practical applications, the algorithm must apply heuristic to yield an estimate of the correct phylogeny. Algorithm 4.4 gives the pseudocode for this approach.

Below, we give an illustration of the quartet puzzling step and how the majority consensus tree is constructed.

Algorithm 4.4 Strimmer, von Haeseler [SvH96]

Input: Taxa $1, \ldots, n$ (all of the same length)

```
for i=1 to (n choose 4) {
    construct and store T_i
       // T_i is i-th maximum likelihood 4-taxon tree
    let L_i be set of leaves of T_i
    derive and store neighbor relation N_i on L_i
    N_i is of the form N_i(a,b;c,d)
       // a,b have common parent, as do c,d
       // results stored in table for use below
}

for k=1 to m {
    randomly permute order of species to produce list
              a_1,...,a_n
    let P be quartet on a_1,a_2,a_3,a_4
    for i=1 to n {
        e = a_i
        let L be set of leaves of P
        if e ∉ L
            // insert taxon e into phylogenetic tree P
            label all edges of P by 0
            for j=1 to (n choose 4)
                if L_j ⊆ L ∪ {e} and e ∈ L_j {
                      let a,b,c,e be leaves of T_j
                      let N_j be of form N_j(a,b;c,e)
                      add 1 to each edge label in path
                          from a to b in P
                }
        choose edge { u,v } of P with minimum edge label
             // if more than one, randomly choose
        adjoin e to tree P by splitting edge { u,v }
             // split edge {u,v} into { u,w}, {w,v},
             // create edge {w,e}, and let w be parent of e
    }
    set P_k = P, and output P_k
}
output majority consensus of P_1,...,P_m
```

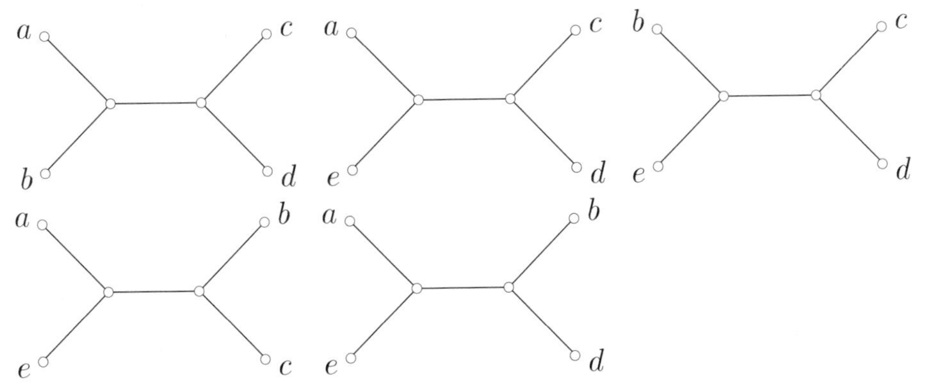

Figure 4.15 $\binom{5}{4}$ quartets for topology of Figure 4.14.

4.5.1 *Quartet Puzzling Step*

For illustrating the quartet puzzling step, we assume that we have the correct but unknown topology as given in Figure 4.14. Note that for the quartet puzzling step, we just need the set of quartet trees. How the quartet trees are inferred (whether by maximum likelihood or another method) does not make a difference. An illustrating example for generating the quartet trees is given in Exercise 5.[15] So suppose that we are given the correct set of induced neighbor relations for this topology, namely $N_1(a, b; c, d)$, $N_2(a, e; c, d)$, $N_3(b, e; c, d)$, $N_4(a, e; b, c)$, $N_5(a, e; b, d)$, as illustrated in Figure 4.15.

Then the quartet puzzling step of Algorithm 4.4 initializes the tree P as the first quartet having neighbor relation $N_1(a, b; c, d)$. We add the next node e as follows. The next neighbor relation to be considered is $N_2(a, e; c, d)$. The principal idea of the quartet puzzling step is to add a penalty of 1 to every edge such that an addition of the new taxon e at this edge would generate a phylogeny that would induce a wrong quartet topology for the quartet $\{a, c, d, e\}$. To see this, suppose one were to add e at the edge starting from c. Then the resulting phylogeny for $\{a, b, c, d, e\}$ would be as shown in Figure 4.16. But this would induce the topology $N(a, d; c, e)$ for the quartet $\{a, c, d, e\}$, which is not the neighbor relation we want to add (which is $N_2(a, e; c, d)$). For this reason, we add the penalty 1 to the edge label for this edge. The same holds for the edge starting from d, which implies that a 1 is also added there. When adding the new taxon e at the edges starting from a or b, then the resulting phylogeny would induce the correct neighbor relation for the quartet $\{a, c, d, e\}$, which implies that no penalty is added for these edges.

To summarize, when we consider the neighbor relation $N_2(a, e; c, d)$ in adding the new taxon e, then the addition of e to any edge on the path between c and d (which are nearest neighbors in $N_2(a, e; c, d)$) would generate a phylogeny that does not induce the correct neighbor relation $N_2(a, e; c, d)$ for the quartet $\{a, c, d, e\}$. Hence, we add the penalty 1 at every edge on the path from the c to d.[16]

Similarly, $N_3(b, e; c, d)$, $N_4(a, e; b, c)$, and $N_5(a, e; b, d)$ add 1 to edges of the path respectively between c, d and between b, c and between b, d. The leaf a then has the minimum

15 Albeit the method used there is not reasonable for practical purposes, since there are better ways to generate quartet trees.

16 Note that the edges on the path between c, d may be only a subset of the edges, where the addition of e would generate a wrong topology.

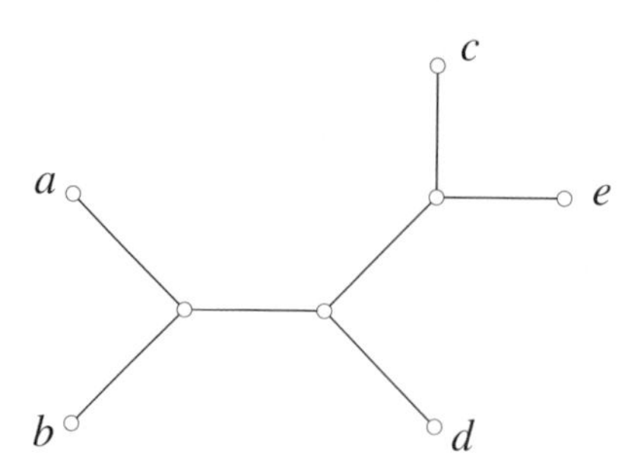

Figure 4.16 Wrong topology generated by adding e at the c-edge.

edge label of 0, which implies that adding e to this edge induces the correct quartet topology for all $\binom{5}{4}$ quartets. So the node e is adjoined at a. These steps appear in Figure 4.17. The final tree coincides with the original topology.

4.5.2 *Majority Consensus Tree*

We now describe how to define the *majority consensus tree*, a notion first investigated by [MM81]. The following definition was adapted from [MM81].

DEFINITION 4.8
An n-tree T is a tree having n leaves, for which every node has a label $\ell \subseteq \{1, \dots, n\}$.

1. *$\emptyset$ is not a label of any node.*
2. *The leaves are labeled by $\{1\}, \dots, \{n\}$.*
3. *If ℓ_1, ℓ_2 are labels of nodes in T, and $\ell_1 \cap \ell_2 \neq \emptyset$, then either $\ell_1 \subseteq \ell_2$ or $\ell_2 \subseteq \ell_1$.*

Note that the tree topology of T is uniquely determined by the subset inclusion relation, because of the following:

- Nodes x, y labeled with ℓ_1, ℓ_2 respectively are immediate neighbors if and only if ℓ_1, ℓ_2 are comparable and there is no intermediate ℓ. This means that either $\ell_1 \subset \ell_2$ or $\ell_2 \subset \ell_1$ and for no other label ℓ is it the case that $\ell_1 \subset \ell \subset \ell_2$ or $\ell_2 \subset \ell \subset \ell_1$.

Now, suppose that $P_1, \dots, P_m$ are the binary trees output in the last step of Algorithm 4.4. Each of the trees P_i has the same leaves – namely, the taxa $1, \dots, n$. Define labels for all nodes of an n-taxon tree P inductively, as follows, where every label ℓ will be a subset of $\{1, \dots, n\}$:

- The label of a leaf node associated with the taxon $i \in \{1, \dots, n\}$ is $\{i\}$.
- Suppose that nodes $x_1, \dots, x_r \in P$ have been labeled respectively by $\ell_1, \dots, \ell_r$ and that y is the parent of $x_1, \dots, x_r$. Then label y by $\ell_1 \cup \dots \cup \ell_r$.

Clearly, the collection of labels on any tree P, as just inductively defined, forms an n-tree, as defined in Definition 4.8.

Assume now that all nodes of $P_1, \dots, P_m$ have been labeled as just described. The *majority consensus tree* M consists of exactly those nodes whose label occurs in more than half of the P_i. It is not immediately apparent that M is a valid n-taxon tree, but this follows from a simple observation. If nodes with labels ℓ_1 and ℓ_2 belong to more than half the trees P_i, then there is

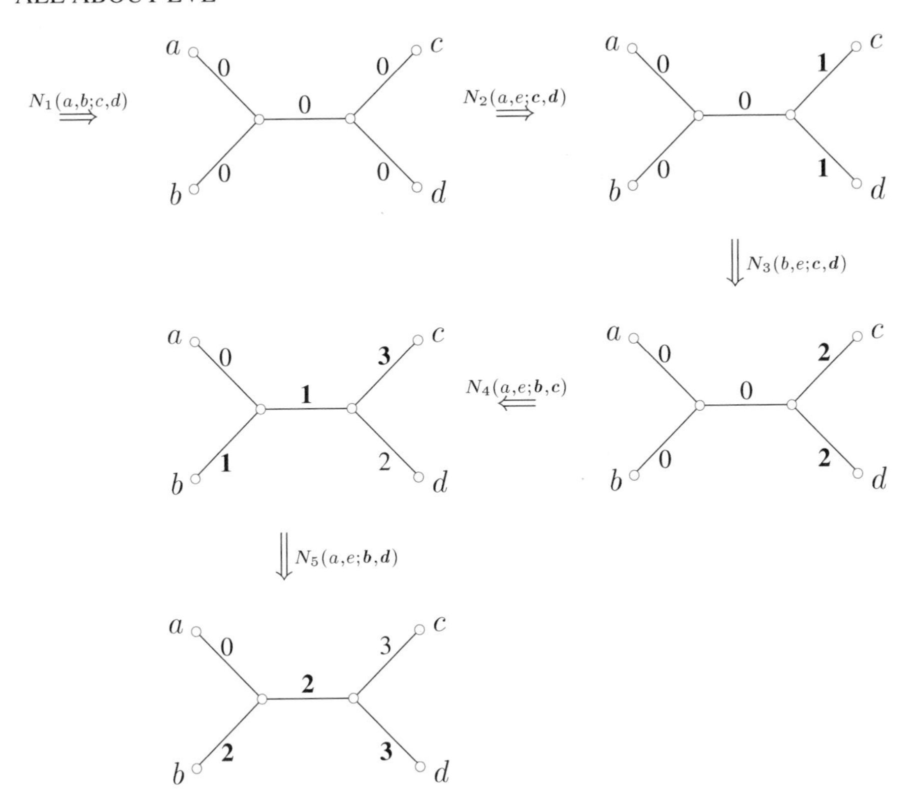

Figure 4.17 Quartet puzzling.

at least one tree P_{i_0} in which both labels appear, and by condition (3) we have that $\ell_1 \subseteq \ell_2$ or $\ell_2 \subseteq \ell_1$. It follows that M is an n-tree.

The method of quartet puzzling is fast, efficient, and appealing, because of its reliance on maximum likelihood quartets. Extensions and refinements of the method can be found in [SGvH97, Kea98].

4.6 Exercises

1. A tree can be defined as a connected acyclic (undirected) graph. Prove that an acyclic graph is connected if and only if the number e of edges is one less than the number n of nodes in the graph.

   Suppose that T is a rooted, binary tree with the property that every internal node has exactly 2 children. Prove that if T has n leaves, then T has $n-1$ internal nodes, hence has $n + (n-1) - 1 = 2n - 2$ many edges.
2. Extend the implementation of the Gotoh algorithm (Exercise 3 in Chapter 3) in order to compute the distance matrix for n nucleotide sequences, rather than only 2. You are *not* asked to perform a multiple sequence alignment, but only to compute the sequence alignment distance $d_{i,j}$ between the ith and the jth sequence.
   The following is an example input:

$s_0 =$	addacdaaa	$s_4 =$	aaacdaaa
$s_1 =$	abca	$s_5 =$	aabbccda
$s_2 =$	abca	$s_6 =$	abbcda
$s_3 =$	aabbcda		

For the previous input file, the output from our implementation is the following:

$$\begin{pmatrix} 0 & 17 & 17 & 15 & 7 & 19 & 16 \\ 17 & 0 & 0 & 10 & 15 & 12 & 8 \\ 17 & 0 & 0 & 10 & 15 & 12 & 8 \\ 15 & 10 & 10 & 0 & 13 & 4 & 4 \\ 7 & 15 & 15 & 13 & 0 & 15 & 12 \\ 19 & 12 & 12 & 4 & 15 & 0 & 8 \\ 16 & 8 & 8 & 4 & 12 & 8 & 0 \end{pmatrix}.$$

3. Experiment with J. Felsenstein's PHYLIP Software Package as well as the demo programs of this book's web site. Construct a phylogenetic tree for the example from the previous exercise.
4. Implement UPGMA, WPGMA and the Farris transformed distance method for both. The output of our implementation of UPGMA for the data from exercise 2 is shown in Figure 4.18

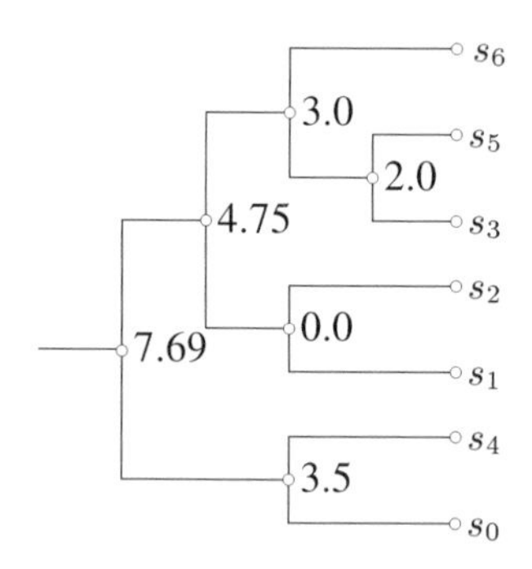

Figure 4.18 UPGMA output for Exercise 4.

5. Implement the quartet puzzling step and use UPGMA to produce the quartet trees. Consider the distance matrix

	a	b	c	d	e
a	0	3	4	4	2
b	3	0	3	3	3
c	4	3	0	2	4
d	4	3	2	0	4
e	2	3	4	4	0

This distance matrix corresponds to the topology given in Figure 4.14, where each edge of the unrooted tree has distance 1. Generate the submatrices for the five quartets $\{a, b, c, d\}$, $\{a, c, d, e\}$, $\{b, c, d, e\}$, $\{a, b, c, e\}$, and $\{a, b, d, e\}$. Then generate a phylogeny using UPGMA for these five submatrices, infer quartet topologies and apply the quartet puzzling step. Is the topology the same as that generated by UPGMA applied to the original distance matrix?

Solution: The five submatrices for the five quartets $\{a, b, c, d\}$, $\{a, c, d, e\}$, $\{b, c, d, e\}$, $\{a, b, c, e\}$, and $\{a, b, d, e\}$ are

	a	b	c	d
a	0	3	4	4
b	3	0	3	3
c	4	3	0	2
d	4	3	2	0

	a	c	d	e
a	0	4	4	2
c	4	0	2	4
d	4	2	0	4
e	2	4	4	0

	b	c	d	e
b	0	3	3	3
c	3	0	2	4
d	3	2	0	4
e	3	4	4	0

	a	b	c	e
a	0	3	4	2
b	3	0	3	3
c	4	3	0	4
e	2	3	4	0

	a	b	d	e
a	0	3	4	2
b	3	0	3	3
d	4	3	0	4
e	2	3	4	0

The corresponding trees produced by UPGMA are shown in Figure 4.19. The induced quartet topologies are the correct one (for the given topology) as shown in Figure 4.15.

Note that UPGMA correctly determines the topology of this example, as shown in Figure 4.20 (UPGMA has added a root and computed different distances, but the topology is correct).

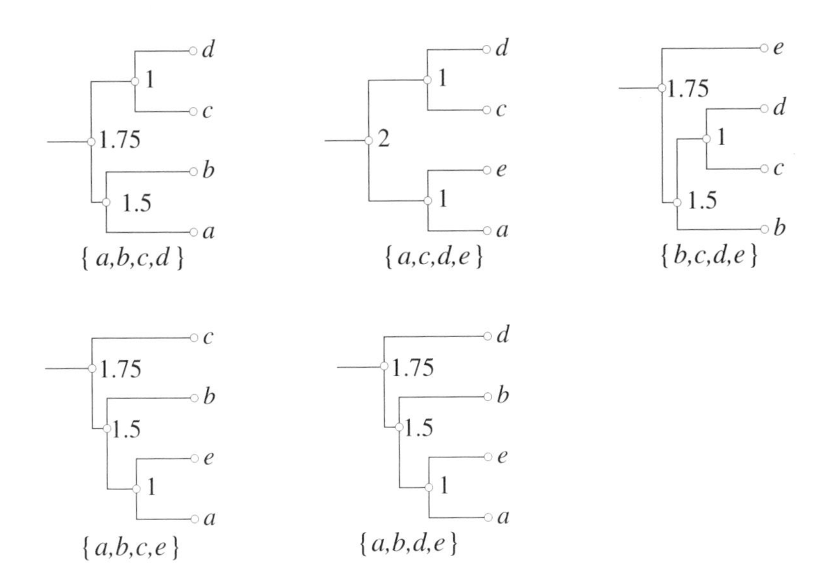

Figure 4.19 Quartet trees produced by UPGMA for Exercise 5.

Acknowledgments and References

Some remarks from the introduction were drawn from the popular book [JJE94] by D. Johanson *et al.* Chapter 11 of [Nei87] and Chapter 14 of [Wat95] survey many techniques of phylogeny construction. Nei's text gives a general overview without many details, and we have attempted to present a more algorithmic approach than Waterman, and have included the relatively new quartet puzzling method of Strimmer and von Haeseler.

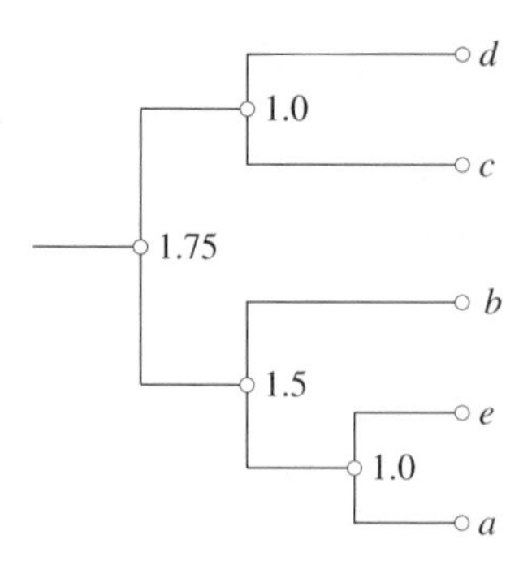

Figure 4.20 Tree produced by UPGMA for Exercise 5.

Much in the section on evolution rates was drawn from [Nei87] and [Wat95]. The articles [DT95] and [GT94] present a modern statistical analysis using *coalescents*, develop a Markov chain model whose state space consists of trees, and suggest a Markov chain Monte Carlo method for phylogeny construction under the assumptions of a constant population size and neutral selection. These are reasonable conditions thought to be satisfied for mitochondrial sequence data in constant population size groups (such as the North American Indian tribe Nuu-Chah-Nulth on Vancouver Island).

The article [Fel83] is a survey of various methods applied for phylogeny construction up to the early 1980s, and [Fel81] is an extremely clear treatment by J. Felsenstein of his application of maximal likelihood in determining phylogenetic trees. Felsenstein went on to develop the software `PHYLIP`, which implements a number of techniques, including certain of those discussed in this chapter. Research in phylogeny, to develop new and more efficient methods for tree construction, is a dynamic and exploding field. Even an overview of the techniques is outside of the scope of this text.

5

Hidden Markov Models

> The conceptual scheme for 'learning' in this context is a machine with an input channel for figures, a pair of YES and NO output indicators, and a *reinforcement* or 'reward' button that the machine's operator can use to indicate his approval or disapproval of the machine's behavior. (M.L. Minsky and S.A. Papert, *Perceptrons* [MP88])

In Chapter 2, we considered (time-homogeneous) Markov chains, where a system in state i goes to state j with transition probability $p_{i,j}$. Implicitly, we have assumed that we can observe the state of the system. Suppose now that the state can no longer be directly observed (the state is *hidden*), but from each state the system emits an output symbol σ from a finite alphabet $\Sigma = \{\sigma_1, \ldots, \sigma_m\}$. Moreover, the symbol $\sigma_k \in \Sigma$ is emitted with probability $b_{i,k}$, provided the system is in state i. If $b_{i,k} = 1$ exactly when $k = i$, then we have (essentially) an ordinary Markov chain.

Algorithms for hidden Markov models infer the probabilities of state transition and output emission from observed data, in order to derive a *statistical model* that generates the observed data (in particular nucleotide data) with high likelihood. One might hope that knowledge of a good statistical model could suggest a physical model, i.e. that the hidden Markov model states actually correspond to physical states of the biological system, but this is not required. There have been applications of hidden Markov models for pattern recognition in a variety of areas: word recognition in speech processing [Rab89], cellular ion channels [CMX$^+$90, FR92a, FR92b, BHH$^+$94, KTH97], multiple sequence alignment, and recognition of protein classes such as globins, immunoglobulins, kinases, HIV membrane proteins, etc. [BCHM94, KBM$^+$94], eukaryotic DNA promotor sequences [Ohl95], isochores [Chu92] (nucleotide segments of more than 3 kb with homogeneous G + C content).

Hidden Markov models are essentially stochastic finite state automata, or equivalently, stochastic regular grammars (actually Moore automata, since outputs are associated with states). There is a natural generalization to stochastic context-free grammars with application to RNA secondary structure, which uses a simple extension of the *forward–backward* algorithm explained in this chapter, along with an efficient parsing algorithm (such as the CYK algorithm or Early parser). See [BB98, DEKM98] for more details on stochastic context-free grammars.

A *discrete Markov model* is a stochastic model with a finite set Q of states, giving rise to a time sequence of states

$$q_0, q_1, q_2, \ldots, q_t, \ldots .$$

The *Markov property* states that the determination of state q_t depends only on

$$q_{t-k}, \ldots, q_{t-1},$$

where k is the *order* of the Markov model. Our earlier definition of Markov chain is a first-order Markov model.

DEFINITION 5.1 (MARKOV CHAIN)
An order-k (time-homogeneous) Markov chain (Q, M) is given by a finite set $Q = \{1, \ldots, n\}$ of states, and an $n^{k+1} = n \times \cdots \times n$ transition matrix of probabilities with the property that for all $1 \leq i_1, \ldots, i_k \leq n$,

$$\sum_{j=1}^{n} M[i_1, \ldots, i_k, j] = 1.$$

The value $M[i_1, \ldots, i_k, j]$ is the probability of transition into state j, provided that the system's last k states were $i_1, \ldots, i_k$; hence

$$M[i_1, \ldots, i_k, j] = Pr[q_t = j | q_{t-k} = i_1, \ldots, q_{t-1} = i_k].$$

Note that an order-0 Markov model does not depend on its current state – for instance, the outcome of a coin flip does not depend on whether the coin was previously heads or tails. More formally, this is given by $(\{1, 2\}, M)$, where $M(1) = \frac{1}{2} = M(2)$, where 1 (resp. 2) represents heads (resp. tails). As another example, [BM93] devised an order-5 Markov model for the recognition of intron/exon splice sites.

As depicted in Figure 5.1, hidden Markov models can be visualized by a collection of n urns

$$U_1, \ldots, U_n,$$

each with a different distribution of colored balls, where an unobserved magician successively draws balls from different urns (with replacement), and we can only observe the sequence of colored balls. Our task is to infer from the observed sequence of balls the succession of urn drawings (states), transition probabilities between states, and emission probabilities.

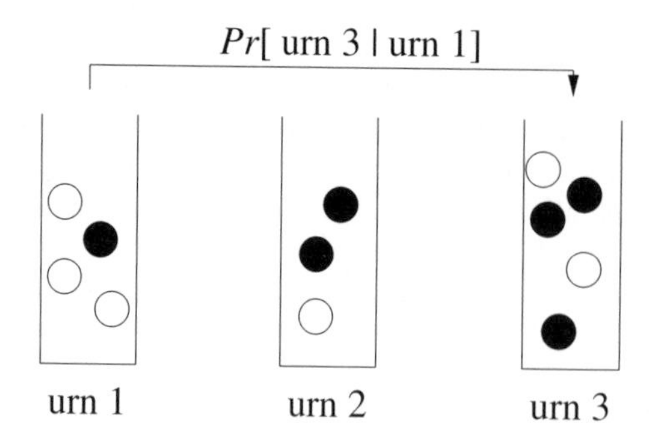

Figure 5.1 HMM depicted as urn model.

DEFINITION 5.2 (MARKOV MODEL)
Let $Q = \{1, \ldots, n\}$ be a set of n states (the urns), $\Sigma = \{\sigma_1, \ldots, \sigma_m\}$ the finite output alphabet, (π_i) the initial state probabilities, $(a_{i,j})$ the transition probabilities, and $(b_{i,k})$ the emission probabilities, so that

$$\begin{aligned} \pi_i &= Pr[q_0 = i], \\ a_{i,j} &= Pr[q_t = j | q_{t-1} = i], \\ b_{i,k} &= Pr[o_t = \sigma_k | q_t = i], \end{aligned}$$

where the system is initially in state q_0, then successively in states $q_1, q_2, q_3, \ldots$, and the length T output sequence is $o_0, o_1, o_2, \ldots, o_{T-1}$. A (first-order, time-homogenous) Markov model is given by

$$M = (Q, \Sigma, \pi, a, b).$$

In applications, we are presented an observation sequence $\mathcal{O} = o_0, \ldots, o_{T-1}$ and, for a fixed state set, would like to compute values for the probability matrices π, a, b for which the resulting model is most likely to generate the observed data. This leads to a discussion of *likelihood* and how to *score a model*. Since we can only see the observation sequence (and not the sequence of states), the Markov model is said to be *hidden*, i.e. a *hidden Markov model* (HMM).

A simple but illustrative example is a variant of the coin flipping model, where three coins are used, each with a different bias for heads. The first coin is fair, with $Pr[H] = 0.5 = Pr[T]$; the second coin has $Pr[H] = 0.75$ and $Pr[T] = 0.25$; the third coin has $Pr[H] = 0.1$ and $Pr[T] = 0.9$. Suppose that a coin is chosen at random at the start, and that if the first coin is flipped, then at the next time instant either the second or third coin will be flipped with equal probability, while if the second or third coin is flipped, then at the next time instant any one of the three coins will be flipped with equal probability. Thus

$$\pi = \begin{pmatrix} 1/3 \\ 1/3 \\ 1/3 \end{pmatrix}, \quad a = \begin{pmatrix} 0 & 1/2 & 1/2 \\ 1/3 & 1/3 & 1/3 \\ 1/3 & 1/3 & 1/3 \end{pmatrix}, \quad b = \begin{pmatrix} 0.5 & 0.5 \\ 0.75 & 0.25 \\ 0.1 & 0.9 \end{pmatrix}.$$

It seems remarkable that one can algorithmically determine these probabilities, simply by observing a series of coin flips.

5.1 Likelihood and Scoring a Model

Suppose that observation sequence $\mathcal{O} = o_0, \ldots, o_{T-1}$ is given, and that $M = (Q, \Sigma, \pi, a, b)$ is a Markov model thought to generate $\mathcal{O}$, and $|Q| = n$. Recalling the definition of likelihood of a model from Chapter 2, we have the following.

DEFINITION 5.3
The likelihood $L_{\mathcal{O}}(M)$ *that model* M *generates* $\mathcal{O}$ *is defined by*

$$\begin{aligned} L_{\mathcal{O}}(M) &= Pr[\mathcal{O}|M] \\ &= \sum_{p \in Q^T} Pr[\mathcal{O}, p|M] \\ &= \sum_{p \in Q^T} Pr[\mathcal{O}|p, M]\, Pr[p|M] \\ &= \sum_{p \in Q^T} A_{p,\mathcal{O}}, \end{aligned}$$

where if $p = p_0, \ldots, p_{T-1}$ *is a path (of length* T*) of states, then*

$$A_{p,\mathcal{O}} = Pr[\mathcal{O}|p, M]\, Pr[p|M]$$

and

$$\begin{aligned} Pr[p|M] &= \pi_{p_0} a_{p_0,p_1} a_{p_1,p_2} \cdots a_{p_{T-2},p_{T-1}}, \\ Pr[\mathcal{O}|p, M] &= b_{p_0,o_0} b_{p_1,o_1} \cdots b_{p_{T-1},o_{T-1}}. \end{aligned}$$

For typographical reasons, we may at times write $p(t)$ in place of p_t for a path $p \in Q^T$. In the case of many different observation sequences

$$\mathcal{O}_1, \ldots, \mathcal{O}_r$$

(for instance, proteins of different lengths that belong to a given protein family), the likelihood of M is taken to be the product of the likelihoods with respect to the observation sequences:

$$L(M) = \prod_s L_{\mathcal{O}_s}(M).$$

The likelihood $L_{\mathcal{O}}(M)$ is sometimes called the *Baum–Welch score*, where often the negative log likelihood, $-\log L_{\mathcal{O}}(M)$, is taken. A likelihood is a probability value between 0 and 1, so the negative log likelihood is a positive *energy* score, where higher likelihood corresponds to lower energy.

The previous definition yields a naive $O(Tn^T)$ exponential time algorithm to compute likelihood. A much better solution uses a dynamic programming technique, called the *forward method.*

DEFINITION 5.4 (FORWARD VARIABLE)

$\alpha_t(i) = Pr[o_0, \ldots, o_t, q_t = i|M]$.

CLAIM 5.5 $\alpha_0(i) = \pi_i b_{i,o_0}$, and $\alpha_{t+1}(j) = \sum_{i=1}^{n} \alpha_t(i) a_{i,j} b_{j,o_{t+1}}$.

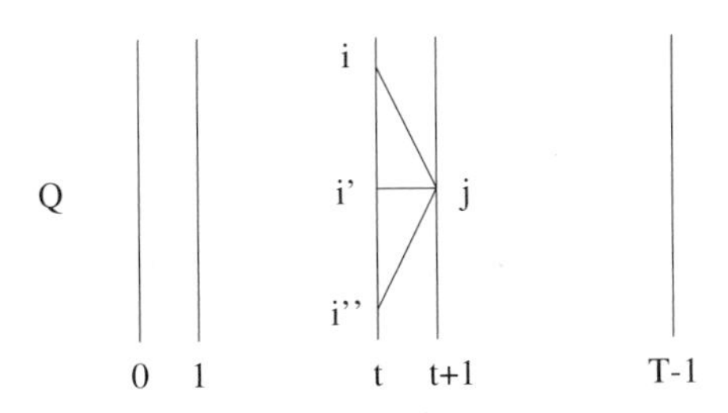

Figure 5.2 $\alpha_{t+1}(j)$ forward method.

PROOF

$$\begin{aligned}
\alpha_{t+1}(j) &= Pr[o_0, \ldots, o_{t+1}, q_{t+1} = j|M] \\
&= \sum_{i \in Q} Pr[o_0, \ldots, o_t, q_t = i, q_{t+1} = j, o_{t+1}|M] \\
&= \sum_{i \in Q} Pr[o_0, \ldots, o_t, q_t = i, q_{t+1} = j|M] b_{j,o_{t+1}} \\
&= \sum_{i \in Q} Pr[o_0, \ldots, o_t, q_t = i|M] a_{i,j} b_{j,o_{t+1}} \\
&= \sum_{i \in Q} \alpha_t(i) a_{i,j} b_{j,o_{t+1}}.
\end{aligned}$$

■

The third and fourth lines follow from the definition of conditional probability: $Pr[A, B] = Pr[B|A]\, Pr[A]$. The recurrence relation for $\alpha_{t+1}(j)$ is illustrated in Figure 5.2.

Clearly, using an array of size nT for the forward variables, the entire computation can be performed in $O(n^2T)$ time. It follows that the likelihood of a model

$$L_{\mathcal{O}}(M) = Pr[\mathcal{O}|M] = \sum_{i \in Q} \alpha_{T-1}(i)$$

can be computed in time $O(n^2T)$.

DEFINITION 5.6

The Viterbi score of a model $M = (Q, \Sigma, \pi, a, b)$ with respect to the observation sequence $\mathcal{O} = (o_0, o_1, \ldots, o_{T-1})$ is defined by

$$Pr[\mathcal{O}|M] = \max_{p \in Q^T} Pr[\mathcal{O}, p|M].$$

Note that the *Viterbi score* is identical to the Baum–Welch score, except with *maximum* in place of *sum* over all paths of states. From the definition, a naive computation of the Viterbi Score uses exponential time $O(Tn^T)$. Using dynamic programming, by a slight modification of the forward method (with maximum in place of sum), we have an $O(n^2T)$ time algorithm for computing the Viterbi score. From the Viterbi score, using *tracebacks*, one can compute

Algorithm 5.1 Viterbi Algorithm for optimal path

(1) INITIALIZATION $\delta_0(i) = \pi_i b_{i,o_0}$, for $i \in Q$.
(2) RECURSION For $0 \leq t < T - 1$, and $j \in Q$,

$$\delta_{t+1}(j) = \max_{i \in Q}\{\delta_t(i) a_{i,j} b_{j,o_{t+1}}\}.$$

(3) PATH DETERMINATION BY TRACEBACKS

$$q^*_{T-1} = \text{argmax}_{i \in Q}\{\delta_{T-1}(i)\}$$

and for $t = T - 2, \ldots, 0$,

$$q^*_t = \text{argmax}_{1 \leq i \leq n}\{\delta_t(i) a_{i,q^*_{t+1}}\}.$$

---

that path of states, which has greatest probability of producing the observation sequence. This is given in Algorithm 5.1.

Letting $q^* = q^*_0, \ldots, q^*_{T-1}$, from the previous considerations, it can be shown that

$$Pr[\mathcal{O}, q^*|M] = \max_{p \in Q^T} Pr[\mathcal{O}, p|M],$$

so that q^* is the most likely path of states to generate the observation sequence, given the model.

5.2 Re-estimation of Parameters

Given an observation sequence $\mathcal{O}$, consider the likelihood landscape as a function of the model parameters π, a, b, for fixed state set Q and alphabet Σ. In principle, one could apply various combinatorial optimization techniques to determine a (local) maximum likelihood. In particular, gradient ascent (hill-climbing) Monte Carlo, genetic algorithms, and especially quasi-Newton numerical methods, etc. come to mind. As reported in [KTH97], quasi-Newton methods applied to maximization problems with many parameters tend to be numerically unstable and dependent on a good initial choice of parameters.[1] In practice, two optimization techniques have successfully be applied: *expectation maximization* from statistics (due to Baum–Welch[2]), and the *gradient descent* method (due to Baldi–Chauvin).

Let $M = (Q, \Sigma, \pi, a, b)$ be a given model for observation sequence $\mathcal{O} = o_0, \ldots, o_{T-1}$. The Baum–Welch method re-estimates the transition probabilities $\overline{a}_{i,j}$ to be the *expected number* of transitions from state i to j along all paths $p \in Q^T$, divided by the *expected number* of transitions out of state i along along all paths $p \in Q^T$. To compute the mathematical expectation, paths are weighted by the quantity $A_{p,\mathcal{O}}$. Similar parameter re-estimations are given for $\overline{\pi}, \overline{b}_{i,k}$. This yields the model $\overline{M} = (Q, \Sigma, \overline{\pi}, \overline{a}, \overline{b})$ whose likelihood is better; i.e.

---

[1] J. Timmer (personal communication) has reported success of the Freiburger Physics group in initially using the EM algorithm, then switching over to quasi-Newton methods, which then more rapidly converge.

[2] Baum and Welch developed their parameter re-estimation technique, before the EM algorithm was first proposed by Demster *et al.* EM is quite general and has been used successfully in many different applications. In this section, we show how EM yields the Baum–Welch parameter re-estimates.

$L_{\mathcal{O}}(\overline{M}) > L_{\mathcal{O}}(M)$. Iterating this, we converge to a (local) likelihood maximum. Simulated annealing may be added to attempt to avoid becoming trapped in a local maximum and improve chances of convergence to a global maximum. This feature is supported by the software HMMER, written by S. Eddy (see [Edd95]).

A newer method, due to P. Baldi and Y. Chauvin [BC94], involves a Boltzmann distribution-like transformation of variables and the application of gradient descent to determine the lowest energy $E = -\log L_{\mathcal{O}}(M)$. Recently, H. Mamitsuka [Mam97] modified both the Baum–Welch and Baldi–Chauvin methods by incorporating a measure of deviation from *target* likelihoods, and thereby was able to incorporate both positive and negative examples in the training set.

5.2.1 *Baum–Welch Method*

Thinking of a hidden Markov model as a physical system, it is natural to ask what is the most probable state at any given time t, given the observation sequence and model.

DEFINITION 5.7 (MOST PROBABLE STATE AT TIME t)
Let

$$\gamma_t(i) = \frac{Pr[q_t = i, \mathcal{O}|M]}{L_{\mathcal{O}}(M)} = \frac{\sum_{p \in Q^T, p(t)=i} A_{p,\mathcal{O}}}{L_{\mathcal{O}}(M)},$$

and for $0 \leq t < T$ *define* $r_t = \text{argmax}_{1 \leq i \leq n} \gamma_t(i)$,

Clearly r_t is the most probable state at time t. To compute $\gamma_t(i)$ efficiently, we introduce the *backward* variable $\beta_t(i)$, in analogy to the forward variable $\alpha_t(i)$.

DEFINITION 5.8 (BACKWARD VARIABLE)
$\beta_t(i) = Pr[o_{t+1}, \ldots, o_{T-1}|q_t = i, M]$.

CLAIM 5.9 $\gamma_t(i) = \frac{\alpha_t(i)\beta_t(i)}{L_{\mathcal{O}}(M)}$.

PROOF Because the state q_{t+1} depends only on q_t (Markov condition), it follows that

$$Pr[o_{t+1}, \ldots, o_{T-1}|q_t = i, M] = Pr[o_{t+1}, \ldots, o_{T-1}|q_t = i, o_0, \ldots, o_t, M]$$

and hence proving the claim is equivalent to showing that

$$\begin{aligned}
&Pr[\mathcal{O}|M]\, Pr[q_t = i|\mathcal{O}, M] \\
&\quad = \alpha_t(i)\beta_t(i) \\
&\quad = Pr[q_t = i, o_0, \ldots, o_t|M]\, Pr[o_{t+1}, \ldots, o_{T-1}|q_t = i, M] \\
&\quad = Pr[q_t = i, o_0, \ldots, o_t|M]\, Pr[o_{t+1}, \ldots, o_{T-1}|q_t = i, o_0, \ldots, o_t, M] \\
&\quad = Pr[q_t = i, \mathcal{O}|M].
\end{aligned}$$

Using the definition of conditional probability,

$$Pr[\mathcal{O}|M]\, Pr[q_t = i|\mathcal{O}, M] = Pr[q_t = i, \mathcal{O}|M].$$

■

The following recurrence relation yields an $O(n^2T)$ time algorithm with $O(nT)$ space for computation of all backward variables.

CLAIM 5.10 $\beta_{T-1}(i) = 1$, and $\beta_t(i) = \sum_{j=1}^{n} a_{i,j}\beta_{t+1}(j)b_{j,o_{t+1}}$, for $t < T-1$.

PROOF

$$\begin{aligned}
\beta_t(i) &= Pr[o_{t+1}, \ldots, o_{T-1} | q_t = i, M] \\
&= \sum_{j \in Q} Pr[q_{t+1} = j, o_{t+1}, \ldots, o_{T-1} | q_t = i, M] \\
&= \sum_{j \in Q} Pr[o_{t+1}, \ldots, o_{T-1} | q_t = i, q_{t+1} = j, M] a_{i,j} \\
&= \sum_{j \in Q} Pr[o_{t+1}, \ldots, o_{T-1} | q_{t+1} = j, M] a_{i,j}.
\end{aligned}$$

Since an observation at time $t+1$ depends only on the state q_{t+1}, we have that

$$Pr[o_{t+1}, \ldots, o_{T-1} | q_{t+1} = j, M] = b_{j,o_{t+1}} Pr[o_{t+2}, \ldots, o_{T-1} | q_{t+1} = j, M]$$

from which we have

$$\begin{aligned}
\beta_t(i) &= Pr[o_{t+1}, \ldots, o_{T-1} | q_t = i, M] \\
&= \sum_{j=1}^{n} a_{i,j}\beta_{t+1}(j)b_{j,o_{t+1}}
\end{aligned}$$

for $0 \leq t < T-1$. Clearly this computation can be performed by dynamic programming in $O(n^2T)$ time and $O(nT)$ space. ■

By the way, note that the likelihood can be efficiently computed from the backward variables, since

$$L_{\mathcal{O}}(M) = \sum_{p \in Q^T} Pr[\mathcal{O}, p | M] = \sum_{i \in Q} \pi_i \beta_0(i) b_{0,o_0}.$$

The following notion is used in determining the *expected* number of transitions from state i to j, which will be applied in the Baum–Welch updating rules. Recall that

$$\begin{aligned}
A_{p,\mathcal{O}} &= \sum_{p \in Q^T} Pr[\mathcal{O} | M, p]\, Pr[p | M] \\
&= \pi_{p_0} \prod_{t < T-2} a_{p_t, p_{t+1}} \prod_{t < T-1} b_{p_t, o_t},
\end{aligned}$$

so that $A_{p,\mathcal{O}}$ is a *weight* for the path p, given the model and observation sequence.

DEFINITION 5.11

$$\begin{aligned}
\eta_t(i,j) &= Pr[q_t = i, q_{t+1} = j | \mathcal{O}, M] \\
&= \frac{\sum_{p \in Q^T, p_t = i, p_{t+1} = j} A_{p,\mathcal{O}}}{L_{\mathcal{O}}(M)}.
\end{aligned}$$

We leave the proof of the following claim to an exercise.

CLAIM 5.12

$$\begin{aligned}\eta_t(i,j) &= Pr[q_t = i, q_{t+1} = j | \mathcal{O}, M] \\ &= \frac{\alpha_t(i) a_{i,j} b_{j,o_{t+1}} \beta_{t+1}(j)}{Pr[\mathcal{O}|M]}.\end{aligned}$$

Note that

$$\begin{aligned}\gamma_t(i) &= Pr[q_t = i | \mathcal{O}, M] \\ &= \frac{\alpha_t(i)\beta_t(i)}{Pr[\mathcal{O}|M]} \\ &= \frac{\alpha_t(i)(\sum_{j\in Q} a_{i,j} b_{j,o_{t+1}})\beta_{t+1}(i)}{Pr[\mathcal{O}|M]} \\ &= \sum_{i\in Q} \eta_t(i,j).\end{aligned}$$

Moreover, since $\gamma_t(i)$ is the probability that the system is in state i at time t, it follows that the expected number of times the system is in state i is $\sum_{t<T} \gamma_t(i)$, and $\sum_{t<T-1} \gamma_t(i)$ is the expected number of times there is a transition from state i. Finally, $\sum_{t<T-1} \eta_t(i,j)$ is the expected number of transitions from state i to state j.

A more graphical manner of seeing the previous relations between $\gamma_t(i)$, $\eta_t(i,j)$ and expected values consists of summing over all paths of states satisfying a certain property, multiplied by the weight of the path. Recall that $A_{p,\mathcal{O}}$ is the expression

$$\pi_{p(0)} \prod_{i<T-2} a_{p(i),p(i+1)} \prod_{i<T-1} b_{p(i),o_i}$$

for a given path $p \in Q^T$, so that $A_{p,\mathcal{O}} = Pr[p,\mathcal{O}|M]$. Let $c_p(i)$ denote the number of times path that p takes on the value i; i.e. $c_p(i) = |\{t < T \mid p(t) = i\}|$. Let $d_p(i) = |\{t < T-1 \mid p(t) = i\}|$ and $c_p(i,j) = |\{t < T-1 \mid p(t) = i, p(t+1) = j\}|$. Then

$$\begin{aligned}\gamma_t(i) &= Pr[q_t = i | \mathcal{O}, M] \\ &= \sum_{p\in Q^T} Pr[q_t = i | p, \mathcal{O}, M]\, Pr[p|\mathcal{O}, M] \\ &= \sum_{p\in Q^T, p(t)=i} Pr[p|\mathcal{O}, M].\end{aligned}$$

Thus

$$\sum_{t<T} \gamma_t(i) = \sum_{p\in Q^T} c_p(i)\, Pr[p|\mathcal{O}, M] = \sum_{p\in Q^T} \frac{c_p(i) A_{p,\mathcal{O}}}{L_{\mathcal{O}}(M)},$$

which is the expected number of times the system is in state i, given the observation sequence and model. Similarly, $\sum_{t<T-1} \gamma_t(i) = \sum_{p\in Q^T} d_p(i)\, Pr[p|\mathcal{O}, M]$ is the expected number of times the system makes a transition from state i, given $\mathcal{O}, M$.

Finally,

$$\begin{aligned}
\eta_t(i,j) &= Pr[q_t = i, q_{t+1} = j | \mathcal{O}, M] \\
&= \sum_{p \in Q^T} Pr[q_t = i, q_{t+1} = j | p, \mathcal{O}, M] \, Pr[p | \mathcal{O}, M] \\
&= \sum_{p \in Q^T, p(t)=i, p(t+1)=j} Pr[p | \mathcal{O}, M] \\
&= \sum_{p \in Q^T, p(t)=i, p(t+1)=j} \frac{A_{p,\mathcal{O}}}{L_{\mathcal{O}}(M)}.
\end{aligned}$$

Letting $\eta_{\mathcal{O}}(i,j) = \sum_{t<T} \eta_t(i,j)$, we have

$$\eta_{\mathcal{O}}(i,j) = \sum_{p \in Q^T} \frac{c_p(i,j) A_{p,\mathcal{O}}}{L_{\mathcal{O}}(M)},$$

which is the expected number times the system makes a transition from i to j, given $\mathcal{O}, M$. With this notation, we have a reformulation of the Baum–Welch parameter re-estimates, given as follows:

$$\begin{aligned}
\overline{\pi}_i &= \frac{E[\text{number of times system begins in state } i | \mathcal{O}, M]}{E[\text{number of times system begins in any state} | \mathcal{O}, M]}, \\
\overline{a}_{i,j} &= \frac{E[\text{number of transitions from } i \text{ to } j | \mathcal{O}, M]}{E[\text{number of transitions from } i | \mathcal{O}, M]}, \\
\overline{b}_{i,k} &= \frac{E[\text{number of emissions of } \sigma_k, \text{ while in state } i | \mathcal{O}, M]}{E[\text{number of times in state } i | \mathcal{O}, M]}.
\end{aligned}$$

Suppose that Q, Σ are fixed, and that $|Q| = n$, $|\Sigma| = m$. Given an observation sequence $\mathcal{O} = o_0, \ldots, o_{T-1}$ of length T, the Baum–Welch method generates parameters π, a, b for which the model $M = (Q, \Sigma, \pi, a, b)$ has a (local) maximum likelihood.

In the updates for $\overline{b}_{i,k}$ in Algorithm 5.2, note that the sum is taken over those indices t, where the observation symbol o_t is σ_k.

It follows from the EM algorithm, about to be discussed, that $Pr[\mathcal{O}|\overline{M}] \geq Pr[\mathcal{O}|M]$ and, unless convergence has occurred, that the likelihood of model $\overline{M}$ is strictly greater than that of M.

5.2.2 EM and Justification of the Baum–Welch Method

In this section, we discuss the *expectation maximization* (EM) algorithm, and show how to derive the Baum–Welch parameter updates from an application of EM. The proof is not particularly difficult, but can certainly by skipped by anyone only interested in implementing HMMs.

Suppose that we have a *many-to-one* map from space $\mathcal{X}$ *onto* space $\mathcal{Y}$, where $\mathcal{X}$ consists of *complete* data, whereas $\mathcal{Y}$ consists of *incomplete* data. Incomplete data is what we can observe, while the complete data consists of observable together with hidden data. This is

Algorithm 5.2 Baum–Welch HMM method

```
initialize using uniform distribution
   for i = 1 to n
      π = 1/n
   for i=1 to n
      for j=1 to n
         a_{i,j} = 1/n
         b_{i,k} = 1/m

repeat
```

$$\overline{\pi}_i = \gamma_0(i)$$
$$\overline{a}_{i,j} = \frac{\sum_{t<T-1} \eta_t(i,j)}{\sum_{t<T-1} \gamma_t(i)}$$
$$\overline{b}_{i,k} = \frac{\sum_{t<T, o_t=\sigma_k} \gamma_t(i)}{\sum_{t<T} \gamma_t(i)}$$

until $L_{\mathcal{O}}(M)$ converges

clearly related to the situation in HMMs, where $y \in \mathcal{Y}$ might consist of an observation sequence $\mathcal{O}$ of length T (e.g. a protein from a particular protein class), which is incomplete data since the states of the Markov model are hidden, while those $x \in \mathcal{X}$ that are mapped to y are of the form $p, \mathcal{O}$, where $p \in Q^T$ is a sequence of states of length T. Letting M denote a model (i.e. fixed model parameters), suppose that $g(y|M)$ is a conditional *probability density* for the space $\mathcal{Y}$, that $f(x|M)$ is a conditional *probability density* for $\mathcal{X}$, and that

$$g(y|M) = \int_{x \in \mathcal{X}(y)} f(x|M)$$

where $\mathcal{X}(y)$ denotes the set of $x \in \mathcal{X}$ that are mapped to y. In the discrete case, we could consider *probabilities* $g(y|M), f(x|M)$ that satisfy $g(y|M) = \sum_{x \in \mathcal{X}(y)} f(x|M)$.[3] Define

$$\begin{aligned} Q(\widetilde{M}|M) &= E[\log f(x|\widetilde{M})|y, M] \\ &= \sum_{x \in \mathcal{X}(y)} Pr[x|y, M] \log f(x|\widetilde{M}). \end{aligned}$$

The EM algorithm (Algorithm 5.3) consists of two steps:

E-STEP: Compute $Q(\widetilde{M}|M)$.

M-STEP: Determine $\text{argmax}_{\widetilde{M}} Q(\widetilde{M}|M)$.

The proof of the following theorem will not be given.

[3] The application of EM to HMMs involves the continuous, non-discrete case, since the collection of all model parameters (π, a, b) is clearly not discrete.

Algorithm 5.3 EM algorithm

```
initialize model parameters M_0
repeat
    compute Q(M|M_t)
    M_{t+1} = argmax_M Q(M|M_t)
until converge
```

THEOREM 5.13 (DEMPSTER *et al.* [DLR77], WU [WU83])
Under reasonable conditions (see [Wu83]), $M_{t+1} \geq M_t$ *for all* t*, with limit*

$$\lim_{t \to \infty} \log L_{\mathcal{O}}(M_t) = \log L_{\mathcal{O}}(M^*)$$

for some model parameters M^**, for which the likelihood is a (local) maximum.*

Note that the *likelihoods* converge, rather than necessarily the model parameters.

Let us perform a computation for the case of HMMs. Assume that $M_t = (Q, \Sigma, \pi^{(t)}, a^{(t)}, b^{(t)})$ and recall that y is an observation sequence $\mathcal{O}$ of length T (incomplete data), while x consists of a path $p \in Q^T$ of states together with $\mathcal{O}$ (complete data). The probability $f(x|M) = Pr[p, \mathcal{O}|M]$, while $g(y|M) = \sum_{x \in \mathcal{X}(y)} f(x|M)$ equals $\sum_{p \in Q^T} Pr[p, \mathcal{O}|M]$. Let $A^{(t)}_{p,\mathcal{O}}$ denote the *constant* expression

$$\pi^{(t)}_{p(0)} \prod_{i<T-2} a^{(t)}_{p(i),p(i+1)} \prod_{i<T-1} b^{(t)}_{p(i),o_i}$$

for path $p \in Q^T$. Then

$$\begin{aligned}
M_{t+1} &= \operatorname*{argmax}_M Q(M|M_t) \\
&= \operatorname*{argmax}_M E[\log f(x|M)|y, M_t] \\
&= \operatorname*{argmax}_M \sum_{p \in Q^T} Pr[p|\mathcal{O}, M_t] \log Pr[p, \mathcal{O}|M] \\
&= \operatorname*{argmax}_M \sum_{p \in Q^T} \frac{Pr[p, \mathcal{O}|M_t]}{L_{\mathcal{O}}(M_t)} \log Pr[p, \mathcal{O}|M] \\
&= \frac{1}{L_{\mathcal{O}}(M_t)} \operatorname*{argmax}_M \sum_{p \in Q^T} Pr[p, \mathcal{O}|M_t] \log Pr[p, \mathcal{O}|M] \\
&= \frac{1}{L_{\mathcal{O}}(M_t)} \operatorname*{argmax}_M \sum_{p \in Q^T} A^{(t)}_{p,\mathcal{O}} \log Pr[p, \mathcal{O}|M].
\end{aligned}$$

Let $\pi_i, a_{i,j}, b_{i,k}$ be new *variables* for $1 \leq i, j \leq n$ and $1 \leq k \leq m$. Let $A_{p,\mathcal{O}}$ denote the *multivariate function*

$$\pi_{p(0)} \prod_{i<T-2} a_{p(i),p(i+1)} \prod_{i<T-1} b_{p(i),o_i}$$

for a given path $p \in Q^T$. It follows that

$$M_{t+1} = \frac{1}{L_{\mathcal{O}}(M_t)} \operatorname*{argmax}_M \sum_{p \in Q^T} A^{(t)}_{p,\mathcal{O}} \log A_{p,\mathcal{O}}.$$

Define the multivariate function

$$F(\pi_1, \ldots, \pi_n, a_{1,1}, \ldots, a_{n,n}, b_{1,1}, \ldots, b_{n,m}) = \frac{1}{L_{\mathcal{O}}(M_t)} \sum_{p \in Q^T} A^{(t)}_{p,\mathcal{O}} \log A_{p,\mathcal{O}}.$$

We wish to determine where F achieves a maximum, subject to the constraint $\sum_{i=1}^n \pi_i = 1$, and for each $1 \leq i \leq n$ the constraints $\sum_{j=1}^n a_{i,j} = 1$ and $\sum_{k=1}^m b_{i,k} = 1$. Using the method of Lagrange multipliers, define

$$G(\pi_1, \ldots, \pi_n, a_{1,1}, \ldots, a_{n,n}, b_{1,1}, \ldots, b_{n,m}, \lambda, \alpha_1, \ldots, \alpha_n, \beta_1, \ldots, \beta_n)$$

to be

$$F + \lambda \left(\sum_{i=1}^n \pi_i - 1 \right) + \sum_{i=1}^n \alpha_i \left(\sum_{j=1}^n a_{i,j} - 1 \right) + \sum_{i=1}^n \beta_i \left(\sum_{k=1}^m b_{i,k} - 1 \right).$$

Computing the partial derivatives, we find

$$\frac{\partial G}{\partial \pi_i} = \frac{1}{L_{\mathcal{O}}(M_t)} \sum_{p \in Q^T, p(0)=i} \frac{A^{(t)}_{p,\mathcal{O}}}{\pi_i} + \lambda,$$

and setting $\frac{\partial G}{\partial \pi_i} = 0$, we obtain $\pi_i = -\frac{1}{L_{\mathcal{O}}(M_t)} \sum_{p \in Q^T, p(0)=i} \frac{A^{(t)}_{p,\mathcal{O}}}{\lambda}$. From the latter together with the constraint $\sum_{i=1}^n \pi_i = 1$, we obtain

$$\lambda = -\frac{1}{L_{\mathcal{O}}(M_t)} \sum_{i=1}^n \sum_{p \in Q^T, p(0)=i} A^{(t)}_{p,\mathcal{O}} = \frac{\sum_{p \in Q^T} A^{(t)}_{p,\mathcal{O}}}{L_{\mathcal{O}}(M_t)} = -\frac{L_{\mathcal{O}}(M_t)}{L_{\mathcal{O}}(M_t)} = -1.$$

It follows that

$$\pi_i = \frac{1}{L_{\mathcal{O}}(M_t)} \sum_{p \in Q^T, p(0)=i} A^{(t)}_{p,\mathcal{O}} = \frac{1}{L_{\mathcal{O}}(M_t)} \sum_{p \in Q^T, p(0)=i} Pr[p, \mathcal{O} | M_t].$$

By the same reasoning, we obtain justifications for the other parameter re-estimations from the Baum–Welch algorithm.

5.2.3 Baldi–Chauvin Gradient Descent

Following [BC94], introduce a variable transformation by defining new variables $\omega_{i,j}$ and $\nu_{i,c}$ so that the following Boltzmann-like distribution holds:

$$a_{i,j} = \frac{e^{\lambda \omega_{i,j}}}{\sum_k e^{\lambda \omega_{i,k}}}, \tag{5.1}$$

$$b_{i,c} = \frac{e^{\lambda \nu_{i,c}}}{\sum_k e^{\lambda \nu_{i,k}}}. \tag{5.2}$$

For a finite training set $\mathcal{O}^1, \ldots, \mathcal{O}^r$, of observation sequences indexed by s, where the length L_s sequence $\mathcal{O}^s = o_0^s, \ldots, o_{L_s-1}^s$, define the increments

$$\begin{aligned} \omega_{i,j}^{new} &= \omega_{i,j}^{old} + \Delta\omega_{i,j}, \\ \nu_{i,c}^{new} &= \nu_{i,c}^{old} + \Delta\nu_{i,c}, \end{aligned}$$

where

$$\begin{aligned} \Delta\omega_{i,j} &= C_a \sum_s \sum_{t<L_s} [\eta_t(i,j) - a_{i,j}\gamma_t(i)], \\ \Delta\nu_{i,c} &= C_b \sum_s \sum_{t<L_s, o_t^s=c} [\gamma_t(i) - b_{i,c}\gamma_t(i)], \end{aligned}$$

and C_a, C_b are appropriate constants.

Unlike the case with the Baum–Welch algorithm, updates can be performed *on-line*, in the sense that the increments $\omega_{i,j}^{\text{new}}$, $\nu_{i,c}^{\text{new}}$ can be computed each time, after reading a new observation sequence $\mathcal{O}$ (i.e. in the sth step, $\mathcal{O} = \mathcal{O}^s$). In the case of on-line parameter re-estimation, the updates are given by

$$\begin{aligned} \Delta\omega_{i,j} &= C_a \frac{L(M)}{L_{\mathcal{O}}(M)} \sum_{t<L_s} [\eta_t(i,j) - a_{i,j}\gamma_t(i)] \\ &= C_a \frac{L(M)}{L_{\mathcal{O}}(M)} [\eta_{\mathcal{O}}(i,j) - a_{i,j}\eta_{\mathcal{O}}(i)] \\ \Delta\nu_{i,c} &= C_b \frac{L(M)}{L_{\mathcal{O}}(M)} \sum_{t<L_s, c=o^s} [\gamma_t(i) - b_{i,c}\gamma_t(i)], \end{aligned}$$

where we let $\eta_{\mathcal{O}}(i)$ denote $\sum_{k\in Q} \eta_{\mathcal{O}}(i,k)$.

A disadvantage of the Baum–Welch method is that in the parameter updates, if any probability is set to 0, then this probability will never later be changed, i.e. 0 is *absorbing*. In the training set of observation sequences, it certainly can happen that certain transitions or emissions have probability 0, yet such transitions or emissions might indeed appear in test sequences, which should test positively. This disadvantage is not present in the Baldi–Chauvin method, since the transition and emission probabilities are defined in a Boltzmann-like manner. Another advantage of the Baldi–Chauvin method is that parameter updates can be made on-line. A disadvantage of this method lies in the choice of constants C_a, C_b, corresponding to step size in the gradient descent method. Gradient descent, unlike the EM algorithm, is a heuristic, which may not converge, depending on the constants.

Justification of Baldi–Chauvin Updates

Let $M = (\Sigma, Q, \pi, a, b)$ be a given HMM having output alphabet Σ, state set $Q = \{1, \ldots, n\}$, state transtition probability matrix a, emission probability matrix b, and initial state probability vector π. Given a sequence of distinct observation sequences $\mathcal{O}_1, \ldots, \mathcal{O}_r$ of possibly varying lengths, where $\mathcal{O}_s = o_0, \ldots, o_{L_{\mathcal{O}_s}-1}$, the likelihood of model M is the product

$$L(M) = \prod_{s=1}^{r} L_{\mathcal{O}_s}(M)$$

of the likelihoods that M generated each individual observation sequence. Define the *energy* $E = -\log L(M)$. Our goal is to find model parameters π, a, b for which energy is a minimum, by using the gradient descent method, an iterative, greedy algorithm, where at each step we move in the direction of steepest descent on the likelihood landscape. If $|Q| = n$ and $|\Sigma| = m$, then thinking of E as a multivariate function of the independent variables $\pi_1, \ldots, \pi_n, a_{1,1}, \ldots, a_{n,n}, b_{1,1}, \ldots, b_{n,m}$ subject to constraints $\sum_{i=1}^{n} \pi_i = 1$, and for $i \in Q$, $\sum_{j=1}^{n} a_{i,j} = 1$, $\sum_{k=1}^{m} b_{i,k} = 1$, the gradient descent method yields increments

$$\begin{aligned} \Delta\pi_i &= -C\frac{\partial E}{\partial \pi_i}, \\ \Delta a_{i,j} &= -C\frac{\partial E}{\partial a_{i,j}}, \\ \Delta b_{i,k} &= -C\frac{\partial E}{\partial b_{i,k}}. \end{aligned}$$

Instead of considering E as a function of model parameters π, a, b, following [BC94] we consider E as a function of π, ω, ν, having made the change of variables as given in equations (5.1) and (5.2). We are thus led to compute the partial derivatives

$$\begin{aligned} \frac{\partial E}{\partial \omega_{i,j}} &= -\frac{1}{L(M)}\frac{\partial L(M)}{\partial \omega_{i,j}}, \\ \frac{\partial E}{\partial \nu_{i,k}} &= -\frac{1}{L(M)}\frac{\partial L(M)}{\partial \nu_{i,k}}. \end{aligned}$$

Concentrating now only on the equation for $\frac{\partial E}{\partial \omega_{i,j}}$, by the chain rule, we have

$$\frac{\partial\, L(M)}{\partial \omega_{i,j}} = \sum_{k=1}^{n} \frac{\partial\, L(M)}{\partial a_{i,k}} \frac{\partial a_{i,k}}{\partial \omega_{i,j}}. \tag{5.3}$$

By the product rule for differentiation, we have

$$\frac{\partial L(M)}{\partial a_{i,k}} = L(M) \sum_{\mathcal{O}} \frac{1}{L_{\mathcal{O}}(M)} \frac{\partial L_{\mathcal{O}}(M)}{\partial a_{i,k}},$$

and, recalling the previously defined *weight* $A_{p,\mathcal{O}}$ of path p with respect to observation sequence $\mathcal{O}$ and model M,

$$A_{p,\mathcal{O}} = \pi_{p(0)} \prod_{t<\ell(\mathcal{O})-1} a_{p(t),p(t+1)} \prod_{t<\ell(\mathcal{O})} b_{p(t),o_t}$$

we have

$$\begin{aligned} \frac{\partial L_{\mathcal{O}}(M)}{\partial a_{i,k}} &= \sum_{p \in Q^{\ell(\mathcal{O})}} \frac{\partial A_{p,\mathcal{O}}}{\partial a_{i,k}} \\ &= \sum_{p \in Q^{\ell(\mathcal{O})}} \frac{c_p(i,k) A_{p,\mathcal{O}}}{a_{i,k}} \\ &= \frac{\eta_{\mathcal{O}}(i,k)}{a_{i,k}} L_{\mathcal{O}}(M). \end{aligned}$$

Here, we use the previously defined notation, where

$$\begin{aligned}\eta_{\mathcal{O}}(i,k) &= \sum_{t<\ell(\mathcal{O})-1} Pr[q_t = i, q_{t+1} = k | \mathcal{O}, M] \\ &= \sum_{p\in Q^{\ell(\mathcal{O})}} \frac{c_p(i,k)A_{p,\mathcal{O}}}{L_{\mathcal{O}}(M)}\end{aligned}$$

and $c_p(i,k)$ is the number of transitions from i to k made in the path p, i.e.

$$c_p(i,k) = |\{t < \ell_{\mathcal{O}} - 1 \mid p(t) = i, p(t+1) = k\}|.$$

Thus

$$\begin{aligned}\frac{\partial L(M)}{\partial a_{i,k}} &= L(M) \sum_{\mathcal{O}} \frac{1}{L_{\mathcal{O}}(M)} \frac{\eta_{\mathcal{O}}(i,k)}{a_{i,k}} L_{\mathcal{O}}(M) \\ &= \frac{L(M)}{a_{i,k}} \sum_{\mathcal{O}} \eta_{\mathcal{O}}(i,k).\end{aligned}$$

We now compute $\frac{\partial a_{i,k}}{\partial \omega_{i,j}}$ by distinguishing whether or not $k = j$.

CASE 1: $k = j$

$$\begin{aligned}\frac{\partial a_{i,j}}{\partial \omega_{i,j}} &= \frac{\partial}{\partial \omega_{i,j}} \left(\frac{e^{\lambda\omega_{i,j}}}{\sum_\ell e^{\lambda\omega_{i,\ell}}} \right) \\ &= \frac{\lambda e^{\lambda\omega_{i,j}}}{\sum_\ell e^{\lambda\omega_{i,\ell}}} - \frac{\lambda e^{\lambda\omega_{i,j}} e^{\lambda\omega_{i,j}}}{(\sum_\ell e^{\lambda\omega_{i,\ell}})^2} \\ &= \left(\frac{\lambda e^{\lambda\omega_{i,j}}}{\sum_\ell e^{\lambda\omega_{i,\ell}}} \right) \left[1 - \frac{e^{\lambda\omega_{i,j}}}{\sum_\ell e^{\lambda\omega_{i,\ell}}} \right] \\ &= \lambda a_{i,j}(1 - a_{i,j}).\end{aligned}$$

CASE 2: $k \neq j$

$$\begin{aligned}\frac{\partial a_{i,k}}{\partial \omega_{i,j}} &= \frac{\partial}{\partial \omega_{i,j}} \left(\frac{e^{\lambda\omega_{i,k}}}{\sum_\ell e^{\lambda\omega_{i,\ell}}} \right) \\ &= \frac{-\lambda e^{\lambda\omega_{i,j}} e^{\lambda\omega_{i,k}}}{(\sum_\ell e^{\lambda\omega_{i,\ell}})^2} \\ &= -\lambda a_{i,j} a_{i,k}.\end{aligned}$$

Thus

$$\begin{aligned}
\frac{\partial E}{\partial \omega_{i,j}} &= -\frac{L(M)}{L(M)}\left\{\left[\sum_{\mathcal{O}} \eta_{\mathcal{O}}(i,j)\right]\left[\frac{\lambda a_{i,j}(1-a_{i,j})}{a_{i,j}}\right] + \left[\sum_{k\neq j}\sum_{\mathcal{O}} \eta_{\mathcal{O}}(i,k)\right]\left[\frac{-\lambda a_{i,j} a_{i,k}}{a_{i,k}}\right]\right\} \\
&= -\lambda(1-a_{i,j})\sum_{\mathcal{O}} \eta_{\mathcal{O}}(i,j) + \lambda \sum_{k\neq j}\sum_{\mathcal{O}} \eta_{\mathcal{O}}(i,k) a_{i,j} \\
&= -\lambda \sum_{\mathcal{O}} \eta_{\mathcal{O}}(i,j) + \lambda \sum_{k\in Q}\sum_{\mathcal{O}} a_{i,j}\eta_{\mathcal{O}}(i,k) \\
&= -\lambda \sum_{\mathcal{O}} \eta_{\mathcal{O}}(i,j) + \lambda \sum_{\mathcal{O}} a_{i,j}\eta_{\mathcal{O}}(i),
\end{aligned}$$

where we let $\eta_{\mathcal{O}}(i)$ abbreviate $\sum_{k\in Q} \eta_{\mathcal{O}}(i,k)$. Finally, using the heuristic of gradient descent,

$$\begin{aligned}
\Delta\omega_{i,j} &= -C\frac{\partial E}{\partial \omega_{i,j}} \\
&= -C\lambda \sum_{\mathcal{O}} \eta_{\mathcal{O}}(i,j) - C\lambda \sum_{\mathcal{O}} a_{i,j}\eta_i(\mathcal{O}),
\end{aligned}$$

and, with a similar derivation,

$$\begin{aligned}
\Delta\nu_{i,k} &= -D\frac{\partial E}{\partial \nu_{i,k}} \\
&= -D\lambda \sum_{\mathcal{O}} \sum_{t<\ell(\mathcal{O}),o_t=k} [\gamma_t(i) - b_{i,k}\gamma_t(i)].
\end{aligned}$$

This concludes the justification of the on-line updates; a similar argument justifies the off-line updates.

5.2.4 Mamitsuka's MA Algorithm

The likelihood of sth sequence with respect to a hidden Markov model M is

$$p_s = Pr[\mathcal{O}^s|M].$$

Let p_s^* be the *target value* likelihood of the sth sequence. Define

$$d_s = \log\left(\frac{p_s^*}{p_s}\right)$$

and

$$d_{\max} = \log\left(\frac{p_{\max}^*}{p_{\min}^*}\right)$$

where $p_{\max}^*$ (resp. $p_{\min}^*$) is the maximum (resp. minimum) of the p_s^*.

Goal

After training, d_s should be 0, so our plan is to minimize the *error distance*

$$\sum_s -\log\left(\frac{d_{\max}^2 - d_s^2}{d_{\max}^2}\right)$$

in parameter re-estimation. To this end, for a finite training set of sequences indexed by s, define

$$\begin{aligned}\Delta\omega_{i,j} &= C_a \sum_s \frac{d_s}{(d_{\max}^2 - d_s^2)} \sum_{t<\ell_s} [\eta_t(i,j) - a_{i,j}\gamma_t(i)],\\ \Delta\nu_{i,c} &= C_b \sum_s \frac{d_s}{(d_{\max}^2 - d_s^2)} \sum_{t<\ell_s, o_t^s=c} [\gamma_t(i) - b_{i,c}\gamma_t(i)],\end{aligned}$$

with the on-line increments for parameter re-estimation given by

$$\begin{aligned}\Delta\omega_{i,j} &= C_a \sum_{t<\ell_s} \frac{d_s}{(d_{\max}^2 - d_s^2)} [\eta_t(i,j) - a_{i,j}\gamma_t(i)],\\ \Delta\nu_{i,c} &= C_b \frac{d_s}{(d_{\max}^2 - d_s^2)} \sum_{t<\ell_s, o_t^s=c} [\gamma_t(i) - b_{i,c}\gamma_t(i)],\end{aligned}$$

where the sequence $\mathcal{O}_s$ has just been read. In the sugar transport protein example given in the next section, Mamitsuka sets

$$p_s^* = \left(\frac{1}{20}\right)^{0.01\ell_s} \approx (0.970487)^{\ell_s}$$

for positive training examples and

$$p_s^* = \left(\frac{1}{20}\right)^{1.99\ell_s} \approx (0.002576)^{\ell_s}$$

for negative training examples, where ℓ_s is the length of the sth sequence.

Justification of Mamitsuko's Updates

Let $L_{\mathcal{O}}^*$ be the *target* likelihood for the observation sequence $\mathcal{O}$, and define

$$\begin{aligned}d_{\mathcal{O}} &= \log\left[\frac{L_{\mathcal{O}}^*}{L_{\mathcal{O}}(M)}\right],\\ g_{\mathcal{O}} &= 1 - \left(\frac{d_{\mathcal{O}}}{d_{\max}}\right)^2\\ &= \frac{d_{\max}^2 - d_{\mathcal{O}}^2}{d_{\max}^2},\end{aligned}$$

where

$$\begin{aligned}d_{\max} &= \log\left(\frac{L_{\max}^*}{L_{\min}^*}\right)\\ &= \log\left(\max\{L_{\mathcal{O}}^* \mid \mathcal{O}\}\right) - \log\left(\min\{L_{\mathcal{O}}^* \mid \mathcal{O}\}\right).\end{aligned}$$

Define the *energy* E by

$$
\begin{aligned}
E &= -\log\left(\prod_{\mathcal{O}} g_{\mathcal{O}}\right) \\
&= \sum_{\mathcal{O}}(-\log g_{\mathcal{O}}).
\end{aligned}
$$

Then

$$
\begin{aligned}
\Delta\omega_{i,j} &= C_a \frac{\partial E}{\partial \omega_{i,j}} \\
&= -C_a \frac{\partial}{\partial \omega_{i,j}}\left(\sum_{\mathcal{O}} -\log g_{\mathcal{O}}\right) \\
&= C_a \sum_{\mathcal{O}} \sum_{k} \frac{\partial \log g_{\mathcal{O}}}{\partial a_{i,k}} \frac{\partial a_{i,k}}{\partial \omega_{i,j}} \\
&= C_a \sum_{\mathcal{O}} \sum_{k} \left(\frac{d_{\max}^2}{d_{\max}^2 - d_{\mathcal{O}}^2}\right)\left(\frac{-2d_{\mathcal{O}}}{d_{\max}^2}\right)\left(\frac{\partial d_{\mathcal{O}}}{\partial a_{i,k}} \frac{\partial a_{i,k}}{\partial \omega_{i,j}}\right) \\
&= C_a \sum_{\mathcal{O}} \sum_{k} \left(\frac{2d_{\mathcal{O}}}{d_{\max}^2 - d_{\mathcal{O}}^2}\right)\left(\frac{1}{L_{\mathcal{O}}(M)}\right)\left(\frac{\partial L_{\mathcal{O}}(M)}{\partial a_{i,k}} \frac{\partial a_{i,k}}{\partial \omega_{i,j}}\right) \\
&= C_a \sum_{\mathcal{O}} \left\{\left(\frac{2d_{\mathcal{O}}}{d_{\max}^2 - d_{\mathcal{O}}^2}\right) \sum_{k} \left[\frac{1}{L_{\mathcal{O}}(M)}\left(\frac{\eta_{\mathcal{O}}(i,k)}{a_{i,k}} L_{\mathcal{O}}(M) \frac{\partial a_{i,k}}{\partial \omega_{i,j}}\right)\right]\right\} \\
&= C_a \sum_{\mathcal{O}} \left\{\left(\frac{2d_{\mathcal{O}}}{d_{\max}^2 - d_{\mathcal{O}}^2}\right) \sum_{k} \left[\frac{\eta_{\mathcal{O}}(i,k)}{a_{i,k}} \frac{\partial a_{i,k}}{\partial \omega_{i,j}}\right]\right\} \\
&= \lambda C_a \sum_{\mathcal{O}} \left(\frac{2d_{\mathcal{O}}}{d_{\max}^2 - d_{\mathcal{O}}^2}\right)\left[\eta_{\mathcal{O}}(i,k) - a_{i,j}\eta_{\mathcal{O}}(i)\right].
\end{aligned}
$$

A similar argument justifies the expression for the increment $\Delta\nu_{i,k}$.

5.3 Applications

5.3.1 Multiple Sequence Alignment

Hidden Markov models have found applications in multiple sequence alignment. The idea is to build a linear hidden Markov model having $3n$ states, where n is the average sequence length in the training set. The training set of sequences to be aligned is treated as a collection of observation sequences. There are n matching states along the backbone, and for each such matching state an additional insertion and deletion state, so that variable-length sequences can be accommodated. Once a hidden Markov model is trained, each sequence from the training set can be scored using the Viterbi algorithm, which then gives rise to an appropriate path of matching, insertion, and deletion states. Thus the sequence is aligned against the stochastic model.

In Figure 5.3 a linear hidden Markov model is depicted for the multiple sequence alignment of sequences whose average sequence length is 4. There are 6 matching states, of which m_0

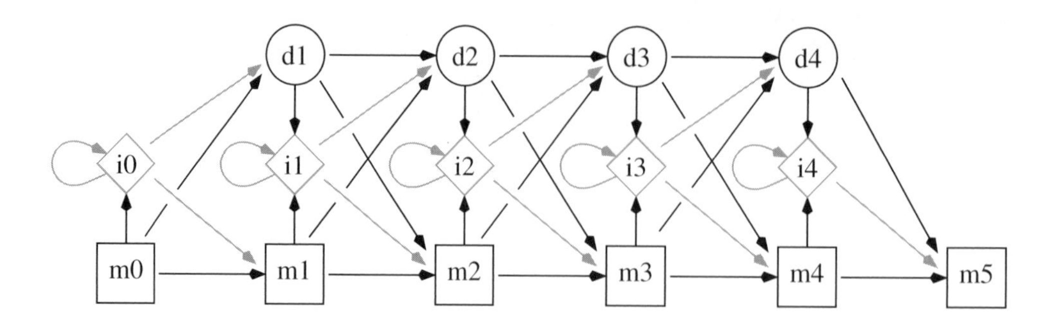

Figure 5.3 Linear HMM for multiple sequence alignment. Adapted from [DEKM98] and [Wat95] with permissions from Cambridge University Press and CRC Press, Boca Raton, Florida

and m_5 are respectively the beginning and end states, 5 insertion states, and 4 delete states. The delete states have no emissions, the insertion states $i_0, \ldots, i_4$ have emissions, m_0 and m_5 have no emissions, while the backbone matching states $m_1, \ldots, m_4$ have emissions. Suppose that we wish to align the sequences GGCT, ACCGAT, and CT. After convergence of the Baum–Welch algorithm, suppose that state transition probabilities and emission probabilities have been computed, and that the Viterbi path for GGCT is $m_0, m_1, m_2, m_3, m_4, m_5$, for ACCGAT it is $m_0, i_0, m_1, d_2, m_3, i_3, i_3, m_4, m_5$, and for CT it is $m_0, m_1, d_2, d_3, m_4, m_5$. Then we obtain the following multiple sequence alignment:

```
a C - C g a T
. G G C . . T
. C - - . . T
```

Multiple sequence alignment software using HMMs has been developed independently by P. Baldi, Y. Chauvin and V. Mittal-Henkle (HMMpro at www.netid.com), S. Eddy (HMMER www.genetics.wustl.edu/ eddy), and K. Karplus *et al.* (SAM www.cse.ucsc.edu/research/compbio/sam.html). Since the application of HMMs to multiple sequence alignment is already well-described in the literature, especially in the monographs [BB98, DEKM98] and in the original papers [KBM$^+$94, Edd95], the topic will not be further discussed here.

5.3.2 Protein Motifs

In [Mam96, Mam97] H. Mamitsuka studied the problem of correctly identifying sugar transport proteins (STP), whose consensus in the database PROSITE is the 12–14 amino acid sequence

$$[LIVMSTA] - [DE] - x - [LIVMFYWA] - G - R - [RK] - x(4,6) - G$$

In the SWISS-PROT release 29.0 database, 49 sugar transport proteins have the previous motif, while 19 non-sugar transport proteins also have this motif (i.e are false positives). Mamitsuka compared the training times and error distributions for four types of hidden Markov models as applied to the prediction problem for STPs, where each model has a different algorithm for parameter re-estimation (Baum–Welch, Baldi–Chauvin, and Mamitsuka's modification of

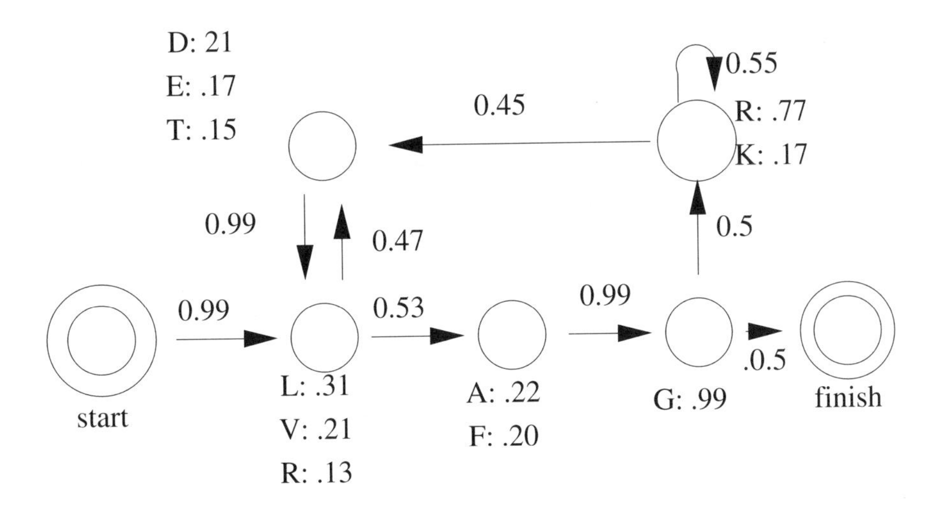

Figure 5.4 Mamitsuka's HMM for sugar transport proteins. Reprinted from [Mam96]. Copyright Mary Ann Liebert, Inc, New York.

each that allowed positive and negative examples to be used in the training set). As earlier mentioned, in applying Mamitsuka's algorithm,

$$p_s^* = \left(\frac{1}{20}\right)^{0.01\ell_s}$$

for positive examples and

$$p_s^* = \left(\frac{1}{20}\right)^{1.99\ell_s}$$

for negative examples. One of the resulting HMMs from [Mam96, Mam97] after training on a randomly selected subset of STPs and non-STPs both having the motif is given in Figure 5.4.

5.3.3 *Eukaryotic DNA Promotor Regions*

Another application was given by U. Ohler [Ohl95], who trained HMMs to recognize promotor sites in eukaryotic DNA. Typically, a promotor contains a binding site for RNA polymerase, the so-called TATA box,[4] as well as binding sites for proteins that control the binding of RNA polymerase, though genes for housekeeping proteins (constantly expressed) have no corresponding TATA box. The promotor region is depicted in Figure 5.5, which is taken from [Ohl95].

Here, one notes the following regions:

- CAP signal: about 8 bp

$$\text{(T)CA(G)T(C)(T)(T)}$$

[4] TATAAT is the consensus sequence for *E. coli*, while TATAT is that for yeast. See Bucher [Buc90] for more information on TATA box structure.

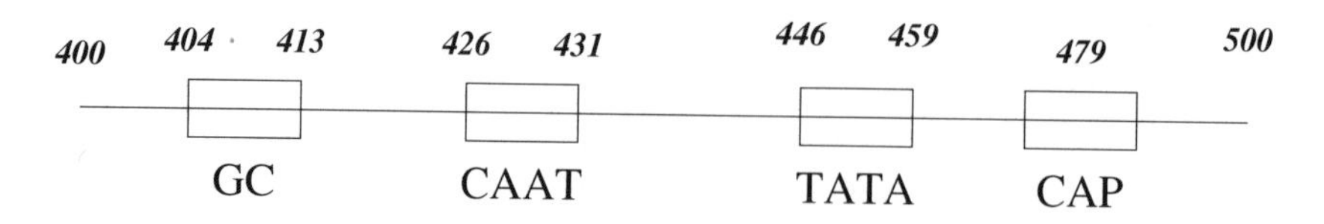

Figure 5.5 Promotor sequence in eukaryotic DNA.

- TATA box: usually between 36 and 20 bp before transcription, length about 15 bp, most significant 8 bp consensus as follows:

$$\text{[G/C]TATA[A/T]A[A/T]}$$

- GC element: 14 bp, no fixed position or orientation, possibly many copies (0 to 5)

$$\text{GGGCGG}$$

- CAAT element: 12 bp, between 120 and 60 bp before transcription

$$\text{[A/G][A/G]CCAAT}$$

In [Ohl95], Ohler defines a regular grammar can be given for (the reversal of) a promotor. With this approach, HMM modules can be defined for the different terminals of the following grammar:

promotor	$\rightarrow$	downstream upstream
downstream	$\rightarrow$	CAP region TATA region
CAP region	$\rightarrow$	CAP signal | CAP waste
TATA region	$\rightarrow$	TATA spacer TATA box | nonTATA spacer TATA waste
upstream	$\rightarrow$	GC element upstream | CAT element upstream | GCrev element upstream | CAATrev element upstream | spacer upstream | ϵ

A large HMM can be connected from HMM modules by serial, parallel and recursive connections, provided one can estimate entrance probabilities for the modules. Let M_1, M_2 be HMMs with initial state probability function π_1, π_2, state transition matrices a_1, a_2, and output symbol probability matrices b_1, b_2. Assume that M_1 has states $1, \ldots, n_1$, while M_2 has states $1, \ldots, n_2$. Assume that M_1 has the designated termination state n_1.

Serial connection Define the serial connection $M = M_1 \& M_2$ by first setting the state set to $\{1, \ldots, n_1 + n_2\}$ and

$$\pi = \begin{pmatrix} \pi_1 \\ 0 \\ \vdots \\ 0 \end{pmatrix},$$

$$a = \begin{pmatrix} A_1 & 0 \\ 0 & A_2 \end{pmatrix},$$

$$b = \begin{pmatrix} B_1 \\ B_2 \end{pmatrix},$$

where A_1 (resp. A_2) is the transition probability matrix for M_1 (resp. M_2), and similarly B_1 (resp. B_2) is the emission probability matrix for M_1 (resp. M_2).

In a similar, but more complicated manner, a parallel $M_1|M_2$ and recursive connection of HMMs is defined. With the resulting connections of HMMs, Ohler successfully trained and tested his program on promotor sequences for eukaryotic DNA, and has used improved versions of his program for promotor sequence prediction on *Drosophilia* in the Berkeley Fly Project (personal communication).

Hidden Markov models are trained to predict whether a protein is likely to belong to a certain class of proteins (as in sugar transport proteins), where a promotor region lies in DNA, and trained Markov models have been used to recognize open reading frames [BM93], etc. An important topic not covered in this text concerns neural networks – see the excellent monograph by P. Baldi and S. Brunak [BB98] for applications of neural networks, hybrid models (HMMs and neural nets), etc. in computational biology.

Almost invariably, mistakes will be made in any machine learning procedure, and one speaks of the *specificity* and *sensitivity* of a method. Taking the example of recognizing intron/exon splice sites, 100% specificity means that only true splice sites are recognized (no *false positives*, i.e. if the method returns YES, then the site is a true splice site), while 100% sensitivity means that *all* splice sites are recognized (no *false negatives*, i.e. if the method returns NO, then the site is not a true splice site[5]).

5.4 Exercises

1. Use the SAM (Sequence Alignment Modeling) software from Santa Cruz to perform a multiple alignment of several tRNAs from *M. jannaschi*.
2. Write a hidden Markov model to find the emission, state transition and initial state probability matrices for the following five sequences of observables. The characters emitted are A, G, C, T, and the observed data consists of the following 5 observation sequences:

   ```
   AGAAAGGTCTAGTGTTTGGTGATGTATCTATAGAGGGACG
   GGTCCTTTCAATATCAGTTGAATATGATGTGAGTGAGTTG
   GGGGGGTGGGGCCTTGATAAGAAGGGCTGTCTTTTGGTAG
   GTACCGGTATAGAAAAGACCGGATTCGAATTAATAATAAG
   TATTACTTGTTCAGCGTTATAAGATTCAGGAGGAGGTGTG
   ```

   To check the output of your program, you should know that the above obervables were generated from the initial state probability matrix

$$\pi = \begin{pmatrix} 0.5 \\ 0.25 \\ 0.25 \end{pmatrix},$$

[5] This assumes the method always returns an answer. *Las Vegas* procedures always return a correct answer, but in certain cases do not return any answer.

the state transition matrix

$$a = \begin{pmatrix} 0.5 & 0.25 & 0.25 \\ 0.25 & 0.35 & 0.40 \\ 0.25 & 0.75 & 0.0 \end{pmatrix},$$

and the emission matrix

$$b = \begin{pmatrix} 0.25 & 0.25 & 0.25 & 0.25 \\ 0.25 & 0.35 & 0.10 & 0.30 \\ 0.25 & 0.25 & 0.0 & 0.50 \end{pmatrix}.$$

Your program should implement the Baum–Welch parameter estimation, and in each re-estimation should print out the model likelihood, so that you see that the likelihood converges to a maximum.

3. Modify your program from the previous problem, by implementing the Baldi–Chauvin method for re-parameterization. Compare accuracy and runtimes for both methods on the data from the previous problem.
4. Prove that if a set $C \subset \Omega$ is partitioned into $\{C_1, \ldots, C_r\}$, where $C_i \cap C_j = \emptyset$ for $i \neq j$, then $Pr[A|C] = \sum_{i=1}^{r}(Pr[A|C_i]\, Pr[C_i|C])$. Use this to give an alternative justification for the recurrence relation $\beta_t(i) = \sum_{j \in Q} \beta_{t+1}(j) a_{i,j} b_{j,o_{t+1}}$.
5. Prove that

$$\begin{aligned} \eta_t(i,j) &= Pr[q_t = i, q_{t+1} = j | \mathcal{O}, M] \\ &= \frac{\alpha_t(i) a_{i,j} b_{j,o_{t+1}} \beta_{t+1}(j)}{Pr[\mathcal{O}|M]}. \end{aligned}$$

6. A context-free *Lindenmayer system* or L system, is specified by a context-free grammar $G = (V, \Sigma, R, S)$, with the unusual requirement that when a grammar rule such as $A \to \alpha$ is applied to a sentential expression $w \in (V \cup \Sigma)$, then *all* occurrences of A must be replaced by α. In contrast, words in a context-free language are generated by replacing (non-deterministically) *one* occurrence of A by α. L systems were originally conceived as a mathematical model to explain plant leaf and stem formation, and have found applications in graphical representations of plants. Stochastic L systems are specified by context-free grammars along with probabilities for rule application. Using ideas from the algorithms for inferring probabilities for stochastic context-free grammars [BB98, DEKM98], describe and implement an algorithm to infer probabilities for stochastic context-free L systems. HINT Extend an efficient parsing algorithm for context-free languages (such as the CYK algorithm or Early parser) to L systems, and generalize the forward–backward algorithms. See [JL87] for more on L systems.

Acknowledgments and References

The single, most lucid introduction to hidden Markov models is [RJ86], from which our exposition is largely influenced. See also [Rab89, KSHG96], and the software SAM (Sequence Alignment and Modeling) from the Santa Cruz group. In [BC94], P. Baldi and Y. Chauvin give a *smooth* algorithm for re-estimation of parameters, which has several advantages over the Baum–Welch method. Using this approach, P. Baldi *et al.* [BCHM94] and A. Krogh *et al.*

[KBM$^+$94] present hidden Markov models for classes of proteins (globins, immunoglobulins, kinases, etc.). In [Mam97] H. Mamitsuka introduced a modification of the Baum–Welch and Baldi–Chauvin methods, which allowed one to train on both positive and negative problem instances. Ohler's [Ohl95] thesis describes a nice application of modularly built HMMs to the problem of DNA promotor detection.

6

Structure Prediction

> In fact, being able to predict a protein's structure from its amino acid sequence is one of the most important unsolved problems of molecular biology and biophysics. Not only would a successful prediction algorithm be a tremendous advance in the understanding of the biochemical mechanisms of proteins, but, since such an algorithm could conceivably be used to *design* proteins to carry out specific functions, it would have profound, far-reaching effects on biotechnology and the treatment of disease. (M. Mitchell, *An Introduction to Genetic Algorithms* 1998 [Mit98])

The *protein structure prediction* problem, also called the *protein folding* problem, is one of the major unsolved problems in computational biology. The problem consists of predicting the *tertiary structure* (in the following also called *conformation*) from the given amino acid sequence. There are three general approaches for attacking protein folding: namely molecular dynamics, protein structure prediction, and homologous modeling. Protein threading can be interpreted as a variant of homologous modeling.

In molecular dynamics, one simulates the actual folding process of a protein, considering mean force fields acting on all atoms in the constituent amino acids, as well as atoms of the solvent (water). Starting with a random initial conformation, one calculates motion vectors for the atoms according to different forces (covalent bonds, electrostatic forces, van der Waals forces, hydrogen bonds between residue atoms and water molecules, etc.). The equations used in calculating the motion vectors are only valid over a short time interval, typically on the order of 10^{-15} seconds. This simulation step is then iterated until a stable conformation is found. Computing motion vectors, while taking into account all atoms of the protein as well as those of the solvent, is itself a time-consuming step. Since proteins typically fold in milliseconds to seconds, this results in a huge number of sampling steps over a large number of atoms before convergence can occur. For this reason, with current computing resources, molecular dynamics cannot be used for *de novo* protein structure prediction.

In the protein structure prediction approach, one considers a fixed energy model $E : \Omega \rightarrow \mathbb{R}$, where Ω is the set of all possible conformations. The native structure is defined to be the structure that has minimal free energy according to the energy model (i.e., that conformation ω where $E(\omega)$ is minimal). There are different approaches in the definition of energy function. Several results strongly indicate that the structure prediction problem is NP-hard (see e.g. [NM92, NMK94]). For this reason, simplified models have been introduced, and used in investigating general properties of protein folding. Our discussion of protein structure prediction is restricted to lattice models.

A third approach is *protein threading*. The idea behind this *knowledge-based* approach is that, since *de novo* protein structure prediction using physical/chemical energy functions appears currently intractable, one instead computes statistical pseudo-energy functions from frequencies of certain amino acids known to lie in proximity to others in conformations of a representative sampling of the protein database. Protein threading is the attempt to align in parallel both the sequence and the structure, by comparing a new protein P with a known protein Q (i.e., one where both the sequence and the structure are known), assuming that P and Q are related, as tested, for example, by using sequence alignment. Though protein threading was originally thought to be a simpler model, following [LS96, AM97] we will show this problem to be NP-complete.

Structure prediction is important not only for proteins, but also for RNA and DNA. RNA molecules perform certain catalytic functions, which are the consequence of their specific three-dimensional structure. This structure is encoded in the nucleotide sequence of the RNA molecule. While RNA tertiary structure prediction appears as difficult as protein structure prediction, there are dynamic programming algorithms for the simpler problem of RNA secondary structure prediction. Even the three-dimensional structure of DNA appears to be important for the functionality of the cell. In particular, the twist and writhe of a DNA molecule dictate in part the site where strand separation is initiated in replication and transcription events. In this chapter, we consider these various aspects of structure prediction.

6.1 RNA Secondary Structure

As mentioned in Chapter 1, ribonucleic acid has hydroxyl groups connected to the $2'$ and $3'$ carbon in the pentose sugar, allowing for more hydrogen bond formation in single-stranded RNA than DNA. From extensive hydrogen bonding, single-stranded RNA is capable of forming complicated three-dimensional structures, capable of certain catalytic functions. *Primary structure* of RNA is the nucleotide sequence, *secondary structure* is the particular planar graph defined by covalent and hydrogen bonds (e.g., the familiar cloverleaf secondary structure of tRNA), and *tertiary structure* is the three-dimensional structure (in tRNA, a three-dimensional L structure). Though reliable prediction of tertiary structure for RNA is not currently possible, for reasons similar to those behind the difficulty of predicting three-dimensional protein structure, there are relatively good RNA secondary structure prediction algorithms.

DEFINITION 6.1
The secondary structure of an RNA sequence of length n is an undirected graph $G = (V, E)$, where $V = \{1, \ldots, n\}$, $E \subseteq V \times V$, such that

1. $(i, j) \in E \iff (j, i) \in E$.
2. $(\forall\, 1 \leq i < n)[(i, i+1) \in E]$.
3. *For $1 \leq i \leq n$, there exists at most one $j \neq i \pm 1$ for which $(i, j) \in E$.*
4. *If $1 \leq i < k < j \leq n$, $(i, j) \in E$ and $(k, \ell) \in E$, then $i \leq \ell \leq j$.*

Condition 1 ensures the graph is undirected, Condition 2 represents the covalent bonds in the sequence, Condition 3 ensures the formation of base pairs, rather than triplets, and Condition 4 disallows knots and pseudoknots.

There is a 1–1 correspondence between RNA secondary structures and well-balanced parenthesis expressions, where the balancing parentheses correspond to base pairings via hydrogen bonds. For example, the following RNA sequence is placed above a parenthesis expression:

```
AGAAACAUCACAU
((...).(...))
```

This notation indicates that there are base pairs $(1, 13)$ with A, U, $(2, 6)$ with G, C, $(8, 12)$ with U, A. Here is another example, where nine possible secondary structures are given for the palindrome ACGUACGU:

```
ACGUACGU
........
....(..)
..(....)
..((..))
(......)
(..(..))
(..)(..)
((....))
(((..)))
```

M. Waterman [Wat78] was one of the first to initiate the study of combinatorics of RNA sequences, and, by using generating functions and a deep theorem due to Bender concerning asymptotic combinatorics, Stein and Waterman [SW78, Wat95] and, building on this, Hofacker, Schuster, and Stadler [HSS98] were able to successfully answer such questions as the following: How many possible secondary structures are there in a sequence of n nucleotides, where stacked base pairs have minimum length of ℓ and hairpin loops minimum size of m? The fact that there are exponentially many such structures points out the necessity for structure prediction algorithms not only to output the optimal structure, but suboptimal ones as well. As well, the combinatorics of RNA secondary structures was the starting point for the development of an interesting theory of *neutral networks* by P. Schuster and co-workers [RSS97, Sch96, SFSH94, BB96, FST$^+$93].

Suppose first that for a sequence of length n, any nucleotide can form a base pair with any other nucleotide, subject to the above rules for secondary structure formation. For this hypothetical example, define $S(n)$ to be the number of secondary structures on $\{1, \ldots, n\}$, and let $S(0) = 1$.

THEOREM 6.2 (WATERMAN)
$S(1) = 1 = S(2)$ and

$$\begin{aligned} S(n+1) &= S(n) + \sum_{k=0}^{n-2} S(k)S(n-k-1) \\ &= S(n) + S(n-1) + \sum_{k=1}^{n-2} S(k)S(n-k-1). \end{aligned}$$

PROOF This is by induction on n. Clearly $S(1) = 1 = S(2)$. For the inductive case, there are two subcases.

CASE 1: $n+1$ is not base-paired. In this case, there are $S(n)$ possible structures.

CASE 2: $n+1$ is base-paired to nucleotide j, for $1 \le j \le n-1$. In this case, secondary structures can independently be formed on the former part $\{1, \ldots, j-1\}$ and latter part $\{j+1, \ldots, n\}$, leading to a contribution $S(j-1)S(n-j)$. Thus

$$S(n+1) = S(n) + \sum_{j=1}^{n-1} S(j-1)S(n-j) \tag{6.1}$$

$$= S(n) + S(n-1) + \sum_{j=2}^{n-1} S(j-1)S(n-j) \tag{6.2}$$

$$= S(n) + S(n-1) + \sum_{k=1}^{n-2} S(k)S(n-k-1). \tag{6.3}$$

■

A small recursive program yields some initial values of $S(n)$:

n	0	1	2	3	4	5	6	7	8	9	10
$S(n)$	1	1	1	2	4	8	17	37	82	185	423

While a recursive program for $S(n)$ takes exponential time, and cannot even compute $S(25)$ (on a 450 MHz Pentium II), a dynamic programming implementation takes quadratic time and immediately yields $S(25) = 226460893$ and $S(40) = 215440028338359$.

Erdös and DeBruijn long ago began looking at closed formulas for such recurrence relations, and in particular for the *convolution* defined by

$$g(n) = \sum_{k=0}^{n-1} g(k)g(n-k)$$

so that $g(n) = g(0)g(n) + g(1)g(n-1) + \ldots + g(n-1)g(1)$, with base case $g(0) = 1 = g(1)$. Their result is that

$$g(n) = \frac{1}{2^{n-1}} \binom{2n}{n} = O(4^n).$$

This should be compared with the *Catalan numbers*

$$c_n = \frac{1}{n+1} \binom{2n}{n} = \Theta\left(\frac{4^n}{n^{3/2}}\right),$$

known to be the number of different binary trees on n nodes. The following simple proposition shows that there are exponentially many secondary structures.

PROPOSITION 6.3 (M. WATERMAN [WAT78])
For $n \ge 2$, $S(n) \ge 2^{n-2}$.

PROOF As before,

$$S(n+1) = S(n) + S(n-1) + \sum_{k=1}^{n-2} S(k)S(n-k-1), \tag{6.4}$$

and so, replacing $n+1$ by n, we have

$$S(n) = S(n-1) + S(n-2) + \sum_{k=1}^{n-3} S(k)S(n-k-2). \tag{6.5}$$

From (6.5), we have

$$\begin{aligned} S(n+1) &= S(n) + S(n-1) + S(n-2)S(1) + \sum_{k=1}^{n-3} S(k)S(n-k-1) \\ &= S(n) + S(n-1) + S(n-2) + \sum_{k=1}^{n-3} S(k)S(n-k-1). \end{aligned} \tag{6.6}$$

Clearly S is a monotonically increasing function, so $S(n-k-1) \geq S(n-k-2)$; so it follows from (6.5) and (6.6) that $S(n+1) \geq S(n) + S(n)$. ■

Thus there are exponentially many secondary possible structures on a sequence of length n, when ignoring the requirement that base pairs must be Watson–Crick or GU. By using generating functions and Bender's Theorem, in [SW78, Wat95] the exact asymptotic solution

$$s(n) \sim \sqrt{\frac{15+7\sqrt{5}}{8\pi}} n^{-3/2} \left(\frac{3+\sqrt{5}}{2} \right)^n$$

is obtained. Recalling that the golden ratio $\alpha = \frac{1+\sqrt{5}}{2} \approx 1.618034$, the Stein–Waterman result states $S(n)$ has growth rate $\Theta(\frac{(1+\alpha)^n}{n^{3/2}})$.

By imposing a restriction on the minimum number of unpaired bases in hairpin loops and in helices (these terms are defined below), one obtains a different asymptotic value for $S(n)$ that is smaller than 2^n. This is an important starting point for the mathematical evolution theory involving *neutral networks* developed by P. Schuster and co-workers. Specifically, define a mapping from the *sequence space* of all 4^n RNA sequences of length n into the *shape space* of all secondary structures for such sequences. This mapping can be construed as a mapping from *genotype* to *phenotype*, and is onto, many–one, even when sequence space consists of all 2^n many purine–pyrimidine sequences of length n, provided that hairpins have at least 3 unpaired bases, and that ladders have a minimum number of stacked bases. One might ask what the expected Hamming distance is between sequences, whose secondary structures are distinct. This has pertinence for evolution theory in that one assumes that the landscape of all secondary structures influences various functional properties of RNA (e.g. catalytic functions in the preprotein RNA world). To this end, for a fixed secondary structure t, consider the set V_t of all RNA sequences s of length n, whose optimal secondary structure is that of t (we shall shortly give an efficient algorithm to compute this optimal structure). Define an edge between sequences $s, s' \in V_t$ if they differ by one unpaired base, or by 2 paired bases, and let E_t be the collection of all such edges. The undirected graph $G_t = (V_t, E_t)$ is called the *neutral network* associated with secondary structure t. Using the theory of random graphs, one can show certain properties about neutral nets: when they are likely to be connected, etc. For more on the fascinating topic of neutral networks, see [RSS97, Sch96, SFSH94, BB96, FST+93].

The previous calculation of $S(n)$ can be made more realistic by allowing base pairings only for the Watson–Crick base pairs and the GU pairs – this is done in our C program that lists all possible secondary structures for a given RNA sequence. Let $R(s_1, \ldots, s_n)$ be the number of secondary structures for the sequence $s_1, \ldots, s_n$, under this more realistic assumption. Then

$$R(s_1, \ldots, s_{n+1}) = R(s_1, \ldots, s_n) + \sum_{j=1}^{n-1} R(s_1, \ldots, s_{j-1}) + R(s_{j+1}, \ldots, s_n) bp(j, n+1),$$

where $bp(j, n+1) = 1$ if s_j, s_{n+1} can base-pair, otherwise 0. Again using Bender's result, assuming that

$$p = Pr[s_i \text{ and } s_j \text{ can base-pair}]$$

where $1 \leq i < j \leq n+1$, Waterman [Wat95] derives the expected number

$$E[R(n)] \approx \frac{c\beta^n}{n^{3/2}}$$

of secondary structures on RNA sequences of length n having probability p of base-pairing, where $\beta = \frac{(1+\sqrt{1+4\sqrt{p}})^2}{4}$ and $c = \frac{\beta(1+4\sqrt{p})^{1/4}}{2\sqrt{\pi p^{3/4}}}$.

An easy modification of the recurrence relation for $S(n)$ yields a recurrence relation for the maximum number $M_{i,j}$ of base pairs in the subsequence $s_i, \ldots, s_j$ of a given RNA sequence $s_1, \ldots, s_n$. First, define the boolean function $bp(x, y)$ that recognizes whether nucleotides x, y can base-pair, by $bp(x, y) = 1$ if x, y can base-pair, else 0. Clearly, we have the following.

THEOREM 6.4

$$\begin{aligned} M_{i,j} &= 0, \text{if } j \leq i+1 \\ M_{i,j+1} &= \max \left\{ M_{i,j}, \max_{i \leq \ell \leq j-1} \{ (M_{i,\ell-1} + 1 + M_{\ell+1,j}) \, bp(s_\ell, s_{j+1}) \} \right\}. \end{aligned}$$

PROOF As before, in considering $M_{i,j+1}$, there are two possible cases: either s_{j+1} does not base-pair with a nucleotide of $s_i, \ldots, s_j$, or s_{j+1} base-pairs with s_ℓ. In the latter case, we have the term $M_{i,\ell-1} + 1 + M_{\ell+1,j}$. This allows a dynamic programming approach to computing the maximum number of base pairings in an RNA sequence, a task first performed by Nussinov and Jacobson [NJ80], and later refined by numerous authors, leading to robust programs of Zuker [ZS84, Zuk] and of Schuster's group in Vienna RNA Package [HFS$^+$]. ■

To refine the algorithm of Nussinov–Jacobson to compute the optimal secondary structure for a given RNA sequence, we need to categorize certain RNA motifs, as given in the following definition. See Figure 6.1 for illustrations of certain motifs.

DEFINITION 6.5 (M. WATERMAN [WAT78])
Let $G = (V, E)$ be a secondary structure on $s_1, \ldots, s_n$.

1. *i is paired if there exists $j \neq i \pm 1$ with $(i, j) \in E$.*
2. *The subsequence $(i+1, \ldots, j-1)$ is a loop if $(i, j) \in E$, and all $i+1, \ldots, j-1$ are unpaired.*
3. *The subsequence $(i+1, \ldots, j-1)$ is a bulge if i, j are paired, $i+1, \ldots, j-1$ are unpaired and $(i, j) \notin E$.*

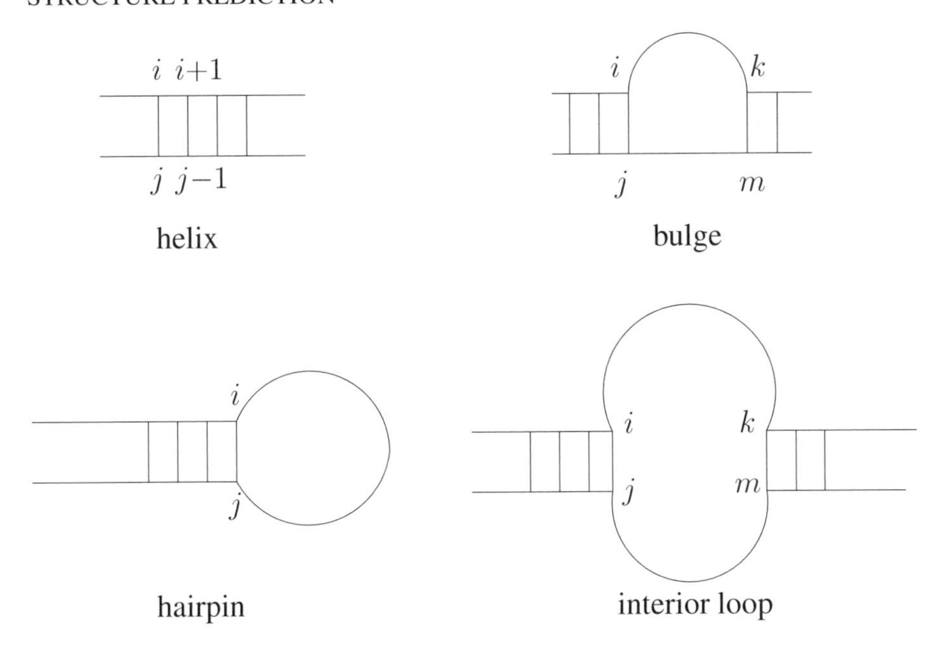

Figure 6.1 RNA secondary structures.

4. *An interior loop is given by two bulges* $(i+1,\ldots,j-1)$ *and* $(k+1,\ldots,\ell-1)$, *such that* $i < j < k < \ell$ *and* $(i,\ell) \in E$ *and* $(j,k) \in E$.
5. *A join is a bulge* $(i,\ldots,j)$ *such that for all* $k < i$, $(k,\ell) \in E$ *implies that* $\ell \leq i$, *and for all* $k > j$, $(k,\ell) \in E$ *implies that* $\ell \geq j$.
6. *A tail is a subsequence of the form* $(1,\ldots,i)$, *where* $1,\ldots,i$ *are unpaired, and either* $i = n$ *or* $i+1$ *paired, or a subsequence of the form* $(i,\ldots,n)$, *where* $i,\ldots,n$ *are unpaired, and* $i = 1$ *or* $i-1$ *is paired.*
7. *A helix or ladder is given by two subsequences* $(i+1,\ldots,i+j)$, $(k+1,\ldots,k+j)$ *such that* $i+j+1 < k$, *and for all* $1 \leq \ell \leq j$, $(i+l,k+j-\ell+1) \in E$. *These latter are the stacked bases.*
8. *A hairpin is the longest subsequence* $(i+1,\ldots,j-1)$ *containing exactly one loop and such that* $(i+1,j-1) \notin E$ *and* $(i,j) \in E$.

It is not difficult to see that every secondary structure can be uniquely decomposed into hairpin loops, interior loops, bulges, ladders (or helical regions), and tails. The following definition is a finite analogue of the notion of Cantor–Bendixson derivative from topology, and is a tool for classifying *multiloops*. This was also the basis for an algorithm developed by Waterman for predicting RNA secondary structure, allowing for arbitrarily complicated multiloops.

DEFINITION 6.6 (M. WATERMAN [WAT78])
Let A *be an adjacency matrix for a given secondary structure* $G = (V,E)$.

1. $A^{(0)} = A$.
2. $A^{(i+1)}$ *obtained from* $A^{(i)}$ *by setting* $a^{(i+1)}_{k,\ell} = 0 = a^{(i+1)}_{\ell,k}$ *wherever* $a^{(i)}_{k,\ell} = 1 = a^{(i)}_{\ell,k}$ *where* k,ℓ *are members of a hairpin, and* $k \neq \ell \pm 1$.

In other words, a multiloop, whose adjacency matrix is given by A, is kth order if k iterations of repeatedly removing stacked base pairs from hairpin loops yields a structure having no loops. In [Wat78] Waterman proved that every secondary structure on $\{1, \ldots, n\}$ has a unique order k for some $k \leq \lfloor \frac{n}{3} \rfloor$.

A first attempt to compute the optimal secondary structure S of a given RNA sequence uses the heuristic of maximizing the number of base pairs, since a structure having many base pairs should be stable. This is the approach of Nussinov and Jacobson [NJ80], where for a given nucleotide sequence $s_1, \ldots, s_n$ an $n \times n$ matrix $M = (m_{i,j})$ is constructed, where $m_{i,j}$ is the maximum number of base pairs occurring in an optimal secondary structure for the subsequence $s_i, \ldots, s_j$. The absence of knots and pseudoknots from condition 4 in Definition 6.1 allows the determination of an optimal secondary structure for the subsequence $s_i, \ldots, s_j$ to be made independently from that for $s_k, \ldots, s_\ell$, provided that $1 \leq i < j < k < \ell \leq n$. This leads to the recurrence relation where $m_{i,j}$ equals the maximum of

$$m_{i,j-1} \tag{6.7}$$

if j is not paired with any k in interval $[i, j]$, and

$$\max_{i \leq k < j} \{1 + m_{i,k-1} + m_{k+1,j-1}\} \tag{6.8}$$

if j is paired with that value k in $[i, j]$ which maximizes $m_{i,k-1} + m_{k+1,j-1}$.

Since the computation of $m_{i,j}$ depends only on values $m_{i',j'}$, where $i \leq i'$ and $j' < j$, this suggests applying the technique of dynamic programming, where the matrix M is filled in, beginning along the diagonal leading from position $(1, 2)$ to $(n-1, n)$, then continuing with the diagonal leading from position $(1, 3)$ to $(n-2, n)$, etc. as shown in Figure 6.2.

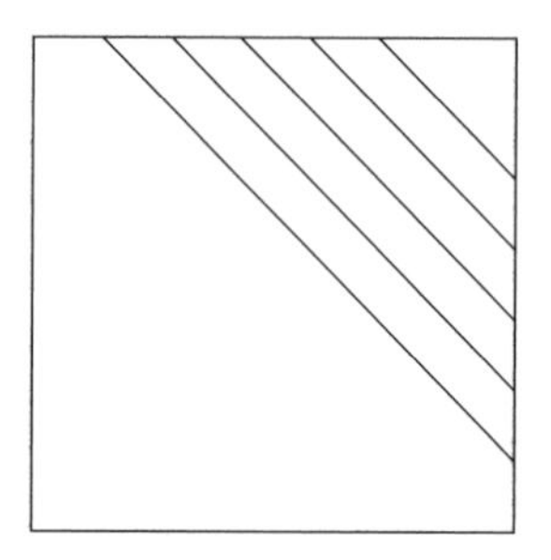

Figure 6.2 Filling in the Nussinov–Jacobson matrix.

Since $m_{i,j}$ is defined as above only if $i < j$, this leaves the lower-triangular part of M unassigned. In the computation of $m_{i,j}$, defined as the maximum of (6.7) and (6.8), we can additionally store in $m_{j,i}$ the index k that is base-paired with j, if j is base-paired, otherwise 0.

Instead of maximizing the number of base pairs, it is biologically more meaningful to determine the minimum energy $E(S)$ associated with an optimal secondary structure for nucleotide sequence $S = s_1, \ldots, s_n$, where

$$E(S) = \sum_{(i,j) \in S} a(i, j)$$

and $a(i,j)$ is the (negative) stabilizing energy from the base pair consisting of the ith and jth nucleotides. Here, one could distinguish between Watson–Crick and GU base pairs, by defining $a(i,j)$ for instance as follows:

$$a(i,j) = \begin{cases} -5 & \text{if } s_i, s_j \text{ is CG or GC,} \\ -4 & \text{if } s_i, s_j \text{ is AU or UA,} \\ -1 & \text{if } s_i, s_j \text{ is GU or UG.} \end{cases}$$

Defining $E(S_{i,j})$ to be the minimum energy of a secondary structure on the sequence $S_{i,j} = s_i, \ldots, s_j$ leads to the recurrence relation

$$E(S_{i,j}) = \min \left\{ \begin{array}{l} E(S_{i,j-1}) \\ a(i,j) + E(S_{i+1,j-1}) \\ \min_{i<k<j} \{a(k,j) + E(S_{i,k-1}) + E(S_{k+1,j-1})\} \end{array} \right\}. \tag{6.9}$$

Using dynamic programming to evaluate $E(S_{i,j})$ yields one of the first algorithms for the determination of RNA secondary structure, developed by Nussinov and Jacobson [NJ80]. Nussinov and Jacobson state that using this simple algorithm predicts most of the base pairs observed in RNA bacteriophage MS2.

Since RNA cannot bend over short regions, it seems reasonable to assume that a base pair between the ith and jth nucleotide cannot form if $|j - i| < \mu$, where μ is a threshold value – for instance $\mu = 3$. This restriction can easily be introduced into the previous equation (6.9).

In Algorithm 6.1, the input RNA nucleotide sequence of length n is given by the string $s_0, \ldots, s_{n-1}$, while $E(i,j)$ denotes the (i,j)th entry in the energy matrix. If $i < j$, then $E(i,j)$ is the energy of the subsequence $s_i, \ldots, s_j$, as computed by the previous recurrence relation (6.9). For $i < j$, $E(j,i)$ stores that index $i \leq k < j$ such that s_k, s_j are base-paired in order to achieve the minimum energy $E(i,j)$, while if s_j is not base-paired, then $E(j,i) = -1$. Variables `min` and `index` are used to determine the minimum energy and the index k.

Using the pointers $E(j,i)$ indicating the index k for which s_k, s_j are base-paired in order to achieve minimum energy $E(i,j)$, we can recursively determine the optimal alignment by backtracking (see Algorithm 6.2). In that algorithm,`paren` is an array of length n, where all entries have been previously initialized to contain '`.`'.

It is more realistic to assume that the unpaired bases in a hairpin loop are destabilizing. This leads to

$$E(S_{i,j}) = \min \left\{ \begin{array}{l} E(S_{i,j-1}) \\ \min_{i<k<j} \{a(k,j) + E(S_{i,k-1}) + E(S_{k+1,j-1})\} \\ E(L_{i,j}) \end{array} \right\}, \tag{6.10}$$

where $L_{i,j}$ is a loop structure on $s_i, \ldots, s_j$ and the ith and jth nucleotides s_i, s_j are paired. Assume we have the following energies from experiments.

1. Let b be the (negative) stabilizing free energy of an additional stacked base pair, in the case that the loop $L_{i,j}$ is a helix (ladder).
2. Let $Ehp(j - i - 1)$ be the (positive) destabilizing free energy of a hairpin loop with $j - i - 1$ unpaired bases, in the case that the loop $L_{i,j}$ is a hairpin loop.
3. Let $Ebu(k)$ be the (positive) destabilizing free energy of a bulge with k unpaired bases, in the case that the loop $L_{i,j}$ is a bulge at i.

Algorithm 6.1 Energy matrix computation

```
for d = μ to n − 1 {
  for i = 0 to n − 1 {
     j=i+d;
     if (j < n){
       min =0; index=-1;
       if ( E(i,j − 1) < min ){
          min = E(i,j − 1);
          index = -1; // j is unpaired
       }
       if ( j − i ≥ μ and a(i,j) + E(i + 1,j − 1) < min ){
          min = a(i,j) + E(i + 1,j − 1);
          index=i;
       }
       for k = i+1 to j-μ {
          val = a(k,j) + E(i,k − 1) + E(k + 1,j − 1);
          if (val < min) {
             min = val;
             index=k;
          }
       }
       E(i,j) = min;
       E(j,i) = index;
     }
  }
}
```

Algorithm 6.2 `backtrack(i,j)`

```
k = E(j,i);
if(k ≠ −1) {
  paren[k] = '(';
  paren[j] = ')';
  if( μ ≤ (j − 1) − (k + 1) + 1 )
     backtrack(k+1,j-1,paren);
  if (μ ≤ k − 1 − i + 1 )
     backtrack(i,k-1,paren);
}
else { // Here k = −1
  if( μ ≤ j − 1 − i + 1 ){
     backtrack(i,j-1,paren);
     }
  else
     return 0;
}
```

4. Let $Ebu(k)$ be the (positive) destabilizing free energy of a bulge with k unpaired bases, in the case that the loop $L_{i,j}$ is a bulge at j.
5. Let $Eil(k+m)$ be the (positive) destabilizing free energy of an interior loop with k unpaired bases in one bulge, and m unpaired bases in the other bulge.

Then

$$E(L_{i,j}) = \min \left\{ \begin{array}{ll} a(i,j) + b + E(S_{i+1,j-1}) & \text{if (1)} \\ a(i,j) + Ehp(j-i-1) & \text{if (2)} \\ a(i,j) + \min_{k \geq 1}\{Ebu(k) + E(S_{i+k+1,j-1})\} & \text{if (3)} \\ a(i,j) + \min_{k \geq 1}\{Ebu(k) + E(S_{i+1,j-k-1})\} & \text{if (4)} \\ a(i,j) + \min_{k,m \geq 1}\{Eil(k+m) + E(S_{i+k+1,j-m-1})\} & \text{if (5)} \end{array} \right\}.$$

Clearly, since for i, j fixed, we must determine a minimum over all k, m in the case of an interior loop, the time of this algorithm is $O(n^4)$.

Following [HFS+94], multiloop energies can be taken into account in the following manner. Let `C[i,j]` be the free energy within the nucleotide subsequence $s_i, \ldots, s_j$ *assuming* that s_i, s_j base-pair with each other, while `F[i,j]` is the free energy for the subsequence $s_i, \ldots, s_j$ without necessarily assuming that s_i, s_j base-pair. `Hairpin(i,j)` is the positive, destabilizing energy of a hairpin loop, where s_i, s_j are base-paired, and $s_{i+1}, \ldots, s_{j-1}$ are unpaired. `Interior(i,j;p,q)` is the positive, destabilizing free energy of an interior loop closed by the base pairs s_i, s_j and s_p, s_q; i.e. s_i, s_j are base-paired with each other, as are s_p, s_q, while $s_{i+1}, \ldots, s_{p-1}$ and $s_{q+1}, \ldots, s_{j-1}$ are unpaired. Note that the case of stacked base pairs is represented by `Interior(i,j;i+1,j-1)` and that the case of a bulge is represented by either `Interior(i,j;p,j-1)` or `Interior(i,j;i+1,q)`. Finally, `FM[i,j]` is the free energy of a multiloop region $s_i, \ldots, s_j$, where the simplifying assumption is made that the multiloop free energy contribution F satisfies

$$F = a(i,j) + bI + cU$$

for a multiloop s where s_i, s_j are base-paired, and there are I interior base pairs and U unpaired bases in the multiloop region (a linearity condition similar to the affine gap penalty in Gotoh's sequence alignment algorithm). Pseudocode for the dynamic programming algorithm from Vienna RNA Package is given in Algorithm 6.3.

The function `Hairpin(i,j)` consists of a positive, destabilizing entropic contribution of `hairpin[j-i-1]`, as given by Figure 6.3, for the $j-i-1$ unpaired bases in the hairpin loop, closed off by the base pair s_i, s_j. There is an additional table `mismatchH` giving negative energies for the base pair s_i, s_j, taking into account the exact nature of the adjacent unpaired bases s_{i+1}, s_{j-1}. Finally, there is a *bonus energy* for certain commonly appearing tetraloops.

The function `Interior(i,j;p,q)` must distinguish between stacked bases, bulges, and true interior loops, as already explained. The positive, destabilizing entropic energy due to $p-i-1+j-q-1$ unpaired bases in the interior loop (combined bulges) is given by `interior[p-i-1+j-q-1]` in Figure 6.3, while the stabilizing energy due to stacked bases is given in `stack[ ][ ]`. For instance, the stacked base pairs CC, as in

$$\begin{array}{c} 5'\text{–CC–}3' \\ 3'\text{–GG–}5' \end{array}$$

Algorithm 6.3 FREEENERGY($s_1, \ldots, s_n$)

```
for (d=1;d<=n;d++)
  for (i=1;i<=d;i++) {
    j=i+d;
    C[i,j] =  MIN ( Hairpin(i,j),
              MIN_{i<p<q<j} (Interior(i,j;p,q) + C[p,q]),
              MIN_{i<k<j} (FM[i+1,k]+FM[k+1,j-1]+a(i,j)) )
    F[i,j] =  MIN ( C[i,j], MIN_{i<k<j} (FM[i,k]+FM[k+1,j]) )
    FM[i,j] = MIN ( b+C[i,j], c+FM[i+1,j], c+FM[i,j-1],
              MIN_{i<k<j} (FM[i,k]+FM[k+1,j]) )
  }
return F[1,n];  // free energy of sequence s_1,...,s_n
```

have energy -2.9 kcal/mol, whereas stacked base pairs

$$\begin{array}{c} 5'-\mathrm{AA}-3' \\ 3'-\mathrm{UU}-5' \end{array}$$

have energy -0.9 kcal/mol, and

$$\begin{array}{c} 5'-\mathrm{CG}-3' \\ 3'-\mathrm{GU}-5' \end{array}$$

have energy -1.2 kcal/mol. Finally, there is a destabilizing energy contribution due to *dangling* ends, which does not appear in the above pseudocode, but is treated in the source code of `Vienna RNA Package`.

All hairpins, bulges, and interior loops are assumed not to exceed 30 unpaired bases, which, according to experimental evidence, is likely. This assumption, along with the affine energy contribution from multiloops, renders an $O(n^3)$ algorithm for RNA secondary structure prediction. The experimentally determined destabilizing energies for hairpins, bulges and interior loops are given in the table below (values at temperature of 37° C and 1 molar sodium chloride concentration are taken from source code from `Vienna RNA Package` [HFS$^+$94, HFS$^+$]) and are given in units of 0.01 kcal/mol).

Waterman [Wat95] follows a different approach in order to reduce the computation time from $O(n^4)$ to $O(n^3)$, sketched as follows. For each i, j, s, store

$$E^*_{i,j,s} = \min\{E(S_{k,m}) : s = (j-i) + (m-k) - 2, k \geq i+2, j-2 \geq m, m-k-1 \geq \mu\},$$

where μ is a threshold value. Then

$$\min_{s \geq 1}\{a(i,j) + Eil(s) + E^*_{i,j,s}\}$$

computes the energy for interior loops at i, j and takes time $O(n^3)$.

As a final remark, we mention that in [HFS$^+$94], pseudocode for a dynamic programming algorithm is given for the computation of the partition function

$$Q = \sum_S e^{-\frac{\Delta G(S)}{RT}},$$

```
int hairpin[31] = {
   INF, INF, INF, 410, 490, 440, 470, 500, 510, 520, 531,
        542, 551, 560, 568, 575, 582, 589, 595, 601, 606,
        611, 616, 621, 626, 630, 634, 638, 642, 646, 650};

int bulge[31] = {
   INF, 390, 310, 350, 420, 480, 500, 516, 531, 543, 555,
        565, 574, 583, 591, 598, 605, 612, 618, 624, 630,
        635, 640, 645, 649, 654, 658, 662, 666, 670, 673};

int interior[31] = {
   INF, INF, 410, 510, 490, 530, 570, 587, 601, 614, 625,
        635, 645, 653, 661, 669, 676, 682, 688, 694, 700,
        705, 710, 715, 720, 724, 728, 732, 736, 740, 744};

int stack[NBPAIRS+1][NBPAIRS+1] =
/*          CG    GC    GU     UG    AU    UA   */
{{ INF,  INF,  INF,  INF,  INF,  INF,  INF, INF},
 { INF, -290, -200, -120, -190, -180, -170, NST},
 { INF, -340, -290, -140, -210, -230, -210, NST},
 { INF, -210, -190,  -40,  150, -110, -100, NST},
 { INF, -140, -120,  -20,  -40,  -80,  -50, NST},
 { INF, -210, -170,  -50, -100,  -90,  -90, NST},
 { INF, -230, -180,  -80, -110, -110,  -90, NST},
 { INF,  NST,  NST,  NST,  NST,  NST,  NST, NST}};
```

Figure 6.3 Some energy functions from `Vienna RNA Package`.

where the sum is over all secondary structures S for a given RNA sequence $s_1, \ldots, s_n$, as well as for the partition functions $Q^b_{i,j}$ of the subsequence $s_i, \ldots, s_j$, assuming that s_i, s_j base-pair. This provides an efficient algorithm to compute the probability $p_{i,j}$ that s_i base-pairs with s_j in a given RNA sequence $s_1, \ldots, s_n$.

6.2 DNA Strand Separation

During the replication of DNA and the transcription of DNA into RNA, the double strands of DNA must separate to allow the formation of a complementary strand of DNA (in the case of replication) and of an RNA transcript (in the case of transcription). In this section, we outline an application of simulated annealing to determine the strand separation sites of double-stranded DNA in transcription and replication events. In vivo, DNA has a negative superhelicity to be explained later, which is a destabilizing factor that lowers the energy required to separate hydrogen-bonded complementary base pairs. It thus seems plausible that those hydrogen bonds that contribute least to the stability of DNA are broken first (e.g. AT rather than GC bonds). Can one compute where the strands of double-stranded DNA first separate when a gene is transcribed? Is it before, in the middle, or after the gene? Are there unique sites where strand separation first occurs when DNA is replicated, and can these sites be predicted by computational methods?

In [Ben90], C. Benham developed a mathematical model and appropriate free energy function to answer such questions. The energy function includes contributions due to *separation* of strands, and to *torsion* resulting from rotation of free strands, as well as the *residual* supercoiled free energy. The energy term for separation of strands concerns the number of separation regions or *runs* r along with the total number n of separated base pairs, and the nucleotide hydrogen bond strengths of the separated base pairs. The torsional and residual energy terms include topological information concerning *linking number*, *twist*, and *writhe* of circular, negatively supercoiled DNA. Note that though topological and secondary structure information is incorporated in the free energy function, no consideration of the exact tertiary structure of DNA is required.

In [SMFB95], Sun, Mezei, Fye, and Benham developed an ergodic, balanced move set for a Monte Carlo algorithm with simulated annealing, and were able to predict those sites where double-strand separation should occur in transcription and replication events. See [FB99] for a recent improvement by Fye and Benham.

Eucaryotic DNA is often found tightly coiled around a core of histones. Indeed, recall from Chapter 4, Section 4.2.1 that histone H4 is one of the most highly conserved proteins. It is known that the mean values for the roll and helical twist angles of a TATA box, a binding site for RNA polymerase, are smaller than those for a random sequence. From such observations, it seems clear that topological considerations might play an important role in any mathematical model for transcription and replication events.

To this end, we discuss the properties of *linking number*, *twist*, and *writhe*, which are topological properties of of circular DNA or of locally constrained linear DNA. Given two oriented closed curves A, B in space, the *linking number* $Lk(A, B)$ is defined as follows (in DNA, the orientation is given by the $5'$ to $3'$ direction). Project the curves onto a plane. For each position p, where there is a crossing in the planar projection of curves A and B, rotate the tangent vector of the *top* curve to coincide with the tangent vector of the *bottom* curve. If the smallest angle of rotation to achieve the coincidence of tangent vectors is counterclockwise then the *index* of p is $+1$, while if it is clockwise then the index is -1. Note that only crossings of the curves A with B are counted, not those crossings of A with itself or B with itself. The *linking number* is defined to be the sum of indices of all crossing points divided by 2. Clearly the linking number is independent of the order of the closed curves; i.e. $Lk(A, B) = Lk(B, A)$. It is known that the linking number does not depend on the projection considered, so that Lk is a *topological invariant* of curves A, B. As an example, the linking number in the right-handed helix of Figure 6.4 is 4, since each of the 8 crossings requires a counterclockwise rotation of the tangent of the top line to be superposed on the tangent of the bottom line.

Given an oriented closed curve C in space, the writhe $Wr(C)$ is defined as follows. Project C onto a plane, and compute the index of all points p where C crosses over itself, as in the linking number index. The *writhe* $Wr(C)$ is the *average*, over all possible projections, of the sum of all crossing point indexes. This sum will usually be the same, except for a few projections (e.g. when the curve is viewed from the side), so the writhe is approximately the sum of indices of crossing points with respect to a canonical projection. (For instance, the writhe in Figure 6.5 is approximately -3.)

The formal definition of *twist* $Tw(A, B)$ is rather complicated, so we only consider the case for a helix, sufficient for our purposes. Let A represent a strand of circular duplex DNA and B the axis of the helix; i.e. consider double-stranded DNA to be a ribbon closed in a circle, and let A be one edge of the ribbon and B be the circle described by the center of the

Figure 6.4 Linking number $Lk = 4$. Adapted with permission from [Whi89]. Copyright CRC Press, Boca Raton, Florida.

Figure 6.5 Writhe $Wr \approx -3$. Adapted with permission from [Whi89]. Copyright CRC Press, Boca Raton, Florida.

ribbon. Then the twist is simply the number of full turns that A makes with respect to B, i.e. the number of times that the helix rotates about its axis. The twist is defined to be *positive* if the helix is right-handed, and *negative* if it is left-handed. The DNA strand in Figure 6.6, for example, has twist $Tw = 1$. *In vivo* B-DNA consists of 10.4 base pairs per turn, so a DNA molecule consisting of n base pairs has positive twist of $\frac{n}{10.4}$.

In [Whi89, Whi95], J.H. White proved the remarkable topological fact that

$$Lk = Tw + Wr. \tag{6.11}$$

White's theorem has applications for computing the twist and writhe of DNA, for analyzing the action of topoisomerases I and II, and, as we shall soon see, for the rotational and residual energy of supercoiled DNA.

DNA is a right-handed helix with close to 10.4 base pairs per turn, so the twist of double stranded DNA having n base pairs is $+\frac{n}{10.4}$. In a non-supercoiled state, if the writhe is 0 then $Tw = \frac{n}{10.4} = Lk$. Presumably for reasons of molecular stability, DNA prefers to maintain its twist without change, though linking number and writhe may be altered – the former by separation of strands and the latter by supercoiling. In this case, by White's invariant (6.11), linking number and writhe change by the same amount.

If the axis of a DNA molecule itself forms a helix, then the DNA is said to form a *superhelix*, and to be *supercoiled*. This happens in particular when DNA is wrapped about a *nucleosome*. Experimentally, it has been determined that there are approximately 146 base pairs of double-stranded DNA, which are wrapped in a left-handed sense about the nucleosome, forming approximately 1.85 superhelical turns. Moreover, when wrapped on a nucleosome, there are roughly 10 base pairs of DNA per helical turn, slightly more than 10.4 bp in the usual B-DNA form. From a calculation in [Whi89, Whi95], using equation (6.11), it follows that the linking

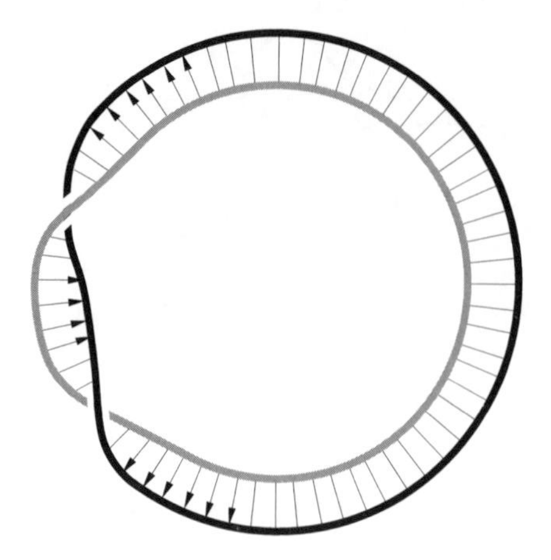

Figure 6.6 Twist $Tw = 1$.Adapted with permission from [Whi89]. Copyright CRC Press, Boca Raton, Florida.

number of DNA *decreases* by 1 for every nucleosome it is so wrapped about, since *in vivo* DNA is negatively supercoiled.

Topoisomerase I cuts one strand of duplex DNA, allows a rotation of the other strand, and then reseals the broken strand. Since the twist is changed by 1, the change of linking number is by 1. Topoisomerase II cuts both strands of duplex DNA, allows the passage of one side of the cut duplex through the break, and then reseals both strands, without any change in twist. Since the writhe is changed by 2, the change in linking number is 2.

With this introduction to linking number, we now turn to the question of determination of DNA strand separation sites. In Benham's model, the free energy G in supercoiled DNA can be written as the sum of three components, G_{sep}, G_{tor}, and G_{res}, as explained below.

Fix L throughout this section as the length of circular or locally constrained linear DNA. Let $G_{\text{sep}}(r, n_{AT}, n_{GC})$ be the free energy necessary to separate double-stranded DNA into r runs or open loops, consisting of n_{AT} base pairs of the form AT and n_{GC} base pairs of the form GC. (Figure 6.7 illustrates the case where $r = 3$.) Then

$$G_{\text{sep}}(r, n_{AT}, n_{GC}) = b_{AT}n_{AT} + b_{GC}n_{GC} + ar, \tag{6.12}$$

where $a > b_{GC} > b_{AT} > 0$; i.e. the cost for *nucleation initiation* is greater than the cost of GC separation (with 3 H bonds), which is greater than the cost of AT separation (with 2 H bonds). We can consider an alternate form of the separation energy $G_{\text{sep}}(r, n)$ in the case of r runs with a total number of n separated base pairs. To this end, let b_i equal b_{AT} if the base pair at the ith position is AT, and equal b_{GC} if the base pair is GC. Let n_i equal 1 if the strands are separated at position i and 0 otherwise. Then the separation energy $G_{\text{sep}}(r, n)$ satisfies

$$G_{\text{sep}}(r, n) = ar + \sum_{i=1}^{L} n_i b_i,$$

which can be rewritten as

$$G_{\text{sep}}(r, n) = \sum_{i=1}^{L} b_i n_i + a \sum_{i=1}^{L} \left(\frac{n_i + n_{i+1}}{2} - n_i n_{i+1} \right) \tag{6.13}$$

by considering the various cases where n_i, n_{i+1} take the values $0, 1$. See Figures 6.8–6.10, and note that since the DNA is circular, one assumes that $n_{L+1} = n_1$. Moreover, the b_i could clearly be modified to allow for near-neighbor effects or for methylated or otherwise modified bases.

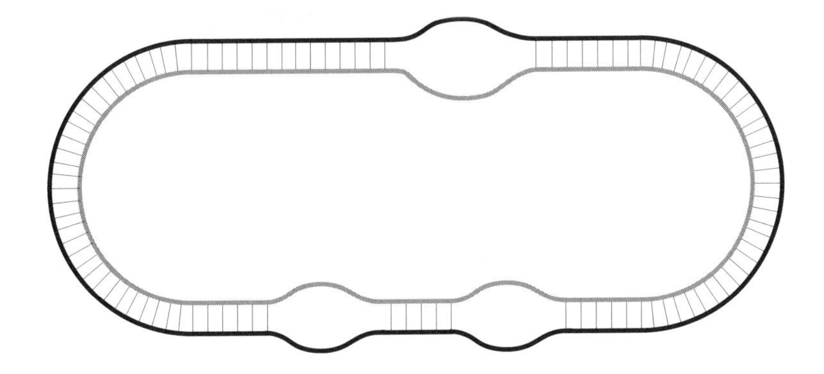

Figure 6.7 Openings in strand separation.

The torsional, or rotational free energy G_{tor} arises because the two single strands of DNA in a separated region may rotate around each other by τ_i radians per base pair at position i. This torsional free energy has been experimentally determined to be quadratic in τ_i, so

$$G_{\text{tor}} = \frac{c}{2} \sum_{i=1}^{L} n_i \tau_i^2, \tag{6.14}$$

where c is a torsional stiffness constant. If the torsion is constant, and independent of position, then $\tau_i = \tau$ and so $G_{\text{tor}} = \frac{cn}{2}\tau^2$.

To compute the residual free energy G_{res}, after strand separation and torsional contributions have been considered, we need to return to topological considerations of equation (6.11) relating linking number, twist, and writhe.

Let Lk_0, Tw_0, and Wr_0 be respectively the linking number, twist, and writhe for DNA in its usual non-negatively supercoiled state. Define the *linking difference* θ to be $Lk - Lk_0$. Assuming that twist $Tw = \frac{n}{10.4} = Tw_0$ remains constant, negatively supercoiled DNA has linking difference $\theta < 0$, where from equation (6.11) the linking difference is the twist plus writhe minus the twist in the unwrithed state; i.e.

$$\theta = (Tw + Wr) - (Tw_0 + Wr_0) = (Tw - Tw_0) + (Wr - Wr_0).$$

When negatively supercoiled double-stranded DNA forms an open region of separated strands, the positive helical twist is reduced, relieving some of the stress, causing the new linking difference to be less negative. If n base pairs are separated, then this contribution from the twist is $\frac{n}{10.4}$. Moreover, the free unpaired strands may twist around each other by an angle of τ_i radians for the ith base pair, or a fixed amount τ, assuming the torsional constants τ_i to be equal, as mentioned in the case of G_{tor}.

It follows that the residual linking difference θ_r satisfies

$$\theta_r + \frac{n\tau}{2\pi} = \theta + \frac{n}{10.4}, \tag{6.15}$$

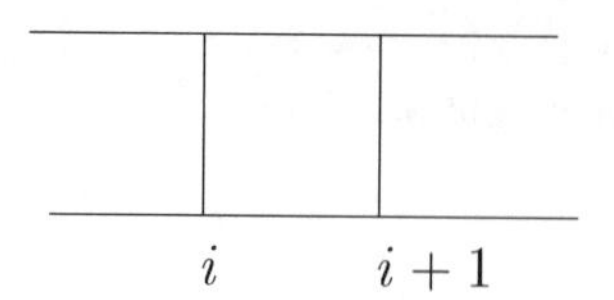

Figure 6.8 Case 1: $\frac{n_i+n_{i+1}}{2} - n_i n_{i+1} = 0$.

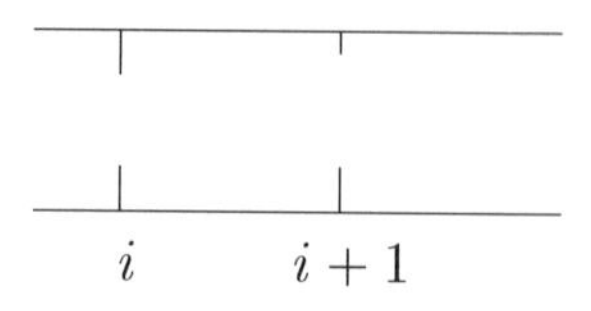

Figure 6.9 Case 2: $\frac{n_i+n_{i+1}}{2} - n_i n_{i+1} = 0$.

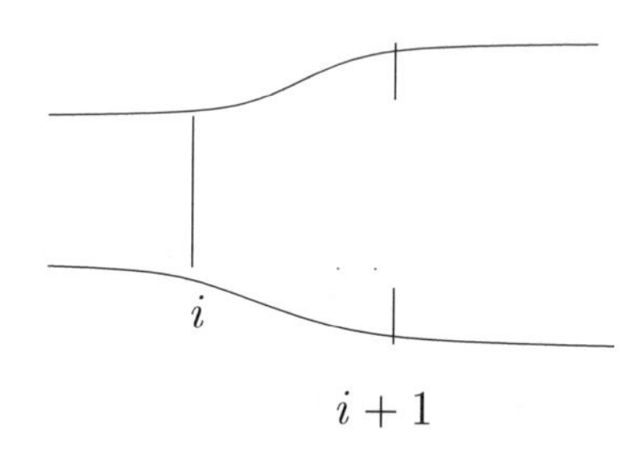

Figure 6.10 Case 3: $\frac{n_i+n_{i+1}}{2} - n_i n_{i+1} = \frac{1}{2}$.

so that the torsion angle τ is negative. According to C. Benham (personal communication), $\tau \approx -15°$ per base pair.

The residual free energy is known to satisfy

$$G_{\text{res}} = \frac{k\theta_r^2}{2}, \tag{6.16}$$

where k is an experimentally measured constant. Thus the total free energy G of supercoiled DNA equals

$$G = G_{\text{sep}} + G_{\text{tor}} + G_{\text{res}}.$$

To find a closed formula for $G(r, n)$, the free energy of supercoiled DNA having r runs with a total number n of separated base pairs, we allow θ_r and τ to equilibrate. Specifically, this means that we minimize the free energy $G_{\text{tor}} + G_{\text{res}}$,

$$G_{\text{tor}} + G_{\text{res}} = \frac{nc\tau^2}{2} + \frac{k\theta_r^2}{2}, \tag{6.17}$$

subject to the constraint

$$\frac{n\tau}{2\pi} + \theta_r = Q, \tag{6.18}$$

where Q is a constant. We apply the method of Lagrange multipliers defining

$$\begin{aligned} H(\tau,\theta_r) &= \frac{n\tau}{2\pi} + \theta_r - Q, \\ F(\tau,\theta_r,\lambda) &= G_{\text{tor}} + G_{\text{res}} + \lambda H(\tau,\theta_r), \end{aligned}$$

hence

$$F(\tau,\theta_r,\lambda) = \frac{nc\tau^2}{2} + \frac{k\theta_r^2}{2} + \lambda\left(\frac{n\tau}{2\pi} + \theta_r - Q\right),$$

and find a solution to

$$\frac{\partial}{\partial\tau}F = \frac{\partial}{\partial\theta_r}F = \frac{\partial}{\partial\lambda}F = 0.$$

Now $\frac{\partial}{\partial\tau}F = nc\tau + \frac{n\lambda}{2\pi} = 0$ implies that

$$\lambda = -2\pi c\tau. \tag{6.19}$$

Also, $\frac{\partial}{\partial\theta_r}F = k\theta_r + \lambda = 0$ implies that

$$\lambda = -k\theta_r. \tag{6.20}$$

Finally, $\frac{\partial}{\partial\lambda}F = \frac{n\tau}{2\pi} + \theta_r - Q = 0$ implies that $\frac{n\tau}{2\pi} + \theta_r = Q$, which is the constraint equation (6.18). Solving equations (6.19) and (6.20) for λ yields

$$2\pi c\tau = k\theta_r. \tag{6.21}$$

From (6.15) and (6.18), we have

$$\theta_r + \frac{n\tau}{2\pi} = Q = \theta + \frac{n}{10.4}, \tag{6.22}$$

where Q is a constant. From (6.21) we have $\theta_r = \frac{2\pi c\tau}{k}$, which when substituted into (6.22) gives

$$\frac{2\pi c\tau}{k} + \frac{n\tau}{2\pi} = \tau\left(\frac{nk + 4\pi^2 c}{2\pi k}\right) = \theta + \frac{n}{10.4},$$

or

$$\tau = \frac{2\pi k}{4\pi^2 c + kn}\left(\theta + \frac{n}{10.4}\right).$$

From (6.21), we have

$$\begin{aligned} \theta_r = \frac{2\pi c}{k}\tau &= \frac{2\pi c}{k}\left(\frac{2\pi k}{4\pi^2 c + kn}\right)\left(\theta + \frac{n}{10.4}\right) \\ &= \frac{4\pi^2 c}{4\pi^2 c + kn}\left(\theta + \frac{n}{10.4}\right). \end{aligned}$$

It now follows that

$$
\begin{aligned}
G_{\text{tor}} + G_{\text{res}} &= \frac{nc\tau^2}{2} + \frac{k\theta_r^2}{2} \\
&= \frac{nc}{2}\left[\frac{2\pi k}{4\pi^2 c + kn}\left(\theta + \frac{n}{10.4}\right)\right]^2 + \frac{k}{2}\left[\frac{4\pi^2 c}{4\pi^2 c + kn}\left(\theta + \frac{n}{10.4}\right)\right]^2 \\
&= \frac{(\theta + n/10.4)^2}{(4\pi^2 c + kn)^2}\left[\frac{(2\pi k)^2 nc}{2} + \frac{k(4\pi^2 c)^2}{2}\right] \\
&= \frac{(\theta + n/10.4)^2}{(4\pi^2 c + kn)^2}\left(\frac{4\pi^2 k^2 nc}{2} + \frac{16\pi^4 c^2 k}{2}\right) \\
&= \frac{(\theta + n/10.4)^2}{(4\pi^2 c + kn)^2}\left(2\pi^2 k^2 nc + 8\pi^4 c^2 k\right) \\
&= \frac{(\theta + n/10.4)^2}{(4\pi^2 c + kn)^2}\left[2\pi^2 kc(kn + 4\pi^2 c)\right] \\
&= \frac{2\pi^2 kc}{4\pi^2 c + kn}\left[\left(\theta + \frac{n}{10.4}\right)^2\right].
\end{aligned}
$$

This completes the derivation of the closed formula for the total free energy

$$G(n, r) = G_{\text{sep}}(r, n) + G_{\text{tor}} + G_{\text{res}}$$

for supercoiled DNA with r runs consisting of a total of n separated base pairs, as given by

$$G(n, r) = ar + \sum_{i=1}^{L} n_i b_i + \frac{2\pi^2 ck}{4\pi^2 c + kn}\left(\theta + \frac{n}{10.4}\right)^2. \tag{6.23}$$

In [SMFB95], calculations were done with the following constants: at 0.01 molar Na^+ concentration, $a = 10.5$ kcal/mol, $b_{AT} = 0.258$ kcal/mol, $b_{GC} = 1.305$ kcal/mol, $c = 3.6$ kcal/rad^2, $k = 2350\frac{RT}{L}$, $R = 8.3146$ J deg^{-1} mole^{-1}, and $T = 310$ K.

To apply Monte Carlo with simulated annealing, we devise a move set for which the underlying Markov chain has stationary probabilities and for which the move set is *balanced*, in that the detailed balance equation (2.5) from Chapter 2,

$$p_i^* p_{i,j} = p_j^* p_{j,i},$$

is satisfied for all states i, j, transition probabilities $p_{i,j}$, $p_{j,i}$, and stationary probabilities p_i^*, p_j^*. By Theorem 2.17, if the transition probability matrix P for the Markov chain corresponding to the move set consists of strictly positive probabilities, then the Markov chain is aperiodic and irreducible, and so has stationary probabilities. Choosing a symmetric move set so that the Markov chain has symmetric transition matrix P clearly satisfies detailed balance. From Chapter 2, it now follows that the Monte Carlo algorithm with simulated annealing converges to a global extremum.

A single move of the Monte Carlo algorithm goes as follows. For $1 \le i \le L$, change n_i to $1 - n_i$ with probability p, for a fixed $0 < p < 1$. In other words, flip each site, either from closed to open, or from open to closed, with probability p. It is clear that this move corresponds to a Markov chain with stationary probabilities satisfying the detailed balance equation. However, even repeating this move a large constant c number of times, in order to decrease the cross-correlation or non-independence of states, the convergence of this Monte Carlo simulated annealing algorithm is reported in [SMFB95] to be unacceptably slow.

For this reason, [SMFB95] devised an additional set of *shuffling* moves, which decrease the cross-correlation between states and allow for reasonable convergence times on phage λ DNA consisting of 48 502 base pairs. The shuffling operations considered are as follows:

1. *Rotation* of all open loops a random distance around the circular DNA without changing their number or the lengths and separations distances of loops.
2. *Shift* the relative starting positions of loops, without changing the number of loops or their lengths.
3. *Squeeze* or redistribute the open pairs among loops, without changing the number of open base pairs or the number of regions.
4. *Exchange* regions by amalgamating or dividing open loops.

It is shown in [SMFB95] that with these shuffling operations detailed balance is satisfied, and stated that the algorithm scales quadratically in the length L of the circular DNA.

We end this section with some concluding remarks about Benham's general statistical mechanical model. Define the partition function Z by

$$Z = \sum_{v \in V} e^{-\beta G(v)}, \tag{6.24}$$

where $\beta = \frac{1}{k_B T}$, T is the absolute temperature, k_B is Boltzmann's constant 13.805×10^{-24} J K^{-1}, and V is the set of all *states*, i.e. all the 2^L possible states of various strand separation resp. non-separation with all possible torsion τ angles. The partition function is given exactly in [Ben90] by

$$e^{-\beta k \theta^2/2} + \sum_{n=1}^{L} \left\{ e^{-\beta(ar+bn)} \int_{-\infty}^{\infty} e^{-\beta\left[\frac{nc\tau^2}{2} + \frac{k}{2}\left(\theta - \frac{n}{10.4} - \frac{n\tau}{2\pi}\right)^2\right]} d\tau \right\},$$

where $\beta = \frac{1}{k_B T}$, and c and k are the constants mentioned earlier.

From the Boltzmann distribution, it follows that the probability of strand separation at site i is

$$p_i = \frac{\sum_{v \in V_i} e^{-\beta G(v)}}{Z}, \tag{6.25}$$

where $V_i = \{v \in V \mid \text{position } i \text{ of strand is separated}\}$. If $p_i = 1$ resp. $p_i = 0$ then the strand at position i is separated resp. not separated.

The ensemble average free energy of all states in V_i is

$$\begin{aligned} \langle G_i \rangle &= E[G(v) : v \in V_i] \\ &= \sum_{v \in V_i} G(v)\, Pr[v] \\ &= \sum_{v \in V_i} G(v) \frac{e^{-\beta G(v)}}{Z_i} \\ &= \frac{\sum_{v \in V_i} G(v) e^{-\beta G(v)}}{\sum_{v \in V_i} e^{-\beta G(v)}}. \end{aligned}$$

Similarly, the average free energy of the equilibrium distribution over all states in V is

$$\langle G \rangle = \frac{\sum_{v \in V} G(v) e^{-\beta G(v)}}{\sum_{v \in V} e^{-\beta G(v)}}.$$

The difference $\Delta G_i = \langle G_i \rangle - \langle G \rangle$ is the *incremental* free energy required to separate the base pair at position i. If $\Delta G_i < 0$, then separation is favored. In [Ben93], ΔG_i is plotted as a function of i for the DNA in pBR322 and ColE1 plasmids of *E. coli*, bacteriophage f1, and the polyoma and bovine papilloma virus genomes. The resulting sites of predicted destabilization in *E. coli* occur both at promotor and at or near terminator regions of certain operons, thus suggesting the existence of a class of prokaryotic transcription units, which are bracketed by destabilized regions. This suggests that the *helix destabilization profile* plot of ΔG_i versus i, a *global* measure, might be combined with hidden Markov models, a *local* measure, in order to determine likely coding regions.

From [Ben93], it was observed that the longest run of AT base pairs in polyoma virus DNA is not contained in any of the 10 most destabilized regions of the molecule. Nevertheless, given enough real strand separation data as well as data output from the Monte Carlo program of [SMFB95], it would be interesting to apply hidden Markov models to determine whether *local* sequence similarity appears to determine where strands separate.

According to Benham (personal communication), easily destabilized sites are strongly correlated with various classes of regulatory regions, where recently found regulation mechanisms specifically involve stress-induced strand separation. A dozen different examples of this phenomenon are now known.

Another application of Benham's statistical mechanical model concerns DNA supercoiling in thermophilic bacteria. We have seen that negative supercoiling can be harnessed for transcription and replication events by lowering the free energy required for strand separation. Similarly, positive supercoiling can prevent denaturation of *in vivo* DNA at temperatures and/or stress conditions, where non-supercoiled DNA would separate. In [Ben96], using the previously developed theoretical model, calculations are given of a *critical* temperature T_c, where DNA denaturation occurs, regardless of amount of positive superhelicity.

Recently, in [FB99], Benham and Fye used the previously derived free energies, but were able to evaluate certain integrals directly, rather than using a Monte Carlo algorithm. With this approach, their method correctly predicted the location and relative frequency of base pair openings in *pBR322* from [KNE88], using an algorithm using $O(L^2)$ operations and $O(L)$ space.

In concluding this section, we mention that it is believed that the determination of strand separation sites depends on *global* information of the entire circular (or locally constrained linear) DNA, rather than on *local* properties of the nucleotide sequence. Protein structure determination, a topic to which we soon turn, is also a global problem, where, in addition to local force contributions, protein folding is known to be influenced by non-local, long-range interactions.

> Which duplex sites are destabilized depends in part on local sequence attributes, with separation energetically favored to occur at A+T-rich sites under normal physiological conditions. But superhelicity globally couples together the secondary structures of every base pair in the molecule. Transition at any one location alters the helicity there, which, by changing the distribution of superhelicity throughout the molecule, alters the level

of stress experienced by every other base pair. [SMFB95]

It remains in our view an interesting open problem to try to quantify to what extent local sequence information is pertinent. With growing genomic database information, by using HMMs along with an implementation of the simulated annealing strand separation algorithm, one might be able to quantify to what extent strand separation is a global versus local phenomenon.

6.3 Amino Acid Pair Potentials

One of the most important open problems in computational biology concerns the computational prediction of the tertiary and quaternary structure of a protein given only the underlying amino acid sequence. Since the enzymatic properties of a protein are determined by its 3-dimensional structure, a feasible computational solution to protein structure prediction would enormously facilitate drug design. During the last few decades, much effort has been made toward solving this problem, with various approaches including

- molecular dynamics,
- secondary structure prediction,
- homology and pattern recognition,
- energy minimization on lattice models using combinatorial optimization methods (Monte Carlo, simulated annealing, genetic algorithms, and constraint programming techniques),
- knowledge-based methods such as amino acid pair potentials and protein threading, etc.

In [Sip90], building on earlier ideas of [MJ85], M. Sippl used the Boltzmann distribution to define an energy function with terms involving *amino acid pair potentials*, computed from a representative database of 3-dimensional coordinates of protein structures. Given an energy function E defined on a finite set V of states, recall that the Boltzmann distribution is defined by

$$p(v) = \frac{e^{\frac{-E(v)}{RT}}}{Z}, \tag{6.26}$$

where the partition function $Z = \sum_{w \in V} e^{\frac{-E(w)}{RT}}$. If the energy function E is not known, but the probabilities $p(v)$ for $v \in V$ can be determined, as in Sippl's case, from the protein database, then by taking the logarithm of (6.26), we can determine $E(v)$:

$$E(v) \quad = \quad -RT \ln p(v) - RT \ln Z. \tag{6.27}$$

It is generally assumed that a protein folds into a unique conformation determined by the global minimum of its free energy. While an exact energy function has yet to be determined, there are contributions from the hydrophobic effect, the electrostatic force between charged amino acids, the Lennard–Jones potential, van der Waals force, etc. A first approximation to such an energy function is to consider only *pairwise* force contributions.

Without knowing an appropriate energy function to minimize, it seems reasonable that the average distance between a given pair of amino acids (say valine–isoleucine) in a representative protein database should correspond to the average energy contribution due to this pair. For example, it is known that proline residues tend to disrupt α-helices, so one expects to find a large average distance between proline and alanine, the latter often found in α-helices. Sippl's idea is to compute a frequency for distances between amino acid pairs, and using (6.26), (6.27) to compute the corresponding *amino acid pair potentials*.

Before proceeding further, a small remark on the choice of a representative protein database is necessary. Since medicinal applications have high priority in biological research, globular proteins such as immunoglobins have received much more attention than other classes. Sippl's work requires the removal of such redundancy from a representative protein database, because the amino acid pair potentials should represent a statistical approximation to a physical energy, rather than an artifact of skewed amino acid pair frequencies occurring in a particular class of proteins. A representative protein database can be produced by using `PDB Select 25`, which selects proteins that share less than 25% sequence homology.

Assume that p is an index varying over all proteins in our selected protein database $\mathcal{P}$. Denote the number of residues in the pth protein by L_p. The amino acid sequence of the pth protein is

$$S_p = \langle S_p(1), S_p(2), \ldots, S_p(L_p) \rangle$$

and the protein's conformation is given by C_p, where $C_p(i,j)$ is the (Euclidean) distance between the α-carbon of the ith and jth residues. Sippl notes that within an α-helix, on average 5.5 Å $< C_p(i, i+4) <$ 6.5 Å, while for extended β-strands, 11.0 Å $< C_p(i, i+4) <$ 14.0 Å.

Define

$$\begin{aligned} lo_k &= \min\{C_p(i,j) \mid k = j - i, p \in \mathcal{P}\}, \\ hi_k &= \max\{C_p(i,j) \mid k = j - i, p \in \mathcal{P}\}, \end{aligned}$$

so that lo_k (resp. hi_k) is the smallest (resp. largest) (Euclidean) distance between residues whose linear distance along the amino acid chain is k units. Let N be the number of intervals into which $[lo_k, hi_k]$ is to be divided (in practice, Sippl sets $N = 20$). For $1 \le s \le N$, define

$$int_k(s) = lo_k + \frac{(s-1)(hi_k - lo_k)}{N}$$

and let $INT_k(s)$ be the interval $(int_k(s-1), int_k(s)]$; i.e.

$$INT_k(s) = \left\{ x \;\middle|\; lo_k + \frac{(s-1)(hi_k - lo_k)}{N} < x \le lo_k + \frac{s(hi_k - lo_k)}{N} \right\}.$$

Let n_k denote the number of amino acid pair observations, regardless of residue; i.e.

$$n_k = |\{(p, i, j) \mid k = j - i, 1 \le i < j \le L_p, p \in \mathcal{P}\}|.$$

For $1 \le s \le N$, define

$$n_k(s) = |\{(p, i, j) \mid k = j - i, 1 \le i < j \le L_p, p \in \mathcal{P}, C_p(i,j) \in INT_k(s)\}|$$

to be the number of amino acid pair observations, regardless of residue, whose Euclidean distance is s. The *reference* amino acid distancy frequency is defined to be

$$f_k(s) = \frac{n_k(s)}{n_k}. \tag{6.28}$$

As indicated above, one can then determine $E_k(s)$, the *reference* potential of mean force of the interaction between two α-carbon atoms of residues at linear distance k. For each possible combination of amino acids a, b (there are 400 such pairs, which cannot be considered symmetric, since there is an amino initial group (NH_2) and a carboxyl terminal group (COOH)), let $n_k^{a,b}$ denote the number of observations of amino acid pairs (a, b) that are located at linear distance k along an amino acid sequence in the protein database; i.e.

$$n_k^{a,b} = |\{(p, i, j) \mid k = j - i, p \in \mathcal{P}, S_p(i) = a, S_p(j) = b\}|.$$

Similarly for $1 \leq s \leq N$, define the number

$$n_k^{a,b}(s) = |\{(p, i, j) \mid k = j - i, p \in \mathcal{P}, S_p(i) = a, S_p(j) = b, C_p(i, j) \in INT_k(s)\}|$$

of observations of amino acid pair (a, b) whose linear distance is k and Euclidean distance is s. Define the *specific* amino acid distance frequencies by

$$g_k^{a,b}(s) = \frac{n_k^{a,b}(s)}{n_k^{a,b}}. \quad (6.29)$$

In Sippl's data set, for $k = 3$, there were 161 distances in the database for the pair alanine–alanine, while there is only one distance for the pair methionine–tryptophan. In general, there may be very few entries of certain pairs (a, b) with respect to other pairs (a', b'), so define the following weighted average of $f_k(s)$ and $g_k^{a,b}(s)$. Let σ be a model-specific parameter ($\sigma = \frac{1}{50}$ in [Sip90]) and define the *normalized* specific amino acid distance frequency by

$$f_k^{a,b}(s) = \frac{f_k(s) + g_k^{a,b}(s) n_k^{a,b} \sigma}{1 + n_k^{a,b} \sigma}. \quad (6.30)$$

Note that $f_k(s)$ and $g_k^{a,b}(s)$ have equal weight after $\frac{1}{\sigma}$ observations. By inverting the Boltzmann distribution, we have

$$E_k^{a,b}(s) = -RT \ln f_k^{a,b}(s) - RT \ln Z_k^{a,b}. \quad (6.31)$$

Define the *net pairwise potential*

$$\begin{aligned}
\Delta E_k^{a,b}(s) &= E_k^{a,b}(s) - E_k(s) \\
&= -RT \left(\ln f_k^{a,b}(s) - \ln f_k(s) + \ln Z_k^{a,b} - \ln Z_k \right) \\
&= -RT \ln \frac{f_k^{a,b}(s)}{f_k(s)} - RT \ln \frac{Z_k^{a,b}}{Z_k}.
\end{aligned}$$

Now, from frequency data extracted from the protein database, there is no possibility of determining the partition function values $Z_k^{a,b}, Z_k$. However, the term $-RT \ln \frac{Z_k^{a,b}}{Z_k}$ is a constant not depending on the state s, and so Sippl assumes that $Z_k^{a,b} \approx Z_k$, and hence the term $-RT \ln \frac{Z_k^{a,b}}{Z_k} \approx 0$. It follows that

$$\Delta E_k^{a,b}(s) = -RT \ln \frac{f_k^{a,b}(s)}{f_k(s)} \quad (6.32)$$

$$= RT \ln(1 + n_k^{a,b} \sigma) - RT \ln \left[1 + n_k^{a,b} \sigma \frac{g_k^{a,b}(s)}{f_k(s)} \right]. \quad (6.33)$$

Note that if the distributions of the $g_k^{a,b}(s)$ are very similar, then $\Delta E_k^{a,b}(s) \approx 0$, and this approach will give no valuable information. However, computations from the protein database indicate quite different distributions.

At this point, one could use Monte Carlo with simulated annealing with the amino acid pair potentials in order to construct a protein folding. To avoid problems of local minima, Sippl introduces an idea now widely used in protein threading. Suppose that S is a (new) amino acid sequence of length L of a protein whose conformation we wish to determine. Let C be a given conformation for an oligopeptide of length L, i.e. $C(i,j)$ is the Euclidean distance between the ith and jth residues. The net potential in folding S in the conformation C is given by

$$\Delta E(S,C) = \sum_{1 \leq i < j \leq L} \Delta E_{j-i}^{S(i),S(j)}(C(i,j)). \tag{6.34}$$

Now, for every L-length subconformation C of a protein in the database $\mathcal{P}$, compute $\Delta E(S,C)$, thus yielding the net potential spectrum for oligopeptide S. From the net potential, one can construct clusters of low-potential conformations.

Nuclear magnetic resonance (NMR) studies have shown that oligopeptides often take on an ensemble of different conformations, rather than having a unique conformation. For instance, Sippl points out that the pentapeptide VNTFV is found in an α-helix in erythrocruorin, but in a β-strand in ribonuclease. Using amino acid pair potentials, Sippl is able to classify small oligopeptides (of length $L \leq 7$) as either *stable* (favoring a particular conformation), *flip-flop* (favoring two or a small number of distinct conformations), *metastable* (favoring a particular conformation, as well as a range of other conformations), and *unstable* (no conformation preference).

In [Krö96], T. Kröger and B. Steipe extend Sippl's approach to compute *vectorial* amino acid pair potentials; e.g. for pair valine–isoleucine, to compute frequencies $f^{V,I}$ with respect to a distance and direction. Their approach is outlined as follows.

For each pair (a,b) of amino acids, measure the 3-diminsional vector $\vec{v}(a,b)$ from a to b in a representative protein database. By translation and rotation, place the α-carbon atom of the amino acid a at the origin, where the nitrogen atom lies in the negative x-axis, and the other carbon atom of the backbone lies on the xy-plane. Define an appropriate 3-dimensional lattice and round the previously calculated vectors to the nearest lattice point, storing the resulting vector frequencies as an oct-tree.

In [Krö96], a number of small modifications of Sippl's approach were taken. For instance, cysteine is a small hydrophilic amino acid with a sulfur atom in its side chain, which often appears in a cysteine–cysteine disulfide bond. Kröger and Steipe distinguish between cysteines depending on whether or not they appear in a disulfide bond – in the former case, cysteine is treated as a 21st amino acid with designation CSS. Another divergence from Sippl's approach lies in the definition of reference frequencies. In our discussion of Sippl's work, the relative frequency $f_k(s)$ is defined by

$$f_k(s) = \frac{n_k(s)}{n_k},$$

as in equation (6.28). This corresponds to the probability that an arbitrary amino acid pair (a,b), at linear distance k in the amino acid chain, lies at Euclidean distance s. By contrast, in [Krö96], for each pair (a,b) of amino acids, the reference frequency $f(\vec{s})$ is defined as follows. Let $\vec{s}$ denote a 3-dimensional vector, varying over grid points. For an amino acid pair

(a, b), the relative frequency

$$f^{a,b}(\vec{s}) = \frac{n^{a,b}(\vec{s})}{n^{a,b}}$$

is compared with the reference frequency for (a, b), given by

$$f(\vec{s}) = \frac{n(\vec{s}) - n^{a,b}(\vec{s})}{n - n^{a,b}}.$$

Thus, in the case of [Krö96], the reference frequency $f^{a,b}(\vec{s})$ measures the probability that in the vector $\vec{s}$ a *different* amino acid pair than (a, b) occurs. In recent work, Kaindl and Steipe have extended this approach to non-contiguous motifs in proteins.

Before closing this section, we make a few remarks about how to perform the translation and rotation required in the amino acid pair vectorial frequencies. First, recall that a rotation by angle θ in the xy-plane can be achieved by application of the rotation matrix

$$\begin{pmatrix} \cos\theta & -\sin\theta & 0 \\ \sin\theta & \cos\theta & 0 \\ 0 & 0 & 1 \end{pmatrix},$$

while a rotation by angle θ in the xz-plane can be achieved by the rotation matrix

$$\begin{pmatrix} \cos\theta & 0 & -\sin\theta \\ 0 & 1 & 0 \\ \sin\theta & 0 & \cos\theta \end{pmatrix}.$$

Recall also that $\cos(-\theta) = \cos\theta$ and $\sin(-\theta) = -\sin\theta$.

With these preliminaries, we now describe the translation and rotations to be made, in order to compute vectors emanating from the α-carbon of a fixed amino acid a in the protein P. First, replace the original coordinates (x, y, z) of each atom of protein P by $(x - x_\alpha, y - y_\alpha, z - z_\alpha)$, where $(x_\alpha, y_\alpha, z_\alpha)$ are the coordinates of the α-carbon of the fixed amino acid in P. This translates the α-carbon to the origin. Now suppose that the coordinates of the nitrogen atom after this translation are (x_0, y_0, z_0). Let $r = \sqrt{x_0^2 + y_0^2}$, and define θ_0 to be the angle such that $\cos\theta_0 = \frac{x_0}{r}$ and $\sin\theta_0 = \frac{y_0}{r}$. Let X be the rotation matrix for angle $-\theta_0$, so that

$$X = \begin{pmatrix} \frac{x_0}{r} & \frac{y_0}{r} & 0 \\ -\frac{y_0}{r} & \frac{x_0}{r} & 0 \\ 0 & 0 & 1 \end{pmatrix}.$$

Let $q = \sqrt{r^2 + z_0^2}$, and define θ_1 to be the angle such that $\cos\theta_1 = \frac{r}{q}$ and $\sin\theta_1 = \frac{z_0}{q}$. Let Y be the rotation matrix for angle $-\theta_1$, so that

$$Y = \begin{pmatrix} \frac{r}{q} & 0 & \frac{z_0}{q} \\ 0 & 1 & 0 \\ -\frac{z_0}{q} & 0 & \frac{r}{q} \end{pmatrix}.$$

Then YX is a rotation matrix that places the nitrogen atom on the positive x-axis. Suppose that after the translation and subsequent rotation YX, the other carbon atom on the backbone has resulting coordinates (x_1, y_1, z_1), and let $s = \sqrt{y_1^2 + z_1^2}$. In a similar fashion to the definition of X and Y, define the rotation matrix Z by

$$Z = \begin{pmatrix} 1 & 0 & 0 \\ 0 & \frac{y_1}{s} & \frac{z_1}{s} \\ 0 & -\frac{z_1}{s} & \frac{y_1}{s} \end{pmatrix}.$$

Then Z rotates the carbon atom with coordinates (x_1, y_1, z_1) to the xy-plane. Finally define U by

$$U = \begin{pmatrix} -1 & 0 & 0 \\ 0 & -1 & 0 \\ 0 & 0 & 1 \end{pmatrix}$$

and note that an application of U rotates the nitrogen, which lay on the positive x-axis, to lie on the negative x-axis.

Summarizing, let $A = UZYX$. Then for each atom b of protein P with coordinates (x, y, z), compute

$$\begin{pmatrix} x' \\ y' \\ z' \end{pmatrix} = A \cdot \begin{pmatrix} x - x_\alpha \\ y - y_\alpha \\ z - z_\alpha \end{pmatrix}.$$

This now allows the determination of vectors $\vec{v}_{a,b}$ from the α-carbon of the fixed amino acid a to the α-carbon of any other amino acid b of the protein P.

In forthcoming work, P. Clote and S. Will have adapted Sippl's approach to develop a genomic motif detection algorithm, making preliminary tests for tRNA detection [Clo98].

6.4 Lattice Models of Proteins

Though experiments on small proteins [Anf73, KS92] suggest that the native state of a protein corresponds to a free energy minimum, this is not yet proven. Nevertheless, this hypothesis is widely accepted, and forms the basis for computational predictions of a protein's conformation from its amino acid sequence.

Molecular dynamics modeling, which simulates the conformational changes of a peptide by taking into account the electrostatic, ionic, van der Waals (dipole–dipole), hydrogen bonding, and other forces considered at the atomic level (for the atoms of the peptide, together with those of the solvent), can currently simulate around 10^{-7} seconds of the folding sequence. This is orders of magnitude less than the time required for a protein to fold (milliseconds to seconds). Moreover, certain studies [Tee86, Tee91] have shown that the energy functions used in molecular dynamics are not fully correct, leading to rather different predictions.

In light of these difficulties, simplified models have been introduced. An important class of simplified models comprises the so-called *lattice models*. The simplifications used in this class of models are as follows: (1) monomers (or residues) are represented using a uniform size, (2) bond length is uniform, and (3) the positions of the monomers are restricted to positions in a regular lattice. In the simplest case, every conformation of a lattice protein is a self-avoiding walk in $\mathbb{Z}^2$ or $\mathbb{Z}^3$ (depending on whether one considers a two-dimensional or three-dimensional lattice). A discussion of lattice proteins can be found in Dill *et al.* [DBY$^+$95].

The most predominant representative of lattice models is the HP model, introduced by Lau and Dill [LD89, LD90]. In this model, the 20-letter alphabet of amino acids (and the corresponding variety of forces between them) is reduced to a two-letter alphabet, namely

6.4.2 *Genetic Algorithm for Folding in the HP Model*

In [UM93], R. Unger and J. Moult described a hybrid genetic algorithm to determine minimal energy conformations on a 2-dimensional square lattice, using the HP model. Building on this, in [BWC00] R. Backofen, S. Will, and P. Clote attempted to quantify the contribution of the hydrophobic force in protein folding, using an extension of the HP model involving Woese's polar requirement (a measure of hydrophobicity) on the 3-dimensional face-centered cubic lattice (FCC). This was done by applying a substantially more efficient version of the Unger-Moult hybrid genetic algorithm to determine a minimal-energy conformation C, where only the hydrophobic force was considered, and then computing the root mean square deviation (RMSD) between conformations C and D, where D is the actual protein conformation, taken from the PDB. The algorithm of [BWC00] uses automorphism groups in handling arbitrary 2- and 3-dimensional lattices, employs *octtrees* for efficient space usage, and performs the energy computation (6.37) in linear, rather than quadratic time.

Given an input length n HP sequence, we maintain a population of P conformations ($P \approx 200$), as represented by a *chromosome*, or relative direction sequences of length $n-1$. The population at time t is denoted $P(t)$. The fitness $F(c)$ of conformation c equals $-E(c)$, where the energy is given by equation (6.36), or the following modification of it using

Algorithm 6.4 Unger–Moult hybrid genetic algorithm

```
t = 0
initialize population P(t) of random coils
best = argmax { F(x) | x in P(t) }
repeat {
     t++
     pointwise mutation
     n=0
     while (n < P) {
          select 2 chromosomes m,f
          produce child c by crossover of m,f
          ave = average( F(m), F(f) )
          if (F(c) >= ave){
               place c in next generation
               n++
               }
          else {
               z = random(0,1)
               if (z < e^(-( ave - F(c) )/T ) ) {
                  place c in next generation
                  n++
                  }
               }
     end while
     update best
until convergence
```

normalized polar requirements:

$$E = - \sum_{1 \leq i < j \leq n} p_i p_j \delta(r_i, r_j) \tag{6.37}$$

where $\delta(r_i, r_j)$ is 1 if if $||r_i - r_j|| = 1$ and $i \neq j \pm 1$, otherwise it is 0, and $p_i = \frac{13-x}{13-4.8}$, where x is the polar requirement value from Table 1.2. This contact energy clearly generalizes that of the HP model.

For each chromosome, choose a site $1 \leq i \leq n-1$ and perform a pointwise mutation at site i with with probability p_m. Each chromosome x is selected for crossover according to its fitness (i.e. with probability $F(x)/\sum_{y \in P} F(y)$), using the roulette wheel technique. It should be noted that this algorithm is not a typical genetic algorithm, but rather a hybrid form that incorporates the Metropolis criterion. The pseudocode is given in Algorithm 6.4.

The pseudocode for the approach in [BWC00] to quantify hydrophobic force in protein folding is given in Algorithm 6.5. The input consists of the α-carbon coordinates from PDB and normalized polar requirement values. This algorithm was run on data from the database of ancient conserved regions drawn from GenBank 101 supplied by W. Gilbert's lab,[1] as well as medium-sized proteins (E. Coli RecA, `2reb`, Erythrocruorin, `1eca`, and Actinidin `2act`). Some sample results are given in the following table (see [BWC00] for more on the methods and the full results):

Name	Energy	RSMD	hyd.meas.1	hyd.meas.2	s.d.(hyd.meas.2)
aat	−12.013371	0.924242	100	61.37	0.4700
acidamy	−10.318862	1.003551	99.94	55.21	0.4756
acyl	−11.396453	0.93119	100	60.70	0.4721
adea	−13.671394	0.880779	100	60.98	0.4701
adh	−10.942641	1.09688	99.87	53.93	0.4754
adk	−15.966104	1.086642	99.23	47.31	0.4694

6.5 Hart and Istrail's Approximation Algorithm

6.5.1 Performance

Recently, the structure prediction problem has been shown to be NP-hard even for the 2-dimensional square lattice and 3-dimensional cubic lattice HP model [BL98, CGP+98] (at present, this problem is open for the 2-dimensional hexagonal and 3-dimensional face-centered cubic lattices). It follows that there cannot be a polynomial-time algorithm to compute the minimal-energy conformation for an arbitrary protein on the HP model (unless P=NP), so that W. Hart and S. Istrail devised an approximation algorithm to determine a conformation provably within a certain bound of the optimal conformation. In the following, for a given HP sequence s, the energy of the minimal-energy conformation (native state) on the 2-dimensional (resp. 3-dimensional) lattice will be denoted as the *minimal energy* $\text{opt}(s)$.

DEFINITION 6.7 (PERFORMANCE)
Let $\mathcal{A}(s)$ denote the energy of that conformation C of the HP sequence s returned by the

[1] We are indebted to W. Gilbert and his lab for generously furnishing data from the database analyzed in their article [dSLK+98].

Algorithm 6.5 Quantifying the Hydrophobic Force in Folding

1. Use GA to determine predicted conformation C as a self-avoiding walk in the FCC lattice.
2. Compute $D_{hp} = (d_{i,j})$, where
$$d_{i,j} = ||r_i - r_j||$$
is the Euclidean distance in conformation C.
3. Compute $D_{pdb} = (e_{i,j})$, where
$$e_{i,j} = \frac{||r_i - r_j||}{ave}$$
in PDB conformation, where `ave` is the average distance between successive α-carbons in the linear chain.
4. Compute $RSMD(D_{hp}, D_{pdb})$:
$$\sqrt{\frac{\sum_{1\leq i<j\leq n}(d_{i,j} - e_{i,j})^2}{\binom{n}{2}}}.$$
5. Generate $M \approx 200$ random coils; compute D_{rc}.

OUTPUT: $RMSD_C$ between conformation C found by GA and conformation from PDB data, and percent contribution of the hydrophobic force.

`hyd.meas.1` is
$$\frac{|\{RC : RSMD_{RC} > RSMD_C\}|}{|\{RC\}|}$$
`hyd.meas.2` is
$$E[\frac{RMSD_{RC} - RMSD_C}{RMSD_{RC}}]$$
and `s.d.(hyd.meas.2)` is the standard deviation for runs this Algorithm.

---

approximation algorithm $\mathcal{A}$. The performance *of $\mathcal{A}$ for a specific sequence s is defined by*

$$\mathcal{R}_{\mathcal{A}}(s) = \frac{\mathcal{A}(s)}{\text{opt}(s)}.$$

Let S_E be the set of all sequences, whose minimal energy is exactly $-E$. The absolute performance *of $\mathcal{A}$ is*

$$\mathcal{R}_{\mathcal{A}} = \inf\{\mathcal{R}_{\mathcal{A}}(s) \mid s \in S_E, E \in \mathbb{N}\},$$

while the asymptotic performance *is*

$$\mathcal{R}_{\mathcal{A}}^{\infty} = \sup_{E\in\mathbb{N}}\{\inf\{\mathcal{R}_{\mathcal{A}}(s) \mid s \in S_E\}\}.$$

Absolute performance is the guaranteed performance that can be achieved for all sequences. Asymptotic performance describes the behavior of $\mathcal{A}$ for long sequences. For instance, suppose that we know that for every sequence s, the minimal energy is $\text{opt}(s) = -4n$, and suppose that the energy of the conformation returned by the algorithm $\mathcal{A}$ is $\mathcal{A}(s) = -2n+1$. Then the performance of $\mathcal{A}$ for the sequence s is

$$\mathcal{R}_{\mathcal{A}}(s) \quad = \quad \frac{-2n+1}{-4n}.$$

The absolute performance is then given by

$$\mathcal{R}_{\mathcal{A}} \quad = \quad \inf_{n\in\mathbb{N}}\left(\frac{-2n+1}{-4n}\right) = \frac{1}{4} \quad \text{(achieved when } n = 1\text{)},$$

and the asymptotic performance is given by

$$\mathcal{R}_{\mathcal{A}}^{\infty} \quad = \quad \lim_{n\to\infty}\left(\frac{-2n+1}{-4n}\right) = \frac{1}{2}.$$

Asymptotic performance seems more appropriate for describing the performance of an algorithm.

6.5.2 Lower Bound

To compute the performance of an algorithm that approximates the minimal energy of a sequence s, it is necessary to have a lower bound for $\text{opt}(s)$. A first bound can be found using the following considerations. For every lattice model L, the number of possible neighbor positions per single monomer is finite and independent of the actual position of the monomer. In the 2-dimensional square lattice, every monomer has 4 possible neighbor sites, and in the 3-dimensional cubic lattice, each monomer has 6 neighbor sites. Let c_L, the *lattice connectivity constant*, denote the maximum number of neighbor positions for a monomer in the lattice L. It follows that for every interior H-monomer, 2 of the c_L possible neighbor positions of an H-monomer are occupied by the neighbors in the HP sequence. For terminal H-monomers, only one of the c_L positions is occupied by a neighbor in the HP-sequence. Thus, every interior (resp. terminal) H-monomer can form at most $c_L - 2$ (resp. $c_L - 1$) contacts. Since we have counted every possible contact twice, it follows that

$$\text{opt}(s) \quad \geq \quad \frac{\mathcal{N}_H(s)(c_l - 2) + \#\text{term}_H(s)}{2}, \tag{6.38}$$

where $\mathcal{N}_H(s)$ is the number of H-monomers in s and $\#\text{term}_H(s)$ is the number of terminal H-monomers in s.

For the square and cubic lattices, an even sharper energy bound can be given. This bound is a consequence of a specific property of the $\mathbb{Z}^d$ lattices, namely that every contact is formed from monomers with different parity, e.g.

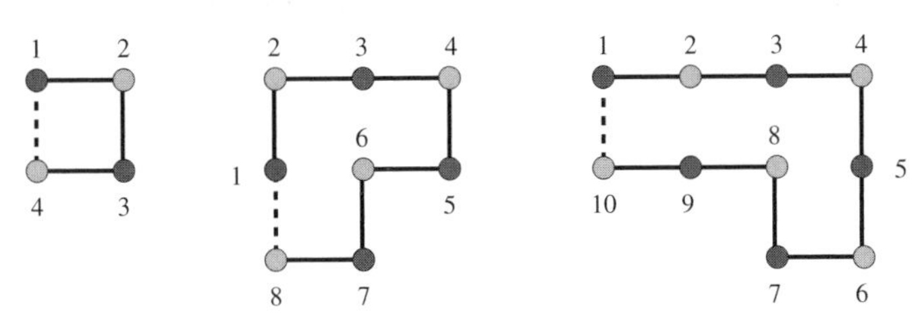

PROPOSITION 6.8
Let s be a sequence and ω be a conformation of s in the 2-dimensional square or 3-dimensional cubic lattice. If the ith and jth monomers form a contact in ω, then i is even and j is odd, or vice versa.

This is a simple consequence of the following proposition.

PROPOSITION 6.9
Let $\vec{x} = (x_1, \ldots, x_d)$ and $\vec{y} = (y_1, \ldots, y_d)$ be two points of $\mathbb{Z}^d$, and let ω be a self-avoiding walk with $|\omega| = n$ such that $\omega(1) = \vec{x}$ and $\omega(n) = \vec{y}$. Then

$$|\omega| \textit{ is even} \quad \Leftrightarrow \quad \left(\sum_{i=1}^{d} x_i \equiv \sum_{i=1}^{d} y_i \right) \mod 2. \tag{6.39}$$

PROOF This is by induction on n. Assume we have proven the claim for all self-avoiding walks ω with $|\omega| \leq n$. Let ω be a walk of length $n+1$. Let $\vec{x} = \omega(1)$ and $\vec{y} = \omega(n+1)$. By the induction hypothesis, we know that

$$|\omega'| \text{ is even} \Leftrightarrow \left(\sum_{i=1}^{d} x_i \equiv \sum_{i=1}^{d} z_i \right) \mod 2$$

where $\omega' = \omega(1) \ldots \omega(n)$ is the subwalk of ω not including $\omega(n+1)$ and $\vec{z} = \omega(n) = (z_1, \ldots, z_d)$. Then $|\omega|$ is even if and only if $|\omega'|$ is odd. Furthermore, $\vec{y} - \vec{z}$ is a unit vector. Hence there exists $1 \leq i \leq k$ such that

$$(y_1, \ldots, y_{i-1}, y_i, y_{i+1}, \ldots, y_d) \quad = \quad (z_1, \ldots, z_{i-1}, z_i \pm 1, z_{i+1}, \ldots, z_d).$$

This implies that $(\sum_{i=1}^{d} x_i \equiv \sum_{i=1}^{d} y_i) \mod 2$ if and only if $(\sum_{i=1}^{d} x_i \not\equiv \sum_{i=1}^{d} z_i) \mod 2$, from which the claim immediately follows. ■

Now, label the H-monomers by X and Y, where all monomers labeled X (resp. Y) have the same parity, and those labeled by X and Y have opposite parity. By the last corollary, every contact is formed between a monomer labeled by X and one labeled by Y. Let $\mathcal{N}_X(s)$ (resp. $\mathcal{N}_Y(s)$) denote the number of H-monomers labeled by X (resp. Y) from s. Since there are two possible labelings, obtained by interchanging X and Y, select the labeling that guarantees that the number of free neighbor positions of monomers labeled by X is at most that of the number of free neighbor positions of monomers labeled Y. In other words, we have either

$$\begin{aligned} &\mathcal{N}_X(s) < \mathcal{N}_Y(s), \text{ or} \\ &\mathcal{N}_X(s) = \mathcal{N}_Y(s) \wedge \#\text{term}_X(s) \leq \#\text{term}_Y(s). \end{aligned} \tag{6.40}$$

Here, $\#\text{term}_X(s)$ (resp. $\#\text{term}_Y(s)$) is the number of terminal monomers labeled X (resp. Y). Since every contact connects an X-monomer with a Y-monomer, and since the number of possible topological neighbors of X-monomers is at most the number of possible neighbors of Y-monomers, it follows that the maximal number of contacts equals the number of possible topological neighbor positions of X-monomers. This yields

$$\text{opt}(s) \quad \geq \quad (2d-2)\mathcal{N}_X(s) + \#\text{term}_X(s), \tag{6.41}$$

where d is the dimension of the lattice.

The main idea of the Hart–Istrail approximation algorithm for the 2-dimensional square lattice is as follows. Consider the sequence

1 3 7 10 12 15 17 19 22 24 26 29 31 33 35 38 42
101000100101001010100101010010101010010001

where we have numbered the H-monomers. Even H-monomers are shown in gray, odd H-monomers in black. By the above convention, there are fewer even monomers, which implies that the monomers $10, 12, 22, 24, 26, 38$, and 42 are labeled by X, while $1, 3, 7, 15, 17, 19, 29, 31, 33$, and 35 are labeled by Y ($\mathcal{N}_X(s) = 7$, $\mathcal{N}_Y(s) = 10$). We now search for a point in the sequence with at least half of the X-monomers on one side, and half of the Y on the other side (in general, we try to maximize the Xs on one side and Ys on the other side, but we always get at least half/half). This is called the *folding point*. In our case, the folding point is between monomer 19 and 22. Then we align the Xs in a column on one side, with the Ys in a column on the other side, such that the aligned X- and Y-monomers form contacts. Proposition 6.8 requires that we have at least distance 2 between X-monomers (and similarly for the Y monomers). One possible alignment is given in Figure 6.13. Using this alignment, our criteria for the selection of the folding point guarantees that we have at least $\frac{\mathcal{N}_X(s)}{2}$ contacts, which implies that

$$\mathcal{R}_A^{\infty} \quad \geq \quad \lim_{\mathcal{N}_X(s) \to \infty} \frac{\frac{\mathcal{N}_X(s)}{2}}{2\mathcal{N}_X(s) + \#\text{term}_X(s)} = \frac{1}{4}. \tag{6.42}$$

Now only the connection between the monomers aligned in the two columns is missing. In principle, we could use an arbitrary self-avoiding walk of the appropriate length connecting two successive monomers in a column. But we have to guarantee that the combination of these walks is a self-avoiding walk itself. The simplest manner of achieving this is to use a U-formed walk for all connections, as in the following diagram:

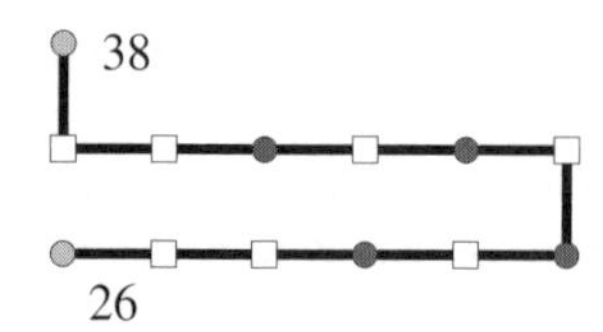

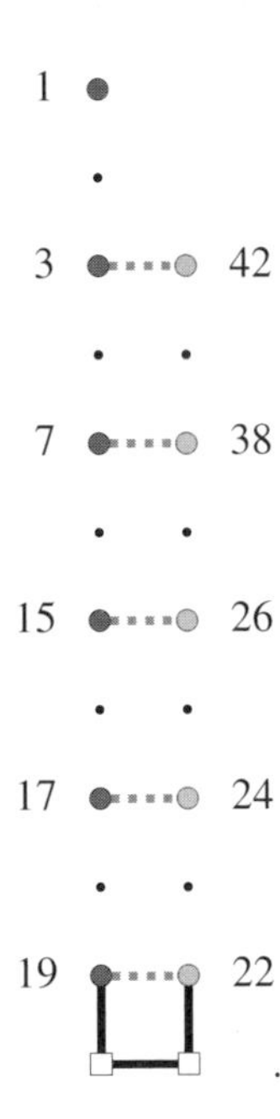

Figure 6.13 Possible alignment with folding point between monomers 19 and 22. The X-monomers are shown in gray, the Ys in black. Note that the distance between two Xs (resp. two Ys) is always 2.

The final conformation is given in the left part of Figure 6.14. The right part is an alternative conformation that has an additional contact. The Hart–Istrail algorithm would produce only the left conformation, since it is easier to calculate. Furthermore, it has the same absolute and asymptotic performance compared with an algorithm that uses a more sophisticated form of connection for the aligned monomers (as in the right conformation). The conformations of the left type are called *basic U-folds*.

The alignment given in Figure 6.13 is not the only possible alignment. An alternative is given in Figure 6.15 together with a possible conformation. But again, this is not used by the approximation algorithm, since we get the same absolute and asymptotic performance.

6.5.3 *Block Structure, Folding Point, and Balanced Cut*

In the following, we assume for a given sequence s a labeling of H-monomers as X- and Y-monomers, as defined by equation (6.40). The labeling of the H-monomers naturally gives rise to a decomposition of a sequence into *blocks* containing only H-monomers having the same label. Such a block always starts and ends with an H-monomer. In the following, we use $\text{subseq}(s, i, j)$ with $i \leq j$ to denote the subsequence of s starting with i and ending with j. For instance, consider the sequence

$$s = 101000100101.$$

Then $\text{subseq}(s, 1, 7) = 1010001$ is one block. The other blocks are $\text{subseq}(s, 1, 3)$, $\text{subseq}(s, 3, 7)$ and $\text{subseq}(s, 10, 12)$, but not, e.g., $\text{subseq}(s, 7, 10)$.

Formally, we say that a sequence $\text{subseq}(s, i, j)$ is a *block* if the sequence starts and ends

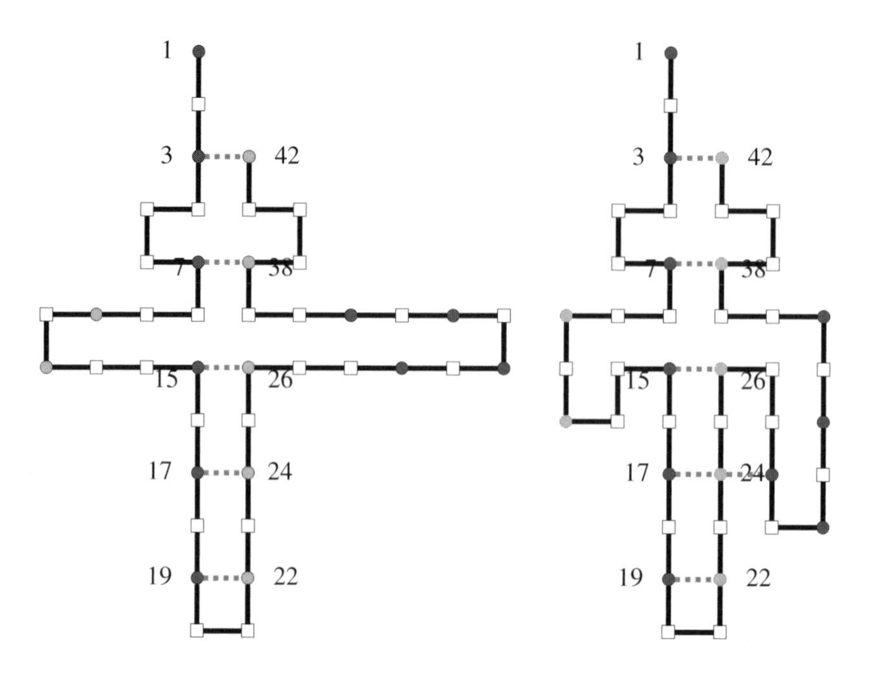

Figure 6.14 Two possible conformations for the alignment given in Figure 6.13. The left conformation is the one output by the approximation algorithm.

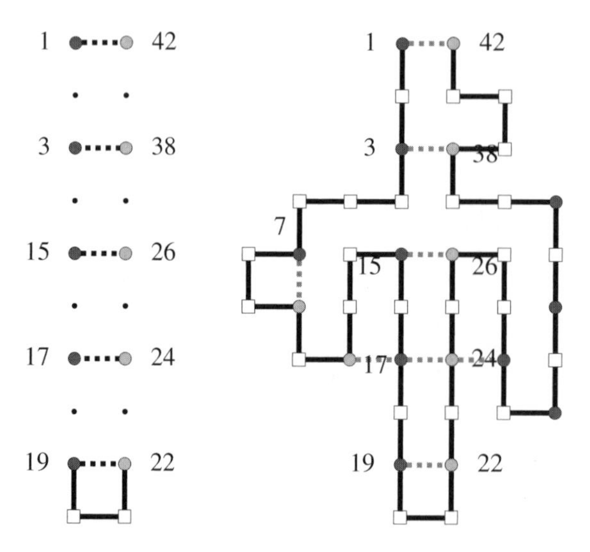

Figure 6.15 An alternative alignment together with a possible final conformation.

with 1, and if the 1s are separated by an odd number of 0s, i.e. if

$$\mathrm{subseq}(s,i,j) \in 1\left(\big(\bigcup_{l\geq 0} 0^{2l+1}\big)1\right)^{*}.$$

Note that every block has odd length. The *label* of a block is the label of its H-monomers. A block $\mathrm{subseq}(s,i,j)$ is called *maximal in* s if it cannot be extended, i.e. there is no $l > j$ (resp. $l < i$) such that $\mathrm{subseq}(s,i,l)$ (resp. $\mathrm{subseq}(s,l,j)$) is a block.

For every sequence, there is a unique decomposition of this sequence into maximal blocks, which are separated by block separators. This is called the *block structure of* s. An example of such a decomposition is

$$s \;=\; \underbrace{1010001}\overset{\overset{a}{\downarrow}}{\overline{0000}}\underbrace{101}\overset{\overset{b}{\downarrow}}{\overline{00}}\underbrace{10101}\overset{\overset{c}{\downarrow}}{\overline{00}}\underbrace{10101}\overset{\overset{d}{\downarrow}}{}\underbrace{1010101}\overset{\overset{e}{\downarrow}}{\overline{00}}\underbrace{1000101}. \tag{6.43}$$

The sequences between consecutive blocks are called *block separators*. In the above example, the sequences marked $a, \ldots, e$ are block separators. Note that block separators consist of a even number of 0s. Furthermore, note that a block separator can be the empty sequence (such as the block separator d). In the following, we fix a sequence s with block structure $s = z_0 b_1 z_1 \ldots b_k z_k$, where the b_is are maximal blocks, and the z_is are the block separators.

We now consider the problem of finding a folding problem that 'balances' the number of X-monomers on the one side with the number of Y-monomers on the other side. This yields the following definition of a folding point. In our discussion below, we consider only sequences that have at least 2 blocks.

DEFINITION 6.10 (CUT)
Let s be a sequence. A cut c for s *is a triple* (s_α, s_l, s_β) *such that there exists* $i \leq k$ *with*

$$\begin{aligned} s_\alpha &= z_0 b_1 z_1 \ldots b_i, \\ s_l &= z_i, \\ s_\beta &= b_{i+1} z_{i+1} \ldots b_k z_k. \end{aligned}$$

The value of c is defined as

$$\mathrm{val}(c) = \max \left\{ \begin{array}{c} \min\{\mathcal{N}_X(s_\alpha), \mathcal{N}_Y(s_\beta)\}, \\ \min\{\mathcal{N}_Y(s_\alpha), \mathcal{N}_X(s_\beta)\} \end{array} \right\}.$$

A cut is of XY type, if

$$\mathrm{val}(c) = \min\{\mathcal{N}_X(s_\alpha), \mathcal{N}_Y(s_\beta)\},$$

and of YX type if

$$\mathrm{val}(c) = \min\{\mathcal{N}_Y(s_\alpha), \mathcal{N}_X(s_\beta)\}.$$

A cut c is called maximal *if its value is maximal.*

A cut is called *balanced* if

$$\mathrm{val}(c) \geq \frac{\mathcal{N}_X(s)}{2}.$$

THEOREM 6.11 (HART AND ISTRAIL)
Every maximal cut is balanced.

A cut $c = (s_\alpha, s_l, s_\beta)$ is a *folding point* if it is a maximal cut and

$$\begin{aligned} c \text{ of } XY\text{-type} &\implies & s_\alpha \text{ ends with an } X\text{-monomer} \\ & & \text{and } s_\beta \text{ begins with a } Y\text{-monomer}, \\ c \text{ of } YX\text{-type} &\implies & s_\alpha \text{ ends with a } Y\text{-monomer} \\ & & \text{and } s_\beta \text{ begins with an } X\text{-monomer}. \end{aligned}$$

PROPOSITION 6.12
Every sequence s has a folding point.

PROOF We consider only sequences whose maximal cut has a value greater than 0. Let s be a sequence, and let $c = (s_\alpha, s_l, s_\beta)$ be a maximal cut for s, where

$$\begin{aligned} s_\alpha &= z_0 b_1 z_1 \ldots b_i, \\ s_l &= z_i, \\ s_\beta &= b_{i+1} z_{i+1} \ldots b_k z_k. \end{aligned}$$

We will consider only the case that c is of XY-type. The other case is analogous. If c is of XY-type, then we distinguish two cases:

1. b_i is labeled by X. Then b_{i+1} is labeled by Y. By the definition of block structure, the last monomer of b_i is an H-monomer labeled by X. Furthermore, the first monomer of b_{i+1} is an H-monomer labeled by Y. Hence, c is a folding point.
2. b_i is labeled Y. Then b_{i+1} is labeled by X. Let c' be the cut $(s'_\alpha, s'_l, s'_\beta)$ with

$$\begin{aligned} s'_\alpha &= z_0 b_1 z_1 \ldots b_i z_i b_{i+1}, \\ s'_l &= z_{i+1}, \\ s'_\beta &= b_{i+2} z_{i+2} \ldots b_k z_k. \end{aligned}$$

Then c' is a cut satisfying

$$\mathcal{N}_X(s'_\alpha) > \mathcal{N}_X(s_\alpha) \quad \text{and} \quad \mathcal{N}_Y(s'_\beta) = \mathcal{N}_Y(s_\beta).$$

Hence, $\mathrm{val}(c') \geq \mathrm{val}(c)$, which implies that c' is maximal. By the previous case, c' is a folding point.

■

Hence, we have only to search through all possible cuts of s to find a folding point, which can be done in linear time. Given the folding point, we can perform the basic U-fold as given by Algorithm 6.6 and its subroutines (Algorithm 6.7). The performance of this algorithm was calculated earlier in equation (6.42).

Algorithm 6.6 Basic U-Fold

```
UFold(s) {
   calculate block structure
   find folding point c = (s_α, s_l, s_β)
   if (c is of XY-type) {
      t_α  =  s_α with all Y-monomers substituted by 0
      t_β  =  s_β with all X-monomers substituted by 0
   }
   else {
      t_α  =  s_α with all X-monomers substituted by 0
      t_β  =  s_β with all Y-monomers substituted by 0
   }
   ω_α  = foldpart(reverse(t_α),(0 0),(-1 0))
   ω_β  = foldpart(t_β,(1 0),(1 0))
   if (|s_l| == 0)
      return(append(reverse(ω_α),ω_β))
   else {
      ω_l  = Uloop(|s_l|/2,(0 -1),(0 -1),(1 0))
      return(append(reverse(ω_α),ω_l,ω_β))
   }
}
```

6.6 Constraint-Based Structure Prediction

Despite the extended use of lattice models, most of the techniques used for protein structure prediction are heuristic methods. Although they may perform well in finding local or perhaps even global optima, one does not know when a global optimum has been found. There are (to the best of our knowledge) only two methods that are able to find global optima and to prove optimality for the HP-model, namely the 'constraint hydrophobic core construction' (CHCC) method by K. Yue and K. Dill [YD93, YD95], and the constraint-based approach by R. Backofen [Bac98]. The later approach has been applied to another lattice model with a more complex energy function in [BWBB99]. We outline the basic constraint formulation that underlies the latter search algorithm. Details of the algorithm in [Bac98] lie outside the scope of this text.

The algorithm is based on constraint optimization, which is the combination of two principles, namely *generate-and-constraint* with *branch-and-bound*. To apply constraint optimization, we have to transform the protein structure prediction problem into a constraint problem. A constraint problem consists of a set of variables together with some constraints on these variables. In the following, we fix a sequence s of length n.

The constraint problem for protein structure prediction consists of variables, which range over finite domains. Additionally, one must also use boolean constraints, entailment constraints, and reified constraints. By reified constraints, we mean constraints of the form

$$(\mathtt{x} = 1) \leftrightarrow (\phi),$$

where $\mathtt{x}$ is a boolean variable and ϕ is a finite domain constraint. The variable $\mathtt{x}$ is 1 if the *constraint store* (i.e. the collection of constraints generated at this point in the execution of the program) entails ϕ, and 0 if the constraint store disentails ϕ. A constraint store *entails*

Algorithm 6.7 Subroutines Uloop and foldpart

```
subroutine Uloop(len,start,loopdir,onestep) {
  if (len == 0)
     return(ε)
  else {
     ω(1) = start
     for (i = 1; i ≤ len/2 − 1; i++)
        ω(i+1) = ω(i) + loopdir
     ω(len/2 + 1) = ω(len/2) + onestep
     for (i = len/2 + 1; i ≤ len − 1; i++)
        ω(i+1) = ω(i) − loopdir
     return(ω)
  }
}

subroutine foldpart(seq,start,dir) {
  ω(1) = start
  i = 2
  while (i ≤ |seq|) {
     if (s_i == 0) {
       zerolen = max{j+1 | i+j ≤ |seq| ∧ s_i s_{i+1} ... s_{i+j} ∈ 0*}
       looplen = zerolen-1
       ω(i)...ω(i+looplen−1) = Uloop(looplen,ω(i−1)+dir,dir,(0 1)ᵀ)
       ω(i+looplen) = ω(i+looplen−1) − dir
       i = i+looplen+1
     }
     else {
       ω(i) = ω(i−1) + (0 1)ᵀ.
       i = i+1
     }
  }
  return(ω)
}
```

a constraint ϕ if every valuation that makes the constraint store valid also makes ϕ valid. It *disentails* ϕ if the conjunction of ϕ with the constraint store is not satisfiable. We use also entailment constraints of the form $\phi \to \psi$, which are interpreted as follows. If a constraint store entails ϕ, then ψ is added to the constraint store. Finite domain constraints and reified constraints can be encoded directly in many modern constraint programming languages.

Given an HP sequence $s = s_1, \ldots, s_n \in \{0,1\}^n$ of length n, we can encode the space of all possible conformations of s as a constraint problem in the following manner. For every $1 \leq i \leq n$, we introduce new variables $\mathtt{X}_i$, $\mathtt{Y}_i$ and $\mathtt{Z}_i$, which denote the x-, y-, and z-coordinates of $c(i)$, the conformation position of the ith monomer s_i of the HP sequence s.

Since we are using the 3-dimensional cubic lattice, we know that these coordinates are all integers. However, we can even restrict the possible values of these variables to the finite domain $[1..2n]$.[2] This is expressed by introducing the constraints

$$\mathtt{X}_i \in [1..(2 \cdot n)] \wedge \mathtt{Y}_i \in [1..(2 \cdot n)] \wedge \mathtt{Z}_i \in [1..(2 \cdot n)] \tag{6.44}$$

for every $1 \leq i \leq n$. The self-avoiding condition is just $(\mathtt{X}_i, \mathtt{Y}_i, \mathtt{Z}_i) \neq (\mathtt{X}_j, \mathtt{Y}_j, \mathtt{Z}_j)$ for $i \neq j$.[3] Next we want to express that the distance between two successive monomers is 1, i.e.

$$||(\mathtt{X}_i, \mathtt{Y}_i, \mathtt{Z}_i) - (\mathtt{X}_{i+1}, \mathtt{Y}_{i+1}, \mathtt{Z}_{i+1})|| = 1.$$

Although this is some sort of constraint on the monomer position variables $\mathtt{X}_i, \mathtt{Y}_i, \mathtt{Z}_i$ and $\mathtt{X}_{i+1}, \mathtt{Y}_{i+1}, \mathtt{Z}_{i+1}$, it cannot be expressed directly in most constraint programming languages. Hence, we must introduce for every monomer i with $1 \leq i < n$ three variables $\mathtt{Xdiff}_i$, $\mathtt{Ydiff}_i$, and $\mathtt{Zdiff}_i$, having values 0 or 1. Then we can express the unit vector distance constraint by

$$\begin{aligned} \mathtt{Xdiff}_i &= |\mathtt{X}_i - \mathtt{X}_{i+1}|, & \mathtt{Zdiff}_i &= |\mathtt{Z}_i - \mathtt{Z}_{i+1}|, \\ \mathtt{Ydiff}_i &= |\mathtt{Y}_i - \mathtt{Y}_{i+1}|, & 1 &= \mathtt{Xdiff}_i + \mathtt{Ydiff}_i + \mathtt{Zdiff}_i. \end{aligned}$$

The constraints described above span the space of all possible conformations. In other words, every valuation of $\mathtt{X}_i, \mathtt{Y}_i, \mathtt{Z}_i$ satisfying the constraints introduced above is an *admissible* conformation for the sequence s, i.e. a self-avoiding walk of s. Given partial information about $\mathtt{X}_i, \mathtt{Y}_i, \mathtt{Z}_i$ (expressed by additional constraints as introduced by the search algorithm), we call a conformation c *compatible* with these constraints on $\mathtt{X}_i, \mathtt{Y}_i, \mathtt{Z}_i$ if c is admissible and c satisfies the additional constraints.

In order to use constraint optimization, we must encode the energy function. For HP-type models, the energy function can be calculated if we know for every pair of monomers (i, j) whether i and j form a contact. To this end, for every pair (i, j) of monomers with $i + 1 < j$, we introduce a variable $\mathtt{Contact}_{i,j}$. The variable $\mathtt{Contact}_{i,j}$ is 1 if i and j have a contact in every conformation which is compatible with the valuations of $\mathtt{X}_i, \mathtt{Y}_i, \mathtt{Z}_i$, and 0 otherwise. We can express this property in constraint programming as follows:

$$\begin{aligned} \mathtt{Xdiff}_{i,j} &= |\mathtt{X}_i - \mathtt{X}_j|, \qquad \mathtt{Zdiff}_{i,j} = |\mathtt{Z}_i - \mathtt{Z}_j|, \\ \mathtt{Ydiff}_{i,j} &= |\mathtt{Y}_i - \mathtt{Y}_j|, \qquad \mathtt{Contact}_{i,j} \in \{0,1\}, \\ (\mathtt{Contact}_{i,j} = 1) &\leftrightarrow (\mathtt{Xdiff}_i + \mathtt{Ydiff}_i + \mathtt{Zdiff}_i = 1), \end{aligned} \tag{6.45}$$

---

[2] We could have used $[1..n]$. However, the domain $[1..2n]$ is more flexible, since we can assign an arbitrary monomer to the vector (n, n, n), and still have the possibility of representing all possible conformations.

[3] This cannot be directly encoded in most constraint programming languages, but we reduce these constraints to difference constraints on integers.

where $\texttt{Xdiff}_{i,j}$, $\texttt{Xdiff}_{i,j}$, and $\texttt{Zdiff}_{i,j}$ are new variables. The constraint (6.45) is an example of a reified constraint.

Using the variables $\texttt{Contact}_{i,j}$, we can easily encode the energy function, which is subject to constraint optimization. For the HP model, we introduce a variable HHContacts that counts the number of contacts between H-monomers. Formally, HHContacts is defined by

$$\texttt{HHContacts} = \sum_{\substack{i+1<j\wedge \\ s(i)=H\wedge s(j)=H}} \texttt{Contact}_{i,j}. \tag{6.46}$$

We now define a variable Energy, and add the constraint

$$\texttt{Energy} \quad = \quad -\texttt{HHContacts}.$$

Summarizing, we have encoded self-avoiding walks by means of the variable Energy.

We can now describe the search procedure, which is a combination of generate-and-constraint and branch-and-bound. In a generate step, select an undetermined variable var from the set of variables $\{\texttt{X}_i, \texttt{Y}_i, \texttt{Z}_i \mid 1 \le i \le n\}$ according to some specified selection strategy. A variable is *determined* if its associated domain consists of only one value, and *undetermined* otherwise. Next, select a value val from the associated domain, and set the variable to this value in the first branch (i.e., the constraint $var = val$ is added to the constraint store), and the search algorithm is called recursively. In the second branch, which is visited after the first branch is completed, the constraint $var \neq val$ is added to the constraint store.

By constraint propagation, each insertion of a constraint leads to a narrowing of some (or many) domains of variables or even to failure, both of which prune the search tree by removing inconsistent alternatives. Thus, the search is done by alternating constraint propagation and branching with constraint insertion. The generate-and-constraint steps are iterated until all variables are determined (which implies that a valid conformation is found). If we have found a valid conformation c, then the constraints will guarantee that Energy is determined. Let E_c be associated value of Energy. Then the additional constraint

$$\texttt{Energy} \quad < \quad E_c \tag{6.47}$$

is added, and the search is continued (via backtracking) in order to find the next best conformation, which must have a smaller energy than the previous ones due to the constraint (6.47). This implies that the algorithm finally finds a conformation with minimal energy.

6.7 Protein Threading

6.7.1 Definition

The protein threading problem is a variant of the protein structure prediction problem, where we have a sequence s with known structure, and we want to determine the structure of a sequence s' that is homologous to s. The fact that s and s' are homologous could be derived using alignment distances, or homology could be known, based on biological reasons. The idea is to use the known structure of s to guide structure prediction for s' by simultaneously aligning s' with s and with the known structure of s. In the following, sequences are always protein sequences, using the one-letter alphabet for amino acids.

DEFINITION 6.13 (CORE MODEL)
Let s be a sequence. A core model *for s is a tuple $(m, \vec{c}, \vec{\lambda}, \vec{l}_{\min}, \vec{l}_{\max})$, where*

$$\begin{aligned}\vec{c} &= (c_1, \dots, c_m),\\ \vec{\lambda} &= (\lambda_0, \dots, \lambda_m),\\ \vec{l}_{\min} &= (l_0^{\min}, \dots, l_m^{\min}),\\ \vec{l}_{\max} &= (l_0^{\max}, \dots, l_m^{\max}),\end{aligned}$$

such that

$$|s| = \lambda_0 + \sum_{1 \le i \le m} (c_i + \lambda_i)$$

and

$$\forall 1 \le i \le m : \; l_i^{\min} \le \lambda_i \le l_i^{\max}.$$

Given a core model $(m, \vec{c}, \vec{\lambda}, \vec{l}_{\min}, \vec{l}_{\max})$ for s, we define the ith core region *of s to be the set of positions*

$$C_i = \left\{ \lambda_0 + \sum_{1 \le j < i} (c_j + \lambda_j) + k \;\middle|\; 1 \le k \le c_i \right\}.$$

The jth position of the ith core is denoted by $C_{i,j}$.

Usually, the core model represents the collection of conserved parts of the secondary structure of s. In this case, the core regions $C_1, \dots, C_m$ correspond to secondary structure elements (α-helices, β-sheets) of s. Let $c_1, \dots, c_m$ denote the lengths of the core regions $C_1, \dots, C_m$. For $1 \le i < m$, the integer λ_i denotes the length of the non-conserved loop or coil region between core region C_i and C_{i+1}. The value λ_0 is the length of the N-terminal loop, while λ_m is the length of the C-terminal loop.

A *threading* of sequence s' through the core model for s is a mapping of the core positions to consecutive positions of s'. Since we are using consecutive regions, a threading is uniquely determined by the mapping of the first position of every core region. Furthermore, this implies that there are no gaps allowed in core regions. All gaps in the alignment must occur in the loop regions. In insertion and deletion positions in the loop regions, one must obey the length restrictions imposed by the core model. Here, $l_i^{\min}$ is the minimal length needed to connect C_i and C_{i+1} according to stereochemical restrictions (depending, e.g., on the distance between the last position in C_i and the first position of C_{i+1} in the structural model of s). The value $l^{\max}$ is the maximal length allowed for the loop region i, and can be used to encode some biological knowledge of the relative distance between core regions C_i and C_{i+1}. If no knowledge is supplied, then $l^{\max}$ is set to ∞.

DEFINITION 6.14
Let $\mathcal{M}_C = (m, \vec{c}, \vec{\lambda}, \vec{l}_{\min}, \vec{l}_{\max})$ be a core model for a sequence s. Let s' be a sequence. A threading *of s' through the core model $\mathcal{M}_C$ for s is a vector*

$$\vec{t} = (t_1, \dots, t_m) \in \mathbb{N}^m$$

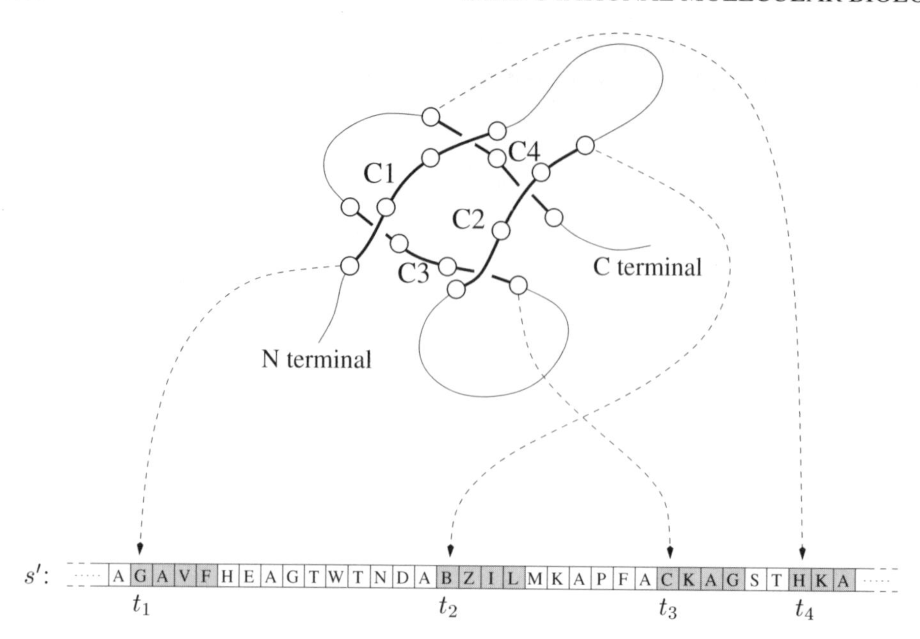

Figure 6.16 Schematic view of a threading. The structure of a sequence s together with its core model is shown. The core model consists of 4 core regions C1...C4 with core lengths 4, 4, 4, and 3. The arrows indicate the threading (t_1, t_2, t_3, t_4) of s' through the core model for s.

such that

$$1 + l_0^{\min} \leq t_1 \leq 1 + l_0^{\max}, \tag{6.48}$$

$$\forall 1 \leq i < m : (t_i + c_i + l_i^{\min} \leq t_{i+1} \leq t_i + c_i + l_i^{\max}), \tag{6.49}$$

and

$$t_m + c_m + l_m^{\min} \leq |s'| + 1 \leq t_m + c_m + l_m^{\max}. \tag{6.50}$$

In the following, we set $c_0 = 0$ for convenience. A schematic view of a threading (for a specific core model) is given in Figure 6.16.

The conditions (6.48)–(6.50) are called *ordering constraints.* These constraints imply so-called *spacing constraints*, which constitute a domain for the ith value of an arbitrary threading

$$\forall 1 \leq i \leq m : \left[1 + \sum_{j<i} (c_j + l_j^{\min}) \leq t_i \leq |s'| + 1 - \sum_{j \geq i} (c_j + l_j^{\min}) \right] \tag{6.51}$$

DEFINITION 6.15 (INTERACTION GRAPH)
Let s be a sequence with a core model $\mathcal{M}_C = (m, \vec{c}, \vec{\lambda}, \vec{l}_{\min}, \vec{l}_{\max})$. An interaction graph $\mathcal{I}$ for $\mathcal{M}_C$ is a graph (V, E), where $E \subseteq V^2$ and V is the set of all core regions, i.e.

$$V = \{C_i \mid 1 \leq i \leq m\}.$$

The interaction graph describes which core regions contain core positions that are 'neighbors' in some biochemical sense, i.e. that are core positions that interact in the folded structure. The corresponding interactions in a threaded structure of s' will be evaluated by the scoring function.

DEFINITION 6.16 (SCORING FUNCTION)
Let s be a sequence with core model $\mathcal{M}_C = (\vec{c}, \vec{\lambda}, \vec{l}_{\min}, \vec{l}_{\max})$ and interaction graph $\mathcal{I}$. A scoring function g for s, $\mathcal{M}_C$, and $\mathcal{I}$ consists of two functions $g_1 \in \mathbb{N}^2$ and $g_2 \in \mathbb{N}^4$ with the property that

$$g_2(i,j,k,l) \neq 0 \Leftrightarrow (C_i, C_j) \in \mathcal{I}. \tag{6.52}$$

Given a threading $\vec{t}$ of s' to s under the core model $\mathcal{M}_C$, the score *of $f(\vec{t})$ of $\vec{t}$ is defined by*

$$f(\vec{t}) = \sum_{i=1}^{m} g_1(i, t_i) + \sum_{i=1}^{m} \sum_{j>i}^{m} g_2(i, j, t_i, t_j).$$

This is the form of scoring function that is most often used in protein threading, where only pairwise interactions are considered. A possible source for the pairwise scoring function g_2 is given by the method of [Sip90, Sip93] described in Section 6.3. Sippl's technique, building on that of Myazawa–Jernigan [MJ85], calculates pairwise potentials from the PDB database of known protein structures. Note that equation (6.52) indicates which core regions interact, and hence could be used as the defining equation for the interaction graph $\mathcal{I}$ given the scoring function g_2.

In general, higher-order interactions could be admitted. To include these, one must extend the definition of interaction graph to that of an interaction hypergraph. Furthermore, one must introduce $2n$-ary functions g_n in order to implement n-ary interactions. If the core model has m regions, then $n \leq m$. Hence, the fully general form of scoring function is

$$\begin{aligned} f(\vec{t}) &= \sum_{i_1} g_1(i_1, t_{i_1}) + \sum_{i_1} \sum_{i_2 > i_1} g_2(i_1, i_2, t_{i_1}, t_{i_2}) + \dots \\ &\quad + \sum_{i_1} \sum_{i_2 > i_1} \cdots \sum_{i_{m-1} > i_m} g_m(i_1, i_2, \dots, i_m, t_{i_1}, t_{i_2}, \dots, t_{i_m}). \end{aligned}$$

To illustrate the principles behind protein threading, we recall the example given in [LS96], which shows two possible threadings of the ancient homologues sperm whale myoglobin ([Wat69], PDB code 1MBN), and the alpha chain of human hemoglobin ([KRA92], PDB code 1DXT). In Figure 6.17, the sequences of these proteins are illustrated together with the core regions, as well as the threads of the core regions. The threading technique can produce an appropriate structure prediction, whereas traditional alignment methods fail due to the low sequence similarity (see [SS91]). In Figures 6.18 and 6.19, we show the structure of these proteins, where additionally the core regions and the threads of the core regions are indicated. One can see that the quality of the threading depends very much on the core model.

6.7.2 A Branch-and-Bound Algorithm

In the next section, we prove that the protein threading problem is NP-complete, and hence, unless P = NP, there cannot be an optimal protein threading algorithm with polynomial

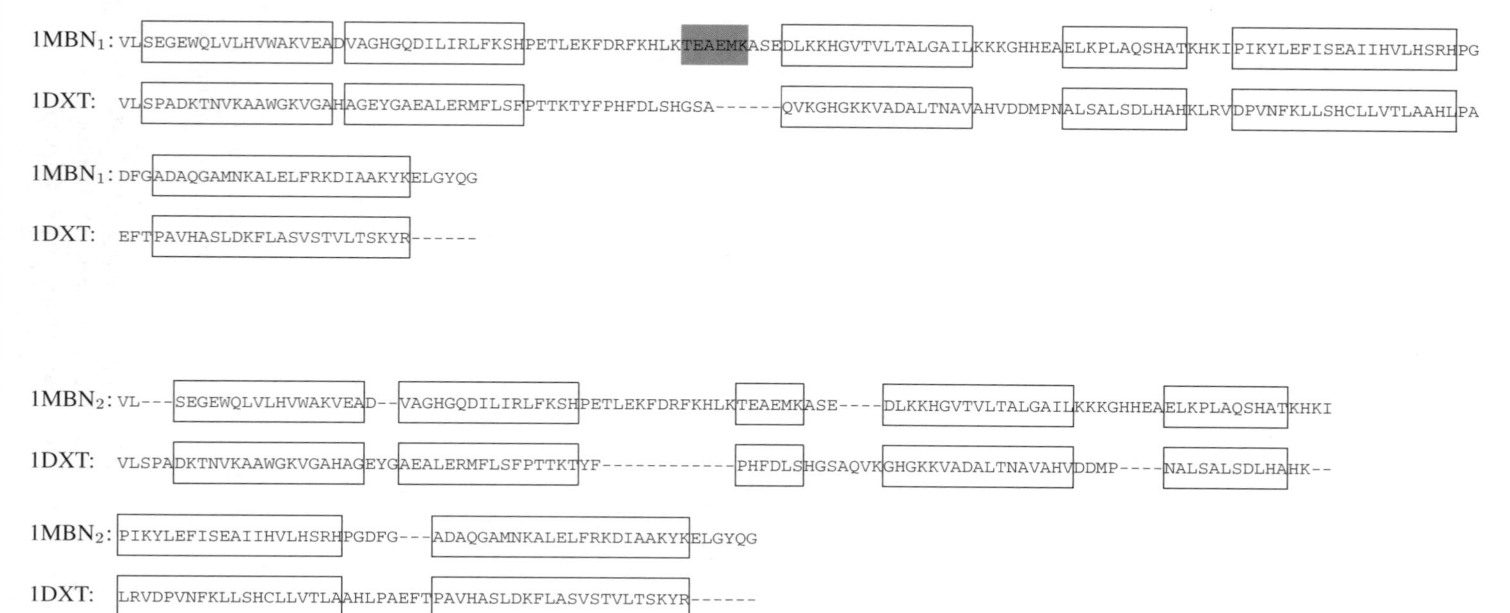

Figure 6.17 Sequences of whale sperm myoglobin ([Wat69], PDB code 1MBN) and human hemoglobin ([KRA92], PDB code 1DXT, chain α) together with the core regions and threadings. The first pair shows the core model for 1MBN without Helix D, together with the optimal threading for 1DXT. The second pair shows the core model for 1MBN including Helix D, and again the corresponding optimal threading (both threadings are taken from [LS96]). Note that a threading predicts only the positions for the core regions, which implies it predicts the number of gaps that have to be introduced in the loop regions; however, it does not predict the exact positions of the gaps. For this reason, we have displayed the gaps always at the end of the loop regions.

runtime. In such cases, often a branch-and-bound approach is used. To do this, we need first a technique to branch on sets of threadings, and second a lower bound on the score for a set of threadings. For this purpose, it is more comfortable to work with relative positions and lengths instead of absolute positions and lengths. In the following, we fix a sequence s together with core model $\mathcal{M}_C = (m, \vec{c}, \vec{\lambda}, \vec{l}_{\min}, \vec{l}_{\max})$ for s.

DEFINITION 6.17 (RELATIVE THREADING)
Let $\vec{t}$ be a threading of s' through the core model $\mathcal{M}_C$ of s. The relative threading *of $\vec{t}$ is the vector $\vec{t^r} = (t_1^r, \ldots, t_m^r)$ with*

$$t_i^r = t_i - \sum_{j<i} (c_j + l_j^{\min}).$$

We define $\vec{l^r} = (l_i^r, \ldots, l_m^r)$ by

$$l_i^r = l_i^{\max} - l_i^{\min},$$

and n^r by

$$n^r = |s'| + 1 - \sum_{i=1}^{m} (c_i + l_i^{\min}).$$

In the following, we assume scoring functions g_1 and g_2 working on relative threadings, obtained by appropriately redefining the original functions g_1 and g_2. Here, l_i^r is the relative length of the ith loop region according to $\mathcal{M}_C$, and n^r is the 'effective' length of the sequence s', relative to the core model $\mathcal{M}_C$. This can be seen from the derived versions of the ordering and spacing constraints. The derived versions of the ordering constraints (6.48), (6.49) and (6.50) are

$$1 \leq t_0^r \leq 1 + l_0^r, \tag{6.53}$$

$$\forall 1 \leq i < m : (t_i^r \leq t_{i+1}^r \leq t_i^r + l_i^r), \tag{6.54}$$

and

$$t_m^r \leq n^r \leq t_m^r + l_m^r, \tag{6.55}$$

respectively. Similarly, the spacing constraint (6.51) becomes

$$1 \leq t_i^r \leq n^r,$$

which justifies our reference to n^r as the 'effective length of s' under $\mathcal{M}_C$'.

DEFINITION 6.18 (THREADING SETS)
*Let s' be a sequence, and $\vec{b}, \vec{d} \in \mathbb{N}^m$ be two vectors (mnemonic for '**b**egin and en**d**'), satisfying*

$$\forall 1 \leq i \leq m : (b_i \leq d_i).$$

We denote by $\mathcal{T}[\vec{b}, \vec{d}]$ the set of all relative threadings of s' through the core model $\mathcal{M}_C$ of s that satisfy the boundaries given by $\vec{b}$ and $\vec{d}$, i.e.

$$\mathcal{T}[\vec{b}, \vec{d}] = \{\vec{t^r} \mid \forall 1 \leq i \leq m : (b_i \leq t_i^r \leq d_i)\}.$$

PROPOSITION 6.19
The set of all possible threadings of s' to s under $\mathcal{M}_C$ is given by $\mathcal{T}[\vec{1}, \vec{n^r}]$, where $\vec{n^r} = (n^r, \ldots, n^r)$. The set $\mathcal{T}[\vec{b}, \vec{d}]$ is empty if there exists $1 \leq i < j \leq m$ with

$$b_i > d_j.$$

The set $\mathcal{T}[\vec{e}, \vec{e}]$ is either empty or contains exactly one element, namely $\vec{t^r} = \vec{e}$.

In the following, we assume for simplicity that $l_i^{\max} = \infty$. Hence, $l_i^r = \infty$, which implies that the right inequality of the relative ordering constraints (6.53)–(6.55) does not restrict the search space. Thus, we get the new relative ordering constraints

$$1 \leq t_0^r$$
$$\forall 1 \leq i < m : (t_i^r \leq t_{i+1}^r)$$

and

$$t_m^r \leq n^r,$$

respectively. For the bound step, we need a lower bound $\mathrm{lb}(\mathcal{T}[\vec{b}, \vec{d}])$ on the score of the threadings contained in set $\mathcal{T}[\vec{b}, \vec{d}]$. A simple version is $\mathrm{lb}_{\mathrm{simpl}}(\mathcal{T}[\vec{b}, \vec{d}])$, which is defined by

$$\sum_{i=1}^{m} \left[\min_{b_i \leq x \leq d_i} g_1(i, x) + \sum_{j>i} \min_{\substack{b_i \leq y \leq d_i \\ b_j \leq z \leq d_j}} g_2(i, j, y, z) \right].$$

Clearly, we have

$$\begin{aligned} \min_{\vec{t^r} \in \mathcal{T}[\vec{b}, \vec{d}]} f(\vec{t^r}) &= \min_{\vec{t^r} \in \mathcal{T}[\vec{b}, \vec{d}]} \sum_{i=1}^{m} \left[g_1(i, t_i^r) + \sum_{j>i} g_2(i, j, t_i^r, t_j^r) \right] \\ &\geq \mathrm{lb}_{\mathrm{simpl}}(\mathcal{T}[\vec{b}, \vec{d}]). \end{aligned}$$

Furthermore, the function $\mathrm{lb}_{\mathrm{simpl}}$ can be calculated with a runtime that is polynomial in m and n^r. Later, we investigate a more sophisticated lower bound. In the following, we assume a selection function

$$\mathrm{sel} : \mathcal{T}[\vec{b}, \vec{d}] \mapsto (i, v),$$

where $(i, v) \in [1..m] \times [b_i..d_i[$ and $b_i < d_i$. The selection function constitutes a search strategy, and usually has an enormous effect on the efficiency. The branch-and-bound method for protein threading is given in Algorithm 6.8.

The function `empty()` checks whether there exists $i < j$ with $b_i > d_j$. Using $\mathrm{lb}_{\mathrm{simpl}}()$ for the lower bound $\mathrm{lb}()$ does not yield a good pruning of the search tree, for which reason we consider a more sophisticated lower bound. The function $\mathrm{lb}_{\mathrm{poly}}$ is defined by

$$\mathrm{lb}_{\mathrm{poly}}(\mathcal{T}[\vec{b}, \vec{d}]) = \min_{\vec{t^r} \in \mathcal{T}[\vec{b}, \vec{d}]} \sum_{i=1}^{m} \left[\begin{array}{l} \quad g_1(i, t_i^r) \\ + \quad g_2(i-1, i, t_{i-1}^r, t_i^r) \\ + \quad \min\limits_{\substack{\vec{u^r} \in \mathcal{T}[\vec{b}, \vec{d}] \\ u_i^r = t_i^r}} \sum\limits_{|i-j|>1} \frac{1}{2} g_2(i, j, t_i^r, u_j^r) \end{array} \right],$$

Algorithm 6.8 Branch-and-Bound for protein threading

opt(fun f, sequence s) {
  queue = $\mathcal{T}[\vec{1}, \vec{n^r}]$
  thread = $\emptyset$;
  opt = ∞;

  while (queue $\neq$ nil) {
    $\mathcal{T}[\vec{b}, \vec{d}]$ = pop(queue);
    if ($\mathrm{lb}(\mathcal{T}[\vec{b}, \vec{d}])$ < opt) {
      if ($\forall i : b_i = d_i$) {
        thread = $\vec{b}$;
        opt = f(thread);
      }
      else {
        (i, v) = $\mathrm{sel}(\vec{b}, \vec{d})$;

        define $\vec{b^v} = (b_1^v, \ldots, b_m^v)$ by $b_j^v = \begin{cases} b_j & \text{if } i < j \\ v + 1 & \text{if } i = j \\ \max(v + 1, b_j) & \text{if } i > j \end{cases}$

        define $\vec{d^v} = (d_1^v, \ldots, d_m^v)$ by $d_j^v = \begin{cases} \min(v, d_j) & \text{if } i < j \\ v & \text{if } i = j \\ d_j & \text{if } i > j \end{cases}$

        TL = $\mathcal{T}[\vec{b}, \vec{d^v}]$;
        TR = $\mathcal{T}[\vec{b^v}, \vec{d}]$:

        if (empty(TL))
          if (NOT (empty(TR)))
            queue = push(TR,queue);
        else
          if (empty(TR))
            queue = push(TL,queue);
          else
            if (lb(TL) $\leq$ lb(TR))
              queue = push(TL,push(TR,queue));
            else
              queue = push(TR,push(TL,queue));
      }
    }
  }
  return(opt,thread);
}

where, by convention, $g_2(i, j, t_i, t_j) = g_2(j, i, t_j, t_i)$ and $g_2(0, 1, t_0, t_1) = 0$. Note that the ordering constraints imply $u_j^r \leq t_i^r$ for $j < i$ and $u_j^r \geq t_i^r$ for $j > i$.

PROPOSITION 6.20
For every $\mathcal{T}[\vec{b}, \vec{d}]$, we have

$$\mathrm{lb}_{\mathrm{poly}}(\mathcal{T}[\vec{b}, \vec{d}]) \leq \min_{\vec{t^r} \in \mathcal{T}[\vec{b}, \vec{d}]} f(\vec{t^r}).$$

Furthermore, if $\mathcal{T}[\vec{b}, \vec{d}] = \{\vec{t^r}\}$, then

$$\mathrm{lb}_{\mathrm{poly}}(\mathcal{T}[\vec{b}, \vec{d}]) = f(\vec{t^r}).$$

Now it remains to be shown that the function $\mathrm{lb}_{\mathrm{poly}}$ can be computed in polynomial time. Note that the definition of $\mathrm{lb}_{\mathrm{poly}}()$ is very similar to the definition of the real lower bound on $\mathcal{T}[\vec{b}, \vec{d}]$. The only difference is the use of a different u_j in the last term of the definition of $\mathrm{lb}_{\mathrm{poly}}()$. This allows a decoupling of the different g_2 terms. Note that if we do not use g_2 terms at all, then the threading problem is no longer NP-hard.

The efficient calculation of $\mathrm{lb}_{\mathrm{simpl}}()$ uses again a dynamic programming technique. We will first concentrate on the g_1 and $g_2(i-1, i, t_{i-1}^r, t_i^r)$ terms, and just count the number of times a calculation of

$$\delta(i, t_i^r, \mathcal{T}[\vec{b}, \vec{d}]) = \min_{\substack{\vec{u^r} \in \mathcal{T}[\vec{b}, \vec{d}] \\ u_i^r = t_i^r}} \sum_{|i-j|>1} \frac{1}{2} g_2(i, j, t_i^r, u_j^r)$$

is called.

LEMMA 6.21
The function $\mathrm{lb}_{\mathrm{poly}}(\mathcal{T}[\vec{b}, \vec{d}])$ can be calculated in polynomial time with polynomially many calls of $\delta(i, t_i, \mathcal{T}[\vec{b}, \vec{d}])$.

PROOF We define the function

$$H(j, x, \mathcal{T}[\vec{b}, \vec{d}]) = \min_{\substack{\vec{t^r} \in \mathcal{T}[\vec{b}, \vec{d}] \\ t_j^r = x}} \sum_{i \leq j} \begin{pmatrix} g_1(i, t_i^r) \\ + g_2(i-1, i, t_{i-1}^r, t_i^r) \\ + \delta(i, t_i^r, \mathcal{T}[\vec{b}, \vec{d}]) \end{pmatrix}.$$

Then

$$\mathrm{lb}_{\mathrm{poly}}(\mathcal{T}[\vec{b}, \vec{d}]) = \min_{b_m \leq x \leq d_m} H(m, x, \mathcal{T}[\vec{b}, \vec{d}]).$$

Furthermore, we have for every $j > 1$,

$$
\begin{aligned}
&H(j, x, \mathcal{T}[\vec{b}, \vec{d}]) \\
&= \min_{b_{j-1} \le y \le d_{j-1}} \left\{ \min_{\substack{\vec{t}^r \in \mathcal{T}[\vec{b}, \vec{d}] \\ t^r_{j-1} = y, t^r_j = x}} \sum_{i \le j} \begin{bmatrix} g_1(i, t^r_i) \\ + g_2(i-1, i, t^r_{i-1}, t^r_i) \\ + \delta(i, t^r_i, \mathcal{T}[\vec{b}, \vec{d}]) \end{bmatrix} \right\} \\
&= \min_{b_{j-1} \le y \le d_{j-1}} \left\{ \begin{array}{l} g_1(j, x) \; + \; g_2(j-1, j, y, x) \; + \; \delta(j, x, \mathcal{T}[\vec{b}, \vec{d}]) \\ + \displaystyle\min_{\substack{\vec{t}^r \in \mathcal{T}[\vec{b}, \vec{d}] \\ t^r_{j-1} = y, t^r_j = x}} \sum_{i \le j-1} \begin{bmatrix} g_1(i, t^r_i) \\ + g_2(i-1, i, t^r_{i-1}, t^r_i) \\ + \delta(i, t^r_i, \mathcal{T}[\vec{b}, \vec{d}]) \end{bmatrix} \end{array} \right\}.
\end{aligned}
$$

In order to substitute the last term by $H(j-1, y, \mathcal{T}[\vec{b}, \vec{d}])$, we need to get rid of the additional condition $t^r_j = x$ in the inner minimization. Since the variables x and $t^r_j, \ldots, t^r_m$ do not occur in the last term, we know that the restriction $t^r_j = x$ only affects the choice of the threadings $\vec{t}^r$. Now, under the assumption of $\vec{l}^r = (\infty, \ldots, \infty)$, we have

$$
\begin{aligned}
&\left\{(t^r_1, \ldots, t^r_{j-1}) \mid \vec{t}^r \in \mathcal{T}[\vec{b}, \vec{d}] \wedge t^r_{j-1} = y \wedge t^r_j = x\right\} \\
&\quad = \begin{cases} \emptyset & \text{if } y > x, \\ \left\{(t^r_1, \ldots, t^r_{j-1}) \mid \vec{t}^r \in \mathcal{T}[\vec{b}, \vec{d}] \wedge t^r_{j-1} = y\right\} & \text{if } y \le x. \end{cases}
\end{aligned}
$$

Hence, we get

$$
\begin{aligned}
&H(j, x, \mathcal{T}[\vec{b}, \vec{d}]) \\
&= \min_{b_{j-1} \le y \le \min(x, d_{j-1})} \left[\begin{array}{l} g_1(j, x) \; + \; g_2(j-1, j, y, x) \; + \; \delta(j, x, \mathcal{T}[\vec{b}, \vec{d}]) \\ + \;\; H(j-1, y, \mathcal{T}[\vec{b}, \vec{d}]) \end{array} \right].
\end{aligned}
$$

■

THEOREM 6.22
The function $\mathrm{lb}_{\mathrm{poly}}$ *can be calculated in polynomial time.*

PROOF By Lemma 6.21, we need only to show that the function $\delta(i, t^r_i, \mathcal{T}[\vec{b}, \vec{d}])$ can be calculated in polynomial time. For this purpose, we define a function H^* by

$$
H^*(i, k, t^r_i, z) \;=\; \min_{\substack{\vec{u}^r \in \mathcal{T}[\vec{b}, \vec{d}] \\ u^r_i = t^r_i, u^r_k \le z}} \sum_{\substack{|j-i|>1 \\ j \le k}} \frac{1}{2} g_2(i, j, t^r_i, u^r_j).
$$

Then

$$
\delta(i, t^r_i, \mathcal{T}[\vec{b}, \vec{d}]) = H^*(i, m, t^r_i, d_m).
$$

Now $H^*(i, 0, t_i^r, x) = 0$. For $k > 0$, we have

$$H^*(i,k,t_i^r,z) = \begin{cases} H^*(i, t_i^r, k-1, z) & \text{if } k \in \{i-1, i, i+1\}, \\ \infty & \text{if } z < b_k \text{ or } z > d_k, \\ \infty & \text{if } k < i-1 \wedge z > t_i^r, \\ \infty & \text{if } k > i+1 \wedge z < t_i^r, \\ \min \left\{ \begin{array}{l} H^*(i, k-1, t_i^r, \min(d_{k-1}, z)) + \frac{1}{2} g_2(i, k, t_i^r, z), \\ H^*(i, k, t_i^r, z-1) \end{array} \right\} & \text{otherwise.} \end{cases}$$

■

6.7.3 NP-hardness

One can even show a stronger result than NP-hardness for protein threading, namely that the protein threading problem is MAX-SNP-hard. A problem is in the class MAX-SNP if it is in NP and there is a constant-size lower bound for the performance of any approximation algorithm for the problem. A well-known problem that is in MAX-SNP is the max cut problem.

DEFINITION 6.23 (MAX CUT)
Let $G = (V, E)$ be an undirected graph. A cut *for G is a subset $V' \subseteq V$. The* cardinality *of the cut V' is the number of edges (v, v') in E such that $v \in V \backslash V'$ and $v' \in V'$. The* max cut problem *is, given G, to find a cut V' whose cardinality is maximal.*

THEOREM 6.24 (AKUTSU AND MIYANO)
The protein threading problem is MAX-SNP-hard.

PROOF By reduction of the max cut problem. Let $G = (V, E)$ be an undirected graph. Let $v_1 \ldots v_n$ be an arbitrary enumeration of V. The sequence s itself is arbitrary. One uses n core regions of length 1 (i.e. every node corresponds to one core region). Thus, we have a core model $(n, \vec{c}, \vec{\lambda}, \vec{l}^{\min}, \vec{l}^{\max})$ with

$$\begin{aligned} \vec{c} &= \underbrace{(1, \ldots, 1)}_{n \text{ times}}, \\ \vec{\lambda} &= (0, \ldots, 0), \\ \vec{l}^{\min} &= (0, \ldots, 0), \\ \vec{l}^{\max} &= (\infty, \ldots, \infty). \end{aligned}$$

The sequence s' that is threaded through the above core model of s has length $2n$ and is of the form

$$s' = \overbrace{01 \ldots 01}^{n \text{ times}}.$$

The scoring function $f(\vec{t})$ is defined by

$$g_1(i, t_i) = 0,$$
$$g_2(i, j, t_i, t_j) = \begin{cases} 1 & \text{if } i < j, (v_{t_i}, v_{t_j}) \in E \text{ and } s'_{t_i} \neq s'_{t_j}, \\ 0 & \text{otherwise.} \end{cases}$$

Under this core model, every threading $\vec{t}$ corresponds to the cut

$$V_{\vec{t}} = \{v_{t_i} \mid s'_{t_i} = 0\}.$$

The scoring function $f(t)$ is nothing other than the cardinality of the cut $V_{\vec{t}}$ (see also Figure 6.20), which implies that the optimal threading corresponds to the maximal cut. ■

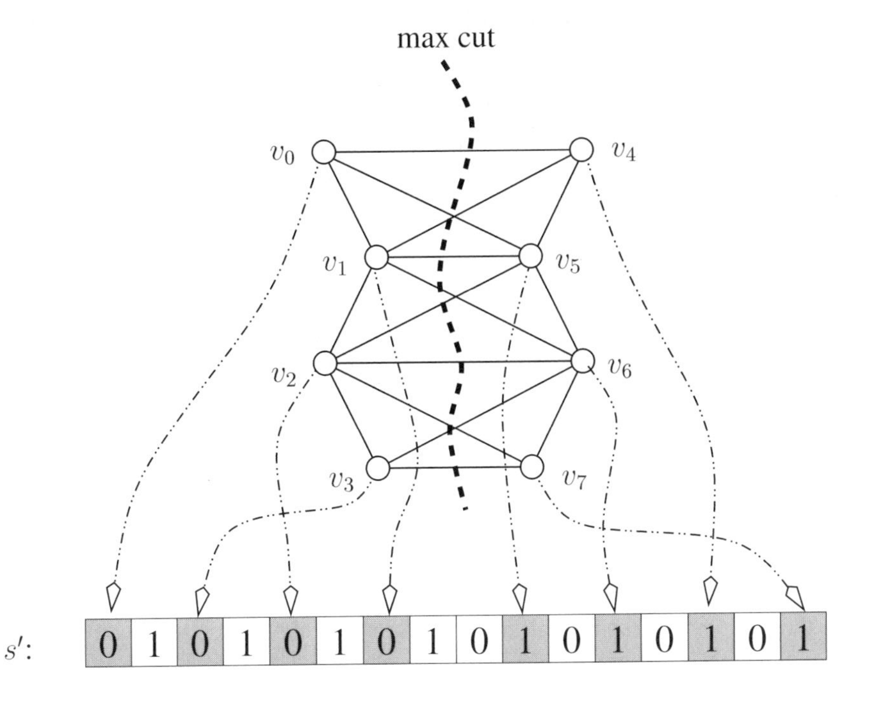

Figure 6.20 Reduction of max cut. Every node is a single core region. The cut indicated by the shown threading is $\{v_0, v_1, v_2, v_3\}$.

6.8 Exercises

1. Sketch all $S(5) = 8$ secondary structures for $[1, 5]$, and all $S(6) = 17$ structures for $[1, 6]$, in both cases with no constraint on base pairing or minimum size for hairpin loops. Sketch all secondary structures for ACGUCG, where base pairs are either Watson–Crick or GU base pairs.
2. Implement a dynamic programming algorithm to compute $S(n)$, the number of secondary structures on $[1, n]$. Using your program, give the values of $S(25)$ and $S(40)$.
   WARNING $S(40)$ leads to integer overflow for 32-bit integers.

3. How many possible secondary structures are there for AAUUAAUUAAUU if you require base pairs to admit hydrogen bonds (i.e. AU, UA, CG, GC, GU, UG)? How many possible secondary structures are there for AAUUAAUUAAUU if you drop this requirement? In other words, how many expressions are there in the symbols '(',')','*' where * means not base-paired, and the parentheses are balanced?
   ANSWER There are 204 secondary structures for AAUUAAUUAAUU with the base pairing condition and 2283 secondary structures without the base pairing condition.
4. Prove the fact, due to M. Waterman [Wat78], that there are $2^{n-2} - 1$ possible hairpin loops on the sequence $(1, \ldots, n)$, provided that base pairs are not required to be Watson–Crick or GU.
   HINT Let $L(n)$ be the number of hairpins on the sequence $(1, \ldots, n)$; i.e. the number of secondary structures with exactly one loop on $(1, \ldots, n)$. Clearly $L(1) = 1 = L(2)$. For the inductive step $L(n+1)$, there are two cases. If $n+1$ does not base-pair, then we have contribution $L(n)$. If $n+1$ is base-paired with $1 \leq j \leq n-1$, then we must consider secondary structures on subsequences $(1, \ldots, j-1)$ and $(j+1, \ldots, n)$. If there were any base pairing in the subsequence $(1, \ldots, j-1)$, then there would be at least two loop structures, which we disallow. Thus we can only consider the subsequence $(j+1, \ldots, n)$, of size $n-(j+1)+1 = n-j$. Thus we have the following:

$$\begin{aligned} L(n+1) &= L(n) + \sum_{j=1}^{n-1} L(n-j) \\ &= L(n) + L(n-1) + \cdots + L(1). \end{aligned}$$

   Hence $L(1) = 1$, $L(2) = 1$, $L(3) = 2$, $L(4) = 4$, $L(5) = 8$, and by induction $L(n) = 2^{n-2}$. Subtracting off the unique structure having no base pairs, we have that the number of hairpins is $2^{n-2} - 1$.
5. Familiarize yourself with the *Vienna RNA Package*. Using this software, determine the secondary structure of three different tRNAs from *M. jannaschii*.
6. Implement the Nussinov–Jacobson algorithm.
7. Write a Monte Carlo program for folding of a 27-mer in a 3-dimensional lattice, using the local moves (end segment rotation, corner, and crankcase) described by Šali–Shakhnovitch–Karplus [ŠSK94b].
8. An alternative to using Monte Carlo (with or without simulated annealing) for determining the minimal-energy conformation of a protein in a lattice model is to employ a genetic algorithm. This has been done by Unger and Moult. Represent a self-avoiding walk of length n in the 2-dimensional rectangular lattice as a sequence of $n-1$ relative directions S, L, R (straight, left, right). Given an HP sequence of length n, develop a genetic algorithm that supports crossover and pointwise mutation of *chromosomes*, where a chromosome is a sequence of $n-1$ relative directions.
9. The main result from [ŠSK94b] is that the speed of folding to the native state depends on the magnitude of the difference between the lowest and next lowest energy levels in the compact cube. Define δ to be the difference between these two energy levels. Is energy continuous in the lattice model? (i.e. if there is a conformation of energy E, then is there a conformation of energy level $E \pm 1$?) Show that there δ is large if and only if there is a pronounced unique minimum.

10. Given a Boltzmann distribution (6.26), write a small program to compute the corresponding energy function; i.e., as in Sippl's work, invert the Boltzmann probability distribution to obtain an estimate for the energy.
11. Assume that for each amino acid pair (a, b), there is a (positive or negative) constant $k(a, b)$, for which the force between residue a and b at distance r is

$$F^{a,b}(r) = \frac{k(a, b)}{r^2}.$$

Assume that the van der Waals force is infinitely repulsive at distances less than or equal to the van der Waals radius, and otherwise 0. Thus the van der Waals force corresponds to sphere packing.

Assume that the free energy is the sum of pair potentials depending on the previously described forces. Prove or disprove the Sippl hypothesis that pair potentials can be obtained from the frequency data. In other words, using the previously described forces, generate sample conformations taking minimal free energy. Now from the frequency data, generate the pair potentials. Do they agree? If so, how much frequency data is required for a certain convergence?
12. Using Sippl's approach, calculate nucleotide pair potentials for nucleosome sequences from the nucleosome database of [IT93]. From the potentials, compute the net potential for a new oligonucleotide, and give a score of likelihood that the sequence is wrapped about a nucleosome. Attempt to discern common nucleosome sequence patterns by using the potentials.

Acknowledgments and References

The combinatorics of RNA secondary structures is an area initiated by M.S. Waterman [Wat78, Wat95]. Theorem 6.2 derives from Theorem 13.1 of [Wat95]. An asymptotic formula for the number of RNA secondary structures was first derived by P.R. Stein and M.S. Waterman in [SW78], and further elaborated in [SFSH94] to count a variety of types of secondary structures. In [SFSH94] neutral networks are studied. The material on dynamic programming algorithms for RNA secondary structure determination is drawn from articles of R. Nussinov, A. Jacobson, M. Zuker, D. Sankhoff, I. Hofacker *et al.* in [NJ80, ZS84, HFS+94]. Thanks are due to I. Hofacker for personal communication concerning the pseudocode for `Vienna RNA Package`. The material on DNA strand separation was drawn from the articles [Ben90, SMFB95, Ben96, Ben93, FB99] by C. Benham and collaborators R. Fye, M. Mezei and H. Sun, along with supplementary material on J.H. White's topological invariant from [Whi89, Whi95]. Special thanks are due to C. Benham for personal communication, especially in the derivation of equation (6.23).

The material concerning the Monte Carlo simulation for folding a heteropolymer 27-bead model into the $3 \times 3 \times 3$ compact cube comes from Šali, Shakhnovich, and Karplus [ŠSK94a, ŠSK94b]. The genetic algorithm for protein folding with the HP model on the 2-dimensional lattice was drawn from the article of Unger and Moult [UM93]. This work was extended to arbitrary lattices, including the 3-dimensional face-centered cubic lattice by R. Backofen, S. Will, and P. Clote [BWC00].

The 2-dimensional square lattice approximation algorithm was taken from [HI96], where one can also find an extension of the algorithm to the case of the 3-dimensional cubic lattice (with an even better approximation ratio of $\frac{3}{8}$). This work was extended to a side-chain lattice model in [HI97]. An approximation algorithm for the HP model on the face-centered cubic lattice can be found in [ABD$^+$97].

The description of the protein threading problem and the branch-and-bound algorithm comes from [LS96], whereas the section on NP-hardness comes from [AM97]. The pictures of the secondary structure of myoglobin and human hemoglobin were produced using the RasMol package [SMW95] (to create the cartoon representation) and the software package Molscript [Kra91], and then rendered using Raster3d [MB97].

Appendix A

Mathematical Background

A.1 Asymptotic complexity

Standard notation for classifying the asymptotic runtime and space requirements for algorithms is given by $O(f)$, $\Omega(f)$ and $\Theta(f)$, whose definitions are given as follows.

Suppose that $f : \mathbb{N} \to \mathbb{N}$ and $g : \mathbb{N} \to \mathbb{N}$. We say that $f = O(g)$, or that f is $O(g)$ (read f is 'big-O' of g), if there exist positive constants c, n_0 such that

$$(\forall n \geq n_0) f(n) \leq c \cdot g(n).$$

In words, $f = O(g)$ means that asymptotically, when neglecting constant multiplicative factors, f is bounded above by g. On the other hand, we say that $f = \Omega(g)$, or that f is $\Omega(g)$, if there exist positive constants c, n_0 such that

$$(\forall n \geq n_0) c \cdot g(n) \leq f(n).$$

In words, $f = \Omega(g)$ means that asymptotically, when neglecting constant multiplicative factors, f is greater than g; i.e. g is a lower bound for f. It follows from the definitions that $f = O(g)$ if and only if $g = \Omega(f)$. Finally, we define $f = \Theta(g)$ to mean that $f = O(g)$ and $g = O(f)$.

A.2 Units of Measurement

Recall the prefixes: milli (10^{-3}), micro (10^{-6}), nano (10^{-9}), pico (10^{-12}), femto (10^{-15}), kilo (10^{3}), mega (10^{6}), giga (10^{9}), tera (10^{12}). Thus, for instance, a femtosecond is 10^{-15} seconds, a time unit of importance in protein folding simulations.

One angstrom, denoted Å, is 10^{-10} meters. Visible light has wavelengths in the micrometre (micron) range (10^{-6} m), while covalent bond lengths are roughly 1 Å. A *mole* is the quantity 6.0229×10^{23} molecules, also known as Avogadro's number. One *dalton* is the molecular weight of one hydrogen atom; sometimes the weight of a molecule is given in kilodaltons, denoted kDa Sometimes rRNA is identified by svedberg units, denoted S. This is a measure of sedimentation rate in an ultracentrifuge.

One *calorie* is roughly the amount of energy required to raise the temperature of 1 gram = 1 ml = 1 cm^3 of water one degree Celsius. Bond energies are often given in kilocalories per mole (kcal/mol), or less often in kilojoules per mole (kJ/mol), where 4.184 J = 1 cal.

Boyle's law for an ideal gas can be generalized by

$$PV = nRT$$

where P is the pressure acting on the gas, V the volume occupied by the gas, n the number of moles of gas, R the *gas constant*, and T the absolute temperature in degrees Kelvin (K). The *gas constant* R is 8.3146 J K^{-1} $mole^{-1}$. *Boltzmann's constant* $k = R/N$, where N is Avogadro's number; thus $k = 13.805 \times 10^{-24}$ J K^{-1}. Boltzmann's constant appears in the Boltzmann probabilities

$$\frac{e^{-E(i)/kT}}{Z}$$

where $Z = \sum_{s \in S} e^{-E(s)/kT}$, and S is the (finite) set of possible states.

A.3 Lagrange Multipliers

In this section, we recall some facts from calculus of several variables, and state Lagrange's *method of undetermined multipliers*. Many problems in computational biology involve determining the maxima or minima of a function of several variables. For instance, determining the tertiary structure of an amino acid sequence can be approached by finding that structure which minimizes an appropriate energy function. Nonlinear optimization problems are in general computationally intractable. However, in certain instances, Lagrange's method yields an answer.

Let us begin with by defining some notions. A *stationary point* $a = a_1, \ldots, a_n$ of the function $f(x_1, \ldots, x_n)$ satisfies

$$\frac{\partial f}{\partial x_i}(a) = 0$$

for $1 \leq i \leq n$. An *extremum* a of f is a local maximum resp. minimum of f; i.e. there is an open neighborhood U of the point a such that $f(x) \leq f(x)$ resp. $f(x) \geq f(a)$ for all $x \in U$. By calculus of several variables, we know that all extrema of a smooth (continuously differentiable) function are stationary, though there are stationary points that are not extrema. For instance, at $x_0 = 0, y_0 = 0$, the function $f(x, y) = x^3 + y^3 - 3xy$ has a non-extremal *saddle point* at the origin $(0, 0)$, where nevertheless partial derivatives $\frac{\partial f}{\partial x}(0, 0) = 0$ and $\frac{\partial f}{\partial y}(0, 0) = 0$.

Suppose we want to determine a local maximum or minimum of the function

$$f(x_1, \ldots, x_n)$$

subject to $m < n$ additional constraints $\phi_1(x_1, \ldots, x_n) = 0, \ldots, \phi_m(x_1, \ldots, x_n) = 0$. Since the variables $x_1, \ldots, x_n$ are not independent, but related by the functions ϕ_i, one might attempt to write f as a function of $n - m$ variables and then determine the stationary points of the resulting function. For linear constraints, this may be feasible, but not in general. Instead of this, extend the function f by defining $F(x_1, \ldots, x_n, \lambda_1, \ldots, \lambda_m)$ to be

$$f(x_1, \ldots, x_n) + \lambda_1 \phi_1(x_1, \ldots, x_n) + \cdots + \lambda_m \phi_m f(x_1, \ldots, x_n).$$

Lagrange's method consists of determining points $a = (a_1, \ldots, a_n, \lambda_1, \ldots, \lambda_m)$ such that all partials of F are zero; i.e. $\frac{\partial F}{\delta x_i}(a) = 0$ for $1 \leq i \leq n$, and $\frac{\partial F}{\delta \lambda_i}(a) = \phi_i(a) = 0$ for $1 \leq i \leq m$. It could be the case that a is a stationary, non-extremal point, but in practical situations one can then check maximality or minimality. For more information on Lagrange multipliers, we refer the reader to [CJ74].

Appendix B

Resources

B.1 Web Sites

Since the URL of a web page can change rapidly, we do not include URLs in this appendix. For links to web pages listed below, please consult the web page for the book, which can be found by following links from `http://www.wiley.co.uk/statistics`. The list of databases presented below is by far incomplete and presents only some of the important databases:

1. Protein Data Bank (PDB) [BKW+77, BWF+00]: database of three-dimensional structures of biological macromolecules (proteins, nucleic acids, ...)
2. The CATH database [OMJ+97]: CATH classifies proteins according to Class (where the different protein classes are *mainly Alpha*, *mainly Beta*, *Alpha and Beta*, and proteins having *few secondary structures*), Architecture (which is the description of the gross overall arrangement of the secondary structure elements), Topology (which takes into account the overall shape as well as the connectivity of the secondary structure elements, using structural comparison algorithms) and Homology (where proteins are grouped together if they are believed to have a common ancestor).
3. The SCOP database [MBHC95]: SCOP has a similar hierarchical structure, where proteins are grouped according to fold (major structural similarity), superfamily (probable common evolutionary origin), and family (clear evolutionary relationship). SCOP is an acronym for Structural Classifcation of Proteins hierarchical.
4. SWISS-PROT [BA00]: database of annotated protein sequences.
5. Unbiased samples of proteins sharing little homology can be retrieved by anonymous FTP from EMBL Heidelberg.
6. For the annotated genome of *Saccharomyces cerevisiae* (yeast), see the MIPS web site (Munich Information Center for Protein Sequences [MHPF98, MHK+99]).
7. The annotated sequence data for the genomes of *M. jannaschii* and certain other bacteria can be found at the web site of TIGR.

B.2 The PDB Format

The Brookhaven Protein Database (PDB) [BKW+77, BWF+00] contains the 3-dimensional coordinates in angstroms of 10310 proteins[1] and 788 nucleic acids (as of February 2000). The database consists of flat Ascii text files, 80 characters per line (due to earlier punch card format).

PDB entries are separated into different sections using keywords in the beginning of the line. There are reserved words such as COMPND, SEQRES, ATOM, etc. For a more detailed description, see the PDB format description at the PDB homepage. We give below two short fragments from 2bna.pdb [DSD82] (PDB code 2BNA). The first fragment is from the title section, and gives general information about the PDB entry:

```
HEADER    DEOXYRIBONUCLEIC ACID                   12-NOV-81   2BNA      2BNA   3
                         :
AUTHOR    H.R.DREW,R.E.DICKERSON                                        2BNA   8
                         :
JRNL        AUTH   H.R.DREW,S.SAMSON,R.E.DICKERSON                      2BNA   9
JRNL        TITL   STRUCTURE OF A B-/DNA$ DODECAMER AT 16 KELVIN        2BNA  10
JRNL        REF    PROC.NAT.ACAD.SCI.USA          V.  79  4040 1982     2BNAA  1
JRNL        REFN   ASTM PNASA6  US ISSN 0027-8424                 040   2BNAA  2
                         :
REMARK   2 RESOLUTION. 2.7 ANGSTROMS.                                   2BNA  46
REMARK   3                                                              2BNA  47
REMARK   3 REFINEMENT. METHOD OF JACK AND LEVITT (ACTA CRYST., A34,      2BNA  48
REMARK   3  931 (1978)), NINETEEN CYCLES STARTING FROM THE NATIVE       2BNA  49
REMARK   3  290 DEGREES K MODEL AND USING UNIFORM TEMPERATURE           2BNA  50
REMARK   3  FACTORS.  THE R VALUE FOR ALL 1836 REFLECTIONS IN THE       2BNA  51
REMARK   3  RANGE 15.0 - 2.7 ANGSTROMS (INCLUDING UNOBSERVED            2BNA  52
REMARK   3  REFLECTIONS) IS 0.21.  FOR THE 1051 REFLECTIONS IN THE      2BNA  53
```

The second fragment is from the coordinate section, where the coordinates of the different atoms are displayed. The following fragment displays one nucleotide base (cytosine) from the DNA Helix:

```
ATOM     13  N1    C A   1      17.117  28.868  24.167  1.00 -1.36      2BNA 100
ATOM     14  C2    C A   1      17.271  27.706  24.834  1.00  3.69      2BNA 101
ATOM     15  O2    C A   1      18.423  27.361  25.122  1.00 -6.94      2BNA 102
ATOM     16  N3    C A   1      16.196  26.968  25.127  1.00 -3.13      2BNA 103
ATOM     17  C4    C A   1      14.974  27.387  24.768  1.00  7.45      2BNA 104
ATOM     18  N4    C A   1      13.880  26.634  25.070  1.00  -.35      2BNA 105
ATOM     19  C5    C A   1      14.779  28.583  24.085  1.00-10.22      2BNA 106
ATOM     20  C6    C A   1      15.901  29.335  23.779  1.00   .77      2BNA 107
```

The result of loading this information in RasMol [SMW95] is shown in Figure B.1. If one clicks on the atom indicated in this figure, then one obtains the following information (in Version 2.6 of RasMol [SMW95]):

```
Atom: O2 15  Group:  C 1  Chain: A
```

[1] This figure includes proteins, peptides and viruses.

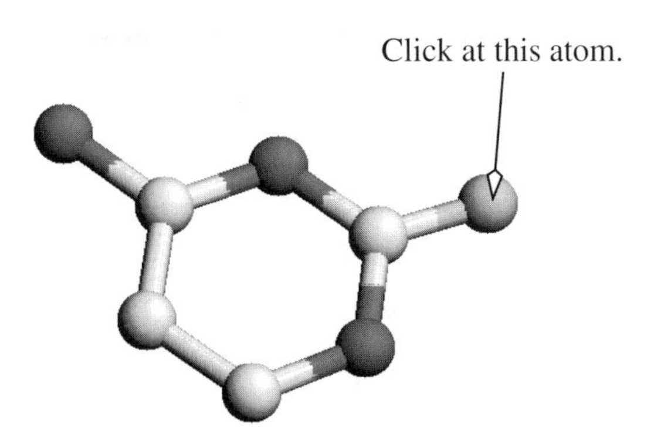

Figure B.1 Result of PDB example. One has to select the BALLS-AND-STICK option in the DISPLAY menu.

References

[ABD+97]Agarwala R., Batzoglou S., Dancik V., Decatur S. E., Farach M., Hannenhalli S., Muthukrishnan S., and Skiena S. (1997) Local rules for protein folding on a triangular lattice and generalized hydrophobicity in the HP model. *Journal of Computational Biology* 4(2): 275–296.

[ABL+94]Alberts B., Bray D., Lewis J., Raff M., Roberts K., and Watson J. D. (1994) *Molecular Biology of the Cell*. Garland, New York, 3rd edition.

[AH94]Adler B. K. and Hajduk S. L. (1994) Mechanisms and origins of RNA editing. *Current Opinion in Genetics and Development* 4: 316–322.

[Alt91]Altschul S. (1991) Amino acid substitution matrices from an information theoretic perspective. *Journal of Molecular Biology* 219: 555–565.

[AM97]Akutsu T. and Miyano S. (1997) On the approximation of protein threading. In *Proceedings of the First Annual International Conferences on Computational Molecular Biology (RECOMB97)*, pp. 3–8. ACM Press.

[Anf73]Anfinsen C. B. (1973) Principles that govern the folding of protein chains. *Science* 181: 223–230.

[BA00]Bairoch A. and Apweiler R. (2000) The SWISS-PROT protein sequence database and its supplement TrEMBL in 2000. *Nucleic Acids Research* 28: 45–48. http://www.expasy.ch/.

[Bac98]Backofen R. (1998) Constraint techniques for solving the protein structure prediction problem. In Maher M. and Puget J.-F. (eds) *Proceedings of 4th International Conference on Principle and Practice of Constraint Programming (CP'98)*, volume 1520 of *Lecture Notes in Computer Science*, pp. 72–86. Springer-Verlag, Berlin.

[BB96]Bornberg-Bauer E. (1996) Random structures and evolution of biopolymers: A computational case study on RNA secondary structures. *Pharmaceutica Acta Helvetiae* 71: 79–85.

[BB97]Bornberg-Bauer E. (1997) How are model protein structures distributed in sequence space? *Biophysical Journal* 73(5): 2393–403.

[BB98]Baldi P. and Brunak S. (1998) *Bioinformatics: The Machine Learning Approach*. MIT Press, Cambridge, MA.

[BC94]Baldi P. and Chauvin Y. (1994) Smooth on-line learning algorithms for hidden Markov models. *Neural Computation* 6.

[BC97]Backofen R. and Clote P. (1997) Evolution as a computational engine. In Nielsen M. and Thomas W. (eds) *Proceedings of the Annual Conference of the European Association for Computer Science Logic (CSL'97)*, volume 1414 of *Lecture Notes in Computer Science*, pp. 35–55. Springer-Verlag, Berlin.

[BCHM94]Baldi P., Chauvin Y., Hunkapiller T., and McClure M. A. (1994) Hidden Markov models of biological primary sequence information. *Proceedings of the National Academy of Sciences USA* 91: 1059–1063.

[Ben90]Benham C. J. (1990) Theoretical analysis of heteropolymeric transitions in superhelical DNA molecules of specified sequence. *Journal of Chemical Physics* 92(10): 6294–6305.

[Ben93]Benham C. J. (1993) Sites of predicted stress-induced DNA duplex destabilization occur preferentially at regulatory loci. *Proceedings of the National Academy of Sciences USA* 90:

2999–3003.

[Ben96]Benham C. J. (1996) Theoretical analysis of the helix–coil transition in positively superhelical DNA at high temperatures. *Physical Review E* 53(3): 2984–2987.

[Ben97]Benson G. (1997) Sequence alignment with tandem repeats. In *Proceedings of the First Annual International Conferences on Computational Molecular Biology (RECOMB97)*, pp. 27–36. ACM Press.

[BGRRO96]Benaola-Galván P., Román-Roldán R., and Oliver J. L. (1996) Compositional segmentation and long-range fractal correlations in DNA sequences. *Physical Review E* 53: 5181–5189.

[BHH+94]Becker J. D., Honerkamp J., Hirsch J., Fröbe U., Schlatter E., and Greger R. (1994) Analysing ion channels with hidden Markov models. *European Journal of Physiology* 426: 328–332.

[BJK+99]Berry V., Jiang T., Kearney P., Li M., and Wareham T. (1999) Quartet cleaning: Improved algorithms and simulations. In Nesetril J. (ed) *Proceedings of the 7th Annual European Symposium on Algorithms (ESA '99)*, volume 1643 of *Lecture Notes in Computer Science*, pp. 313–324. Springer-Verlag, Berlin.

[BKW+77]Bernstein F. C., Koetzle T. F., Williams G. J., Meyer Jr. E. E., Brice M. D., Rodgers J. R., Kennard O., Shimanouchi T., and Tasumi M. (1977) The protein data bank: a computer-based archival file for macromolecular structures. *Journal of Molecular Biology* 112: 535.

[BL98]Berger B. and Leighton T. (1998) Protein folding in the hydrophobic–hydrophilic (HP) model is NP-complete. In Istrail S., Pevzner P., and Waterman M. (eds) *Proceedings of the Second Annual International Conferences on Computational Molecular Biology (RECOMB98)*, pp. 30–39. ACM Press.

[BM93]Borodovsky M. and McIninch J. (1993) Genmark: Parallel gene recognition for both DNA strands. *Computers and Chemistry* 17(2): 123–133.

[Buc90]Bucher P. (1990) Weight matrix descriptions of four eukaryotic RNA polymerase II promoter elements derived from 502 unrelated promoter sequences. *Journal of Molecular Biology* 212.

[Bun71]Buneman P. (1971) The recovery of trees from measures of dissimilarity. In Hodson F. R., Kendall D. G., and Tautu P. (eds) *Mathematics in the Archaeological and Historical Sciences*. Edinburgh University Press.

[BWBB99]Backofen R., Will S., and Bornberg-Bauer E. (1999) Application of constraint programming techniques for structure prediction of lattice proteins with extended alphabets. *Bioinformatics* 15(3): 234–242.

[BWC00]Backofen R., Will S., and Clote P. (2000) Algorithmic approach to quantifying the hydrophobic force contribution in protein folding. In Altman R. B., Lauderdale K., Dunker A. K., Hunter L., and Klein T. E. (eds) *Proceedings of the Pacific Symposium on Biocomputing 2000*, pp. 92–103. World Scientific, Singapore.

[BWF+00]Berman H. M., Westbrook J., Feng Z., Gilliland G., Bhat T. N., Weissig H., Shindyalov I. N., and Bourne P. E. (2000) The protein data bank. *Nucleic Acids Research* 28: 235–242. http://www.rcsb.org/pdb/.

[BWO+96]Bult C. J., White O., Olsen G. J., Zhou L., Fleischmann R. D., Sutton G. G., Blake J. A., FitzGerald L. M., Clayton R. A., Gocayne J. D., Kerlavage A. R., Dougherty B. A., Tomb J.-F., Adams M. D., Glodek A., Scott J. L., Geoghagen N. S. M., Weidman J. F., Fuhrmann J. L., Nguyen D., Utterback T. R., Kelley J. M., Peterson J. D., Sadow P. W., Hanna M. C., Cottn M. D., Roberts K. M., Hurst M. A., Kaine B. P., Borodovsky M., Klenk H.-P., Fraser C. M., Smith H. O., Woese C. R., and Venter J. C. (1996) Complete genome sequence of the methanogenic archaeon, *Methanococcus jannaschii*. *Science* 273: 1053–1072.

[CD89]Chan H. S. and Dill K. A. (1989) Compact polymers. *Macromolecules* 22: 4559–4573.

[CD91]Chan H. S. and Dill K. A. (1991) Compact polymers. In Nall B. T. and Dill K. A. (eds) *Conformations and Forces in Protein Folding*, pp. 43–66. American Association for the Advancement of Science, Washington, DC. Reprint of [CD89].

[CGP+98]Crescenzi P., Goldman D., Papadimitriou C., Piccolboni A., and Yannakakis M.

(1998) On the complexity of protein folding. *Journal of Computational Biology* 5(3): 523–466.
[Cha95]Chan H. S. (1995) Kinetics of protein folding. *Nature* 373: 664–665. Criticism to [ŠSK94a].
[Chu92]Churchill G. A. (1992) Hidden Markov chains and the analysis of genome structure. *Computers and Chemistry* 16(2): 107–115.
[CJ74]Courant R. and John F. (1974) *Introduction to Calculus and Analysis*, volume II. John Wiley and Sons, New York.
[CL88]Carrillo H. and Lipman D. (1988) The multiple sequence alignment problem in biology. *SIAM Journal of Applied Mathematics* 48: 1073–1082.
[Clo98]Clote P. (1998) A boolean circuit generation language and an appliction to computational biology. Talk given at the Complexity Workshop, *IEEE Symposium on Functional Programming*. Workshop organizer D. Leivant.
[Clo99]Clote P. (1999) Protein folding, the Levinthal paradox and rapidly mixing Markov chains. In Wiedermann J., van Emde Boas P., and Nielsen M. (eds) *Automata, Languages and Programming, Proceedings of the 26th International Colloquium, ICALP'99*, volume 1644 of *Lecture Notes in Computer Science*. Springer-Verlag, Berlin.
[CMX+90]Chung S. H., Moore J. B., Xia L., Premkumar L. S., and Gage P. W. (1990) Characterization of single channel currents using digital signal processing techniques based on hidden Markov models. *Philosophical Transactions of the Royal Society of London B* 329: 265–285.
[Cre92]Creighton T. E. (ed) (1992) *Protein Folding*. W. H. Freeman, New York.
[Cri58]Crick F. H. C. (1958) On protein synthesis. *Symposium of the Society of Experimental Biology* 12: 138–167.
[CSE67]Cavalli-Sforza L. L. and Edwards A. W. F. (1967) Phylogenetic analysis: Models and estimation procedures. *American Journal of Human Genetics* 19: 233–257.
[CSW87]Cann R. M., Stoneking M., and Wilson A. (1987) Mitochondrial DNA and human evolution. *Nature* 325: 31–36.
[CV99]Crochemore M. and Vérin R. (1999) Zones of low entropy in genomic sequences. *Computers and Chemistry* 23: 275–282.
[Dar58]Darwin C. (1958) *The Origin of Species*. New American Library. Second printing 1986.
[Daw89]Dawkins R. (1989) *The Selfish Gene*. Oxford University Press, new edition.
[DBY+95]Dill K. A., Bromberg S., Yue K., Fiebig K. M., Yee D. P., Thomas P. D., and Chan H. S. (1995) Principles of protein folding – A perspective of simple exact models. *Protein Science* 4: 561–602.
[DEKM98]Durbin R., Eddy S., Krogh A., and Mitchison G. (1998) *Biological Sequence Analysis: Probabalistic Models of Proteins and Nucleic Acids*. Cambridge University Press.
[DLR77]Dempster A. P., Laird N. M., and Rubin D. B. (1977) Maximum likelihood from incomplete data via the EM algorithm. *Journal of the Royal Statistical Society B* 39: 1–22.
[DSD82]Drew H. R., Samson S., and Dickerson R. E. (1982) Structure of a B-DNA dodecamer at 16 kelvin. *Proceedings of the National Academy of Sciences USA* 79(13): 4040–4044.
[dSLK+98]de Souza S. J., Long M., Klein R. J., Roy S., Lin S., and Gilbert W. (1998) Toward a resolution of the introns early/late debate: Only phase zero introns are correlated with the structure of ancient proteins. *Proceedings of the National Academy of Sciences USA* 95(9): 5094–5099.
[DSO78]Dayhoff M. O., Schwartz R. M., and Orcutt B. C. (1978) A model of evolutionary change in proteins. In Dayhoff M. O. (ed) *Atlas of Protein Sequence and Structure*, volume 5, supplement 3, pp. 345–352. National Biomedical Research Foundation, Washington, DC.
[DT95]Donnelly P. and Tavaré S. (1995) Coalescents and genealogical structure under neutrality. *Annual Reviews of Genetics* 29: 401–421.
[Due92]Dueck G. (1992) New optimisation heuristics: The great deluge algorithm and the record-to-record travel. *Journal of Computational Physics* 104: 86–92.
[Edd95]Eddy S. R. (1995) Multiple alignment using hidden Markov models. In Rawlings C.,

Clark D., Altman R., Hunter L., Lengauer T., and Wodak S. (eds) *Proceedings of the 3rd International Conference on Intelligent Systems for Molecular Biology (ISMB'95)*, pp. 114–120. AAAI Press, Menlo Park, CA.

[ELT$^+$89]Eigen M., Lindemann B. F., Tietze M., Winkler-Oswatitsch R., Dress A., and von Haeseler A. (1989) How old is the genetic code? Statistical geometry of tRNA provides an answer. *Science* 244(4905): 673–679.

[EMD95]Eddy S. R., Mitchison G., and Durbin R. (1995) Maximum discrimination hidden Markov models of sequence consensus. *Journal of Computational Biology* 2(1): 9–24.

[ES99]Estévez A. M. and Simpson L. (1999) Uridine insertion/deletion RNA editing in trypanosome mitochondria – A review. *Gene* 240(2): 247–260.

[Far77]Farris J. S. (1977) On the phenetic approach to vertebrate classification. In Hecht M. K., Goody P. C., and Hecht B. M. (eds) *Major Patterns in Vertebrate Evolution*, pp. 823–850. Plenum Press, New York.

[FAW$^+$95]Fleischmann R. D., Adams M. D., White O., Clayton R. A., Kirkness E. F., Kerlavage A. R., Bult C. J., Tomb J.-F., Dougherty B. A., Merrick J. M., McKenney K., Sutton G., FitzHugh W., Fields C., Gocayne J. D., Scott J., Shirley R., Liu L.-I., Glodek A., Kelley J. M., Weidman J. F., Phillips C. A., Spriggs T., Hedblom E., Cotton M. D., Utterback T. R., Hanna M. C., Nguyen D. T., Saudek D. M., Brandon R. C., Fine L. D., Fritchman J. L., Fuhrmann J. L., Geoghagen N. S. M., Gnehm C. L., McDonald L. A., Small K. V., Fraser C. M., Smith H. O., and Venter J. C. (1995) Whole-genome random sequencing and assembly of *Haemophilus influenzae*. *Science* 269(5223): 496–513.

[FB99]Fye R. M. and Benham C. J. (1999) Exact method for numerically analyzing a model of local denaturation in superhelically stressed DNA. *Physical Review E* 59: 3408–3426.

[Fel68a]Feller W. (1968) *An Introduction to Probability Theory and its Applications*, volume 2. John Wiley and Sons, New York, 3rd edition.

[Fel68b]Feller W. (1968) *An Introduction to Probability Theory and its Applications*, volume 1. John Wiley and Sons, New York, 3rd edition.

[Fel81]Felsenstein J. (1981) Evolutionary trees from DNA sequences: A maximum likelihood approach. *Journal of Molecular Evolution* 17: 368–378.

[Fel83]Felsenstein J. (1983) Statistical inference of phylogenies. *Journal of the Royal Statistical Society A* 146, part 3: 246–272.

[FGW$^+$95]Fraser C. M., Gocayne J. D., White O., Adams M. D., Clayton R. A., Fleischmann R. D., Bult C. J., Kerlavage A. R., Sutton G., Kelley J. M., Fritchman J. L., Weidman J. F., Small K. V., Sandusky M., Fuhrmann J., Nguyen D., Utterback T. R., Saudek D. M., Phillips C. A., Merrick J. M., Tomb J.-F., Dougherty B. A., Bott K. F., Hu P.-C., Lucier T. S., Peterson S. N., Smith H. O., Hutchison III C. A., and Venter J. C. (1995) The minimal gene complement of *Mycoplasma genitalium*. *Science* 270(5235): 397.

[FH98]Freeland S. J. and Hurst L. D. (1998) The genetic code is one in a million. *Journal of Molecular Evolution* 47: 238–248.

[FLSS92]Fischetti V., Landau G., Schmidt J., and Sellers P. (1992) Identifying periodic occurrences of a template with applications to a protein structure. In *Proceedings of the 3rd Symposium on Combinatorial Pattern Matching*, volume 644 of *Lecture Notes in Computer Science*, pp. 111–120. Springer-Verlag, Berlin.

[FM67]Fitch W. M. and Margoliash E. (1967) Construction of phylogenetic trees. *Science* 155: 279–284.

[FNS96]Ferretti V., Nadeau J. H., and Sankoff D. (1996) Original synteny. In *Proceedings of 7th Annual Symposium on Combinatorial Pattern Matching*, volume 1075 of *Lecture Notes in Computer Science*, pp. 159–167. Springer-Verlag, Berlin.

[FR92a]Fredkin D. R. and Rice J. A. (1992) Bayesian restoration of single channel patch clamp recordings. *Biometrics* 48: 427–448.

[FR92b]Fredkin D. R. and Rice J. A. (1992) Maximum likelihood estimation and identification directly from single-channel recordings. *Proceedings of the Royal Society of London B* 249: 125–132.

[FST+93]Fontana W., Stadler P. F., Tarazona P., Weinberger E., and Schuster P. (1993) RNA folding and combinatory landscapes. *Physical Review E* 47(3): 2083–2099.

[Gar86]Garland S. J. (1986) *Introduction to Computer Science with Applications in Pascal.* Addison-Wesley, Reading, MA.

[Gat72]Gatlin L. L. (1972) *Information Theory and the Living System.* Columbia University Press, New York.

[GG84]Geman S. and Geman D. (1984) Stochastic relaxation, Gibbs distributions, and the Bayesian restoration of images. *IEEE Transactions in Pattern Analysis and Machine Intelligence* PAMI 6(9): 721–741.

[GG95]Govindarajan S. and Goldstein R. A. (1995) Searching for foldable protein structures using optimized energy functions. *Biopolymers* 36: 43–51.

[GG96]Govindarajan S. and Goldstein R. A. (1996) Why are some protein structures so common? *Proceedings of the National Academy of Sciences USA* 93: 3341–3345.

[Giu89]Giulio M. D. (1989) The extension reached by the minimization of the polarity distances during the evolution of the genetic code. *Journal of Molecular Evolution* 29: 288–293.

[GJO+95]Georgiadis M. M., Jessen S. M., Ogata C. M., Telesnitsky A., Goff S. P., and Hendrickson W. A. (1995) Mechanistic implications from the structure of a catalytic fragment of moloney murine leukemia virus reverse transcriptase. *Structure (London)* 3: 879.

[Gol93]Goldman N. (1993) Further results on error minimization in the genetic code. *Journal of Molecular Evolution* 37: 662–664.

[Got82]Gotoh O. (1982) An improved algorithm for matching biological sequences. *Journal of Molecular Biology* 162: 705–708.

[GT94]Griffiths R. C. and Tavaré S. (1994) Ancestral inference in population genetics. *Statistical Science* 9(3): 307–319.

[HFS+]Hofacker I., Fontana W., Stadler P. F., Bonhoeffer L. S., Tucker M., and Schuster P. Vienna RNA package. http://www.tbi.univie.ac.at/ ivo/RNA.

[HFS+94]Hofacker I., Fontana W., Stadler P. F., Bonhoeffer L. S., Tucker M., and Schuster P. (1994) Fast folding and comparison of RNA secondary structures. *Monatshefte für Chemie* 125: 167–188.

[HH91]Haig D. and Hurst L. (1991) A quantitative measure of error minimization in the genetic code. *Journal of Molecular Evolution* 33: 412–417.

[HH92]Henikoff S. and Henikoff J. G. (1992) Amino acid substitution matrices from protein blocks. *Proceedings of the National Academy of Sciences USA* 89: 10915–10919.

[HH93]Henikoff S. and Henikoff J. G. (1993) Performance evaluation of amino acid substitution matrices. *Proteins* 17: 49–61.

[HI96]Hart W. E. and Istrail S. C. (1996) Fast protein folding in the hydrophobic-hydrophilic model within three-eighths of optimal. *Journal of Computational Biology* 3(1): 53–96.

[HI97]Hart W. E. and Istrail S. C. (1997) Lattice and off-lattice side chain models of protein folding: Linear time structure prediction better than 86% of optimal. *Journal of Computational Biology* 4(3): 241–259.

[HP95]Hannenhalli S. and Pevzner P. (1995) Transforming cabbage into turnip (polynomial algorithm for sorting signed permutations by reversals). In *Proceedings of the 27th Annual ACM Symposium on Theory of Computing, Las Vegas*, pp. 178–189.

[HSS98]Hofacker I., Schuster P., and Stadler P. F. (1998) Combinatorics of RNA secondary structures. *Journal of Discrete Applied Mathematics* 88: 207–237.

[IT93]Ioshikhes I. and Trifonov E. N. (1993) Nucleosomal DNA sequence database. *Nucleic Acids Research* 21: 4857–4859. EMBL deposit on FTP server `ftp.ebi.ac.uk` in directory *nucleosome_dna*.

[Jay57]Jaynes E. T. (1957) Information theory and statistical mechanics. *Physical Review* 106(4): 620–630.

[JC69]Jukes T. H. and Cantor C. R. (1969) Evolution of protein molecules. In Munro H. N. (ed) *Mammalian Protein Metabolism*, pp. 21–132. Academic Press, New York.

[JJE94]Johanson D., Johanson L., and Edgar B. (1994) *Ancestors: In Search of Human Origins.*

Villard Books, New York.

[JL87]Jurgensen H. and Lindenmayer A. (1987) Inference algorithms for developmental systems with cell lineages. *Bulletin of Mathematical Biology* 49: 93–123.

[Kar97a]Karlin S. (1997) Assessing inhomogeneities in bacterial long genomic sequences. In *Proceedings of the First Annual International Conferences on Computational Molecular Biology (RECOMB97)*, pp. 164–171. ACM Press.

[Kar97b]Karplus M. (1997) Protein dynamics: From the native to the unfolded state and back again. In *Proceedings of the First Annual International Conferences on Computational Molecular Biology (RECOMB97)*, p. 172. ACM Press. Invited Lecture.

[Kau70]Kauffman S. A. (1970) Behavior of randomly constructed genetic nets: binary element nets. In Waddington C. H. (ed) *Towards a Theoretical Biology*, pp. 18–37. Aldine, Chicago.

[KBM+94]Krogh A., Brown M., Mian I. S., Sjölander K., and Haussler D. (1994) Hidden Markov models in computational biology: Applications to protein modeling. *Journal of Molecular Biology* 235: 1501–1531.

[Kea98]Kearney P. E. (1998) The ordinal quartet method. In Istrail S., Pevzner P., and Waterman M. (eds) *Proceedings of the Second Annual International Conferences on Computational Molecular Biology (RECOMB98)*, pp. 125–134. ACM Press.

[Kec91]Kececioglu J. (1991) *Exact and Approximation Algorithms for DNA Sequence Reconstruction.* PhD dissertation, Department of Computer Science, The University of Arizona. Technical Report 91-26.

[Kec93]Kececioglu J. D. (1993) The maximum weight trace problem in multiple sequence alignment. In *Proceedings of the 4th Symposium on Combinatorial Pattern Matching*, volume 684 of *Lecture Notes in Computer Science*, pp. 106–119. Springer-Verlag, Berlin.

[KGV83]Kirkpatrick S., Gelatt Jr. C. D., and Vecchi M. P. (1983) Optimization by simulated annealing. *Science* 220: 671–680.

[Khi57]Khinchin A. I. (1957) *Mathematical Foundations of Information Theory.* Dover Publications, New York.

[KKBM79]Klotz L. C., Komar N., Blanken R. L., and Mitchell R. M. (1979) Calculation of evolutionary trees from sequence data. *Proceedings of the National Academy of Sciences USA* 76: 4516–20.

[KLJ98]Kearney P., Li M., and Jiang T. (1998) Orchestrating quartets: Approximation and data correction. In *Proceedings of the 39th Annual Symposium on Foundations of Computer Science (FOCS '98)*, pp. 416–425.

[KLM+99]Kececioglu J., Lenhof H.-P., Mehlhorn K., Mutzel P., Reinert K., and Vingron M. (1999) A polyhedral approach to sequence alignment problems. *Discrete Applied Mathematics* submitted.

[KLT97]Kececioglu J., Li M., and Tromp J. (1997) Inferring a DNA sequence from erroneous copies. *Theoretical Computer Science* 185: 3–13.

[KNE88]Kowalski D., Natale D. A., and Eddy M. J. (1988) Stable DNA unwinding, not "breathing", accounts for single-strand-specific nuclease hypersensitivity of specific A+T-rich sequences. *Proceedings of the National Academy of Sciences USA* 85(24).

[Knu73]Knuth D. (1973) *The Art of Computer Programming*, volume 1. Addison-Wesley, Reading, MA, 2nd edition.

[Knu81]Knuth D. (1981) *The Art of Computer Programming*, volume 2. Addison-Wesley, Reading, MA, 2nd edition.

[Kra91]Kraulis P. J. (1991) MOLSCRIPT: a program to produce both detailed and schematic plots of protein structures,. *Journal of Applied Crystallography* 24: 946–950.

[KRA92]Kavanaugh J. S., Rogers P. H., and Arnone A. (1992) High-resolution x-ray study of deoxy recombinant human hemoglobins synthesized from beta-globins having mutated amino termini. *Biochemistry* 31: 8640–8647.

[Krö96]Kröger T. (1996) *Untersuchung von Aminosäurepaarpotentialen aus Proteinstrukturdaten.* Diplomarbeit, Technische Universität München.

[KS60]Kemeny J. G. and Snell J. L. (1960) *Finite Markov Chains.* Van Nostrand, New York.

[KS80]Kinderman R. and Snell J. L. (1980) *Markov Random Fields and Their Applications*. American Mthematics Society, Providence, RI.

[KS92]Karplus M. and Shakhnovich E. (1992) Protein folding: theoretical studies of thermodynamics and dynamics. In Creighton T. E. (ed) *Protein Folding*, pp. 237–196. W. H. Freeman, New York.

[KS95]Kececioglu J. and Sankoff D. (1995) Exact and approximation algorithms for sorting by reversals, with application to genome rearrangement. *Algorithmica* 13: 180–210.

[KSH95]Konecny J., Schöniger M., and Hofacker G. L. (1995) Complementary coding conforms to the primeval comma-less code. *Journal of Theoretical Biology* 173: 263–270.

[KSHG96]Karplus K., Sjölander K., Hughey R., and Grate L. (1996) Hidden markov models (Part 1 and 2). Tutorial held at the *4th International Conference on Intelligent Systems in Molecular Biology (ISMB'96)*.

[KŠS95]Karplus M., Šali A., and Shakhnovich E. (1995) Kinetics of protein folding. *Nature* 373: 665. Reply to [Cha95].

[KT75]Karlin S. and Taylor H. M. (1975) *A First Course in Stochastic Processes*. Academic Press, New York, 2nd edition.

[KTH97]Klein S., Timmer J., and Honerkamp J. (1997) Analysis of multichannel patch clamp recordings by hidden Markov models. *Biometrics* 53: 870–884.

[LAB$^+$93]Lawrence C. E., Altschul S. F., Boguski M. S., Liu J. S., Neuwald A. F., and Wootton J. C. (1993) Detecting subtle sequence signals: A Gibbs sampling strategy for multiple alignment. *Science* 262: 208–214.

[LAK89]Lipman D. J., Altschul S. F., and Kececioglu J. D. (1989) A tool for multiple sequence alignment. *Proceedings of the National Academy of Sciences USA* 86: 4412–4415.

[LC91]Luczak T. and Cohen J. E. (1991) Stability of vertices in random boolean cellular automata. *Random Structures and Algorithms* 2: 237–334.

[LD89]Lau K. F. and Dill K. A. (1989) A lattice statistical mechanics model of the conformational and sequence spaces of proteins. *Macromolecules* 22: 3986–3997.

[LD90]Lau K. F. and Dill K. A. (1990) Theory for protein mutability and biogenesis. *Proceedings of the National Academy of Sciences USA* 87: 638–642.

[Lev68]Levinthal C. (1968) Are there pathways for protein folding? *Journal de Chimie Physique* 65: 44–45.

[LHTW96]Li H., Helling R., Tang C., and Wingreen N. (1996) Emergence of preferred structures in a simple model of protein folding. *Science* 273: 666–669.

[LMR99]Lenhof H.-P., Morgenstern B., and Reinert K. (1999) An exact solution for the segment-to-segment multiple sequence alignment problem. *Bioinformatics* 15(3): 203–210.

[LS96]Lathrop R. H. and Smith T. F. (1996) Global optimum protein threading with gapped alignment and empirical pair score functions. *Journal of Molecular Biology* 255: 641–665.

[Lyn93]Lynch J. F. (1993) A criterion for stability in random boolean cellular automata. *Ulam Quarterly* 2: 32–44.

[Lyn95]Lynch J. F. (1995) On the threshold of chaos in random boolean cellular automata. *Random Structures and Algorithms* 6: 239–260.

[Mam96]Mamitsuka H. (1996) A learning method of hidden Markov models for sequence discrimination. *Journal of Computational Biology* 3(3): 361–373.

[Mam97]Mamitsuka H. (1997) Supervised learning of hidden Markov models for sequence discrimination. In *Proceedings of the First Annual International Conferences on Computational Molecular Biology (RECOMB97)*, pp. 202–208. ACM Press.

[MB97]Merritt E. A. and Bacon D. J. (1997) Raster3d photorealistic molecular graphics. *Methods in Enzymology* 277: 505–524.

[MBHC95]Murzin A. G., Brenner S. E., Hubbard T., and Chothia C. (1995) SCOP: A structural classification of proteins database for the investigation of sequences and structures. *Journal of Molecular Biology* 247: 536–540. http://scop.mrc-lmb.cam.ac.uk/scop/.

[MHK$^+$99]Mewes H. W., Heumann K., Kaps A., Mayer K., Pfeiffer F., Stocker S., and Frishman D. (1999) MIPS: A database for protein sequences and complete genomes. *Nucleic Acids*

Research 27: 44–48.

[MHPF98]Mewes H. W., Hani J., Pfeiffer F., and Frishman D. (1998) MIPS: A database for genomes and protein sequences. *Nucleic Acids Research* 26: 33–37.

[Mic96]Michalewicz Z. (1996) *Genetic Algorithms + Data Structures = Evolution Programs*. Springer-Verlag, Berlin, 3rd edition.

[Mit98]Mitchell M. (1998) *An Introduction to Genetic Algorithms*. MIT Press, Cambridge, MA.

[MJ85]Myazawa S. and Jernigan R. (1985) Estimation of effective interresidue contact energies from protein crystal structures: Quasi-chemical approximation. *Macromolecules* 18: 534–552.

[MM81]Margush T. and McMorris F. R. (1981) Consensus n-trees. *Bulletin of Mathematical Biology* 43(2): 239–244.

[MM89]Miller W. and Meyers E. (1989) Approximate matching of regular expressions. *Bulletin of Mathematical Biology* 51: 5–37.

[MP80]Masek W. J. and Paterson M. S. (1980) A faster algorithm for computing string-edit distances. *Journal of Computer and System Sciences* 20(1): 18–31.

[MP88]Minsky M. L. and Papert S. A. (1988) *Perceptrons*. MIT Press, Cambridge, MA. Expanded edition, 4th printing.

[MRR+53]Metropolis N., Rosenbluth A. W., Rosenbluth M. N., Teller A. H., and Teller E. (1953) Equation of state calculations by fast computing machines. *Journal of Chemical Physics* 21: 1087–1092.

[MS87]Madras N. and Sokol A. D. (1987) Nonergodicity of local, length-conserving Monte–Carlo algorithms for the self-avoiding walk. *Journal of Statistical Physics* 47: 573–595.

[MSH+94]Moran L. A., Scrimgeour K. G., Horton H. R., Ochs R. S., and Rawn J. D. (1994) *Biochemistry*. Neil Patterson/Prentice Hall, Englewood Cliffs, NJ, 2nd edition.

[Nei87]Nei M. (1987) *Molecular Evolutionary Genetics*. Columbia University Press.

[NJ80]Nussinov R. and Jacobson A. B. (1980) Fast algorithm for predicting the secondary structure of single stranded RNA. *Proceedings of the National Academy of Sciences USA* 77(11): 6309–6313.

[NM92]Ngo J. T. and Marks J. (1992) Computational complexity of a problem in molecular structure prediction. *Protein Engineering* 5: 313–321.

[NMK94]Ngo J. T., Marks J., and Karplus M. (1994) Computational complexity, protein structure prediction and the levinthal paradox. In Merz K. and Grand S. L. (eds) *The Protein Folding Problem and Tertiary Structure Prediction*, pp. 433–505. Birkhäuser, Boston.

[NW70]Needleman S. B. and Wunsch C. D. (1970) A general method applicable to the search for similarities in the amino acid sequence of two proteins. *Journal of Molecular Biology* 48: 443–453.

[Ohl95]Ohler U. (1995) *Polygramme und Hidden Markov Modelle zur DNA-Sequenzanalyse: Identifikation und Lokalisation eukaryontischer Promotoren*. Studienarbeit, Friedrich-Alexander-Universität Erlangen-Nürnberg.

[OMJ+97]Orengo C. A., Michie A. D., Jones S., Swindells M. B., Hutchinson G., Martin A., Jones D. T., and Thornton J. M. (1997) CATH – A hierarchic classification of protein domain structures. *Structure* 5(8): 1093–1108. http://www.biochem.ucl.ac.uk/bsm/cath/.

[Pau70]Pauling L. (1970) *General Chemistry*. Dover Publications, New York.

[PJP98]Poole A. M., Jeffares D. C., and Penny D. (1998) The path from the RNA world. *Journal of Molecular Evolution* 46: 1–17.

[PL88]Pearson W. R. and Lipman D. J. (1988) Improved tools for biological sequence comparison. *Proceedings of the National Academy of Sciences USA* 85(8): 2444–2448.

[PP97]Pudlák P. and Pudlák P. (1997) Does dominance–recessiveness have a survival value? Unpublished preprint.

[Rab89]Rabiner L. (1989) A tutorial on hidden Markov models and selected applications in speech recognition. *Proceedings of IEEE* 77(2): 257–286.

[REG+95]Ren J., Esnouf R., Garman E., Somers D., Ross C., Kirby I., Keeling J., Darby G., Jones Y., Stuart D., and Stammers D. (1995) High resolution structures of HIV-1 RT from four RT-inhibitor complexes. *Nature Structural Biology* 2: 293.

[Rei99]Reinert K. (1999) *A Polyhedral Approach to Sequence Alignment Problems*. Dissertation, Universität des Saarlandes, Saarbrücken.

[RJ86]Rabiner L. and Juang B. H. (1986) An introduction to hidden Markov models. *IEEE ASSP Magazine* pp. 4–16.

[RLM+97]Reinert K., Lenhof H.-P., Mutzel P., Melhorn K., and Kececioglu J. P. (1997) A branch-and-cut algorithm for multiple sequence alignment. In *Proceedings of the First Annual International Conferences on Computational Molecular Biology (RECOMB97)*, pp. 241–249. ACM Press.

[RM99]Roberts R. J. and Macelis D. (1999) REBASE – Restriction enzymes and methylases. *Nucleic Acids Research* 27: 312–313. http://rebase.neb.com.

[Rob97]Roberts R. (1997) Hunting for new restriction enzymes in GenBank. In *Proceedings of the First Annual International Conferences on Computational Molecular Biology (RECOMB97)*, p. 251. ACM Press.

[Roz77]Rozanov Y. A. (1977) *Probability Theory: A Concise Course*. Dover Publications, New York.

[RRBGO98]Román-Roldán R., Benaola-Galván P., and Oliver J. L. (1998) Sequence compositional complexity of DNA through an entropic segmentation method. *Physical Review Letters* 80(6): 1344–1347.

[RSS97]Reidys C., Stadler P. F., and Schuster P. (1997) Generic properites of combinatory maps: Neutral networks of RNA secondary structures. *Bulletin of Mathematical Biology* 59(2): 339–397.

[RT98]Rocke E. and Tompa M. (1998) An algorithm for finding novel gapped motifs in DNA sequences. In Istrail S., Pevzner P., and Waterman M. (eds) *Proceedings of the Second Annual International Conferences on Computational Molecular Biology (RECOMB98)*, pp. 228–233. ACM Press.

[SC97]Schönauer M. and Clote P. (1997) How optimal is the genetic code? In Frishman D. and Mewes H.-W. (eds) *Proceedings of German Conference on Bioinformatics (GCB'97)*, pp. 65–67.

[Sch96]Schuster P. (1996) Landscapes and molecular evolution. Technical Report 96-07-047, Santa Fe Institute.

[SFSH94]Schuster P., Fontana W., Stadler P. F., and Hofacker I. L. (1994) From sequences to shapes and back: A case study in RNA secondary structures. *Proceedings of the Royal Society of London B* 255: 279–284.

[SGvH97]Strimmer K., Goldman N., and von Haeseler A. (1997) Bayesian probabilities and quartet puzzling. *Molecular Biology and Evolution* 14(2): 210–211.

[Sin93]Sinclair A. (1993) *Algorithms for Random Generation and Counting: A Markov Chain Approach*. Birkhäuser.

[Sip90]Sippl M. (1990) Calculation of conformation ensembles from potentials of mean force. *Journal of Molecular Biology* 213: 859–883.

[Sip93]Sippl M. J. (1993) Boltzmann's principle, knowledge-based mean fields and protein folding. *Computer-Aided Molecular Design* 7: 473–501.

[SM97]Setubal J. C. and Meidanis J. (1997) *Introduction to Computational Molecular Biology*. PWS Publishing, Boston.

[SMFB95]Sun H., Mezei M., Fye R., and Benham C. J. (1995) Monte Carlo analysis of conformational transitions in superhelical DNA. *Journal of Chemical Physics* 103: 8653–8665.

[SMW95]Sayle R. and Milner-White E. J. (1995) RasMol: Biomolecular graphics for all. *Trends in Biochemical Sciences (TIBS)* 20(9): 374.

[SS91]Sander C. and Schneider R. (1991) Database of homology-derived protein structures and the structural meaning of sequence alignment. *Proteins: Structure Function Genetics* 9: 56–68.

[ŠSK94a]Šali A., Shakhnovich E., and Karplus M. (1994) How does a protein fold? *Nature* 369: 248–251.

[ŠSK94b]Šali A., Shakhnovich E., and Karplus M. (1994) Kinetics of protein folding: A lattice model study of the requirements for folding to the native state. *Journal of Molecular Biology* 235: 1614–1636.

[SvH96]Strimmer K. and von Haeseler A. (1996) Quartet puzzling: A quartet maximum-likelihood method for reconstructing tree topologies. *Molecular Biology and Evolution* 13(7): 964–969.

[SW78]Stein P. R. and Waterman M. S. (1978) On some new sequences generalizing the Catalan and Motzkin numbers. *Discrete Mathematics* 26: 261–272.

[SW81]Smith T. F. and Waterman M. S. (1981) Identification of common molecular subsequences. *Journal of Molecular Biology* 147: 195–197.

[Tee86]Teeter M. (1986) An empirical examination of potential energy minimization using the well-determined structure of the protein crambin. *Journal of the American Chemical Society* 108: 7163–7172.

[Tee91]Teeter M. (1991) Water-protein interactions: Theory and experiment. *Annual Reviews of Biophysics and Biophysical Chemistry* 20: 577–600.

[UM93]Unger R. and Moult J. (1993) Genetic algorithms for protein folding simulations. *Journal of Molecular Biology* 231: 75–81.

[vHBS+92]von Haeseler A., Blum B., Simpson L., Sturm N., and Waterman M. S. (1992) Computer methods for locating kinetoplastid cryptogenes. *Nucleic Acids Research* 20(11): 2717–2724.

[VKBS95]Vieth M., Kolinski A., Brooks III C. L., and Skolnick J. (1995) Prediction of the quaternary structure of coiled coils. Application to mutants of the GCN4 leucine zipper. *Journal of Molecular Biology* 251: 448–467.

[WAC97]Wei L., Altman R. B., and Chang J. T. (1997) Using the radial distribution of physical features to compare amino acid environments and align amino acid sequences. In Altman R. B., Dunker A. K., Hunter L., and Klein T. E. (eds) *Proceedings of the Pacific Symposium on Biocomputing'97*, pp. 465–476. World Scientific, Singapore.

[Wat69]Watson H. C. (1969) The stereochemistry of the protein myoglobin. *Progress in Stereochemistry* 4: 299–333.

[Wat78]Waterman M. S. (1978) Secondary structure of single-stranded nucleic acids. *Studies in Foundations and Combinatorics, Advances in Mathematics Supplementary Studies* 1: 167–212.

[Wat95]Waterman M. S. (1995) *Introduction to Computational Biology - Maps, Sequences and Genomes*. Chapman & Hall.

[WDD+66]Woese C. R., Durge D. H., Dugre S. A., Condo M., and Saxinger W. C. (1966) On the fundamental nature and evolution of the genetic code. *Cold Spring Harbor Symposium on Quantitative Biology* 31: 723–736.

[Whi89]White J. H. (1989) An introduction to the geometry and topology of DNA structure. In Waterman M. S. (ed) *Mathematical Methods for DNA Sequences*, pp. 225–253. CRC Press, Boca Raton, FL.

[Whi95]White J. H. (1995) Winding the double helix: Using geometry, topology, and mechanics of DNA. In Lander E. S. and Waterman M. S. (eds) *Calculating the Secrets of Life*, pp. 153–178. National Academy Press.

[WHR+87]Watson J. D., Hopkins N. H., Roberts J. W., Steitz J. A., and Weiner A. M. (1987) *Molecular Biology of the Gene*. Benjamin/Cummings, Menlo Park, CA, 4th edition.

[WSB76]Waterman M., Smith T. F., and Beyer W. A. (1976) Some biological sequence metrics. *Advances in Mathematics* 20: 367–387.

[WTK83]Watson J. D., Tooze J., and Kurtz D. T. (1983) *Recombinant DNA: A Short Course*. Scientific American Books, New York.

[Wu83]Wu C. F. J. (1983) On the convergence properties of the EM algorithm. *Annals of Statistics* 11(1): 95–103.

[YD93]Yue K. and Dill K. A. (1993) Sequence-structure relationships in proteins and copolymers. *Physical Review E* 48(3): 2267–2278.

[YD95]Yue K. and Dill K. A. (1995) Forces of tertiary structural organization in globular proteins. *Proceedings of the National Academy of Sciences USA* 92: 146–150.

[YFT^{+}95]Yue K., Fiebig K. M., Thomas P. D., Chan H. S., Shakhnovich E. I., and Dill K. A. (1995) A test of lattice protein folding algorithms. *Proceedings of the National Academy of Sciences USA* 92(1): 325–9.

[ZS84]Zuker M. and Sankhoff D. (1984) RNA secondary structures and their prediction. *Bulletin of Mathematical Biology* 46(4): 591–621.

[Zuk]Zuker M. RNA resource web page. http://www.ibc.wustl.edu/ zuker/rna/.

Index